ANIMAL SCIENCE, ISSUES AND PROFESSIONS

INVERTEBRATES

CLASSIFICATION, EVOLUTION AND BIODIVERSITY

ANIMAL SCIENCE, ISSUES AND PROFESSIONS

Additional books in this series can be found on Nova's website
under the Series tab.

Additional e-books in this series can be found on Nova's website
under the e-book tab.

ANIMAL SCIENCE, ISSUES AND PROFESSIONS

INVERTEBRATES

CLASSIFICATION, EVOLUTION AND BIODIVERSITY

RAFAEL RIOSMENA-RODRIGUEZ
EDITOR

NOVA BIOMEDICAL

New York

For permission to use material from this book please contact us:
Telephone 631-231-7269; Fax 631-231-8175
Web Site: http://www.novapublishers.com

Library of Congress Cataloging-in-Publication Data

Invertebrates : classification, evolution and biodiversity / editor, Rafael Riosmena-Rodriguez.
pages cm
Includes index.
ISBN: 978-1-62618-789-4 (hardcover)
1. Invertebrates--Classification. 2. Invertebrates--Evolution. 3. Invertebrates--Ecology. 4. Biodiversity. I. Riosmena-Rodrmguez, Rafael.
QL362.I59 2013
592--dc23

2013015675

Published by Nova Science Publishers, Inc. † New York

Contents

Preface

Invertebrates are common in marine and freshwater ecosystems and key elements in processes such as trophic dynamics and nutrient recycling. At the present time we have a limited knowledge of their diversity and how they have evolved over time. A key element of study in this book, are the current efforts to produce revised classificatory systems which include modern approaches and an update of the current taxonomical system. Another topic discussed in great detail is the relation of invertebrates and their contribution to biodiversity in terms of unique species per habitat. "Invertebrates: Classification, Evolution and Biodiversity" represents an effort to show the relevance of this unique group of living species and the current efforts to better understand them. This book presents a well balanced set of chapters in which a detailed review of the classification of poorly known groups (such as freshwater arthropods, oligochaetes, and marine ascidians) are studied. Another set of chapters are devoted to understanding the associated biodiversity of invertebrates in marine ecosystems and how they respond to environmental variables. Biogeographical comparisons and evolutionary fractal analysis close the contributions and provide a high standard of knowledge towards the goal of this book. This book also discusses the role of invertebrates in coastal systems and provides key investigation for undergraduate and graduate students and researchers and professional researchers who are interested in ecology and evolution.

Chapter 1 - Peatlands are typically reported to sustain low diverse soil fauna communities due to cold temperatures and waterlogging conditions which restrict biological activities. However, very few studies have undertaken a comprehensive survey of soil invertebrate communities in these systems and in most cases the information available is only based on very few records taken on few sampling occasions. In the case of arthropods, difficulties in obtaining accurate identifications and quantifications due to the fact that some groups only stay in the soil for short periods of time, add complications to their study. Their ecological role as litter transformers and predators is determinant in the decomposition and turnover of soil organic matter. Therefore, in this chapter the authors provide a detailed study of the Arthropod community from two contrasting climatic and vegetational peatland sites to evaluate their habitat selection and their responses to changes in environmental conditions. The Arthropoda community at the two investigated upland peatlands was represented by three subphyla (Chelicerata, Myriapoda and Hexapoda), but macroarthropods only accounted for 1% of the total number of invertebrates collected. The author's results indicate that habitat characteristics can have a strong influence on the structure of the Arthropoda community within a particular peatland area, both in terms of population numbers and diversity. Overall, significant higher abundances of most groups (exceptions were Aranea, Pseudoescorpionida, Opilionida and Insecta) were observed at the drier locality (Chan do Eume) than at the most oceanic and wettest site (Chan do Lamoso). Although at both localities Oribatid mites dominated the arthropod community and with Entomobryomorpha collembolans and Prostigmata mites being the two following more abundant groups, their percentages differed slightly between sites and thus, at Chan do Eume oribatids represented 58% of the total numbers, Entomobryomorpha 17% and Prostigmata 11%, whereas at Chan do Lamoso they followed the order Oribatida (37%) > Entomobryomorpha (29%) > Prostigmata (7%). Interestingly, 15 out of the 48 Acari families and half of the insect orders identified were only recorded at one site; in addition, proturans and symphylans were only found at the most continental locality but pseudoscorpions and opilions at the most oceanic one. These differences in the Arthropod community could be partly related to the particular microclimatic conditions at each site, with a strong effect on soil moisture, but also to the different vegetation. The author's results suggest that more frequent waterlogged conditions and marine proximity have negative effects on the two dominant groups in these systems (mites and collembolans) and therefore, alterations in the nutrient cycling dynamics associated to climatic changes are anticipated.

Chapter 2 - Unlike in other European countries, peatland soils have a limited presence in Spain; in the Galicia region (NW Spain) they survive in upland areas where cold temperatures and continuous moisture supply allow their presence. Very little is known about the invertebrate diversity in these areas, but previous research has highlighted the dominance, in terms of biomass, of oligochaete worms (families Lumbricidae and Enchytraeidae). As a result of their burrowing and casting activities, these two groups play an important role in ecosystem functioning by promoting microbial growth, soil structure and porosity as well as soil organic matter incorporation and mixing and nutrient cycling. Therefore, any changes in their presence and/or abundance in response to climatic variations (namely, temperature and soil moisture) could have important consequences on the amount of carbon being stored in these soils. In this chapter, the authors will describe the Oligochaeta community composition of two peatland sites with different microclimate and dominant vegetation to determine the relative importance of abiotic and biotic factors in the size and diversity of their populations. The Oligochaeta community from the upland peatlands within Serra do Xistral was mainly represented by two families, Lumbricidae and Enchytraeidae. However, only two species of lumbricids were collected in very low densities (about 50 individuals per m^2) at both localities, whereas enchytraeids showed a far greater diversity (up to 11 species were identified) and their population numbers reached, on average, more than 60,000 individuals per m^2. In addition, although at both sites the enchytraeid community was dominated by one species (*Cognettia sphagnetorum*), its structure showed marked differences between localities. Overall, enchytraeid numbers were significantly higher at the oceanic site (Chan do Lamoso) due to the greater abundance of *C. sphagnetorum, Cernosvitoviella atrata, Marionina filiformis* and *Mesenchytraeus sanguineus*, whereas drier conditions at Chan do Eume significantly favoured higher population numbers of five species and in particular those of *C. hibernica* and *Achaeta* spp. which reached similar densities to those of the dominant *C. sphagnetorum* at this site. Finally, *M. flavus* and *M. lusitanicus* showed similar abundances at both sites. The author's results highlight the close link between enchytraeids and soil water content in these Galician peatland habitats. Due to the key role these oligochaetes play on soil organic matter decomposition and nutrient cycling, any alterations in the abiotic factors are expected to have important implications in the functioning of these systems.

Chapter 3 - The first work in Mexico on macrozooplankton grazing was published in 1989 and that on microzooplankton appeared in 1991. Only six reports exist on microzooplankton grazing processes; in all of them the dilution method was used. Seven works have been published on macrozooplankton grazing; three of them were carried out in Todos Santos Bay, B.C., at sites with different dynamic characteristics.In two of them, ^{14}C was used as a tracer, one in an upwelling area and another in an estuarine environment. The method of change in chlorophyll-a concentration was used to develop a third survey in a stable hydrographic area with oceanic influence. The fourth report, carried out in the central region of the Gulf of California, where ^{14}C was also used as a tracer. Results on daily variations indicate that grazing rates are up to 75% higher during the night-time that during the day. The perspective of the future scenario is not encouraging for the development of this research line; only a reduced group of researchers are working currently on this topic in Mexico. It is necessary to realise the importance of having information on *in situ* consumption rates of the zooplankton; it is only by having this information that the authors will be able to estimate the ecological role of the zooplankton and its function in the ecosystem. This situation is changing due to the efforts several authors or co-authors of the above papers who are actively working on this topic or are in the stage of preparation of articles on specific grazing rates and filtration for micro- and macrozooplankton, using information derived from previous work. Moreover, the number of working groups that address this issue are increasing gradually.

Chapter 4 - One of the most exceptional and multifunctional habitats for cryptic and epibenthic fauna has been established inthe costal zones of Magdalena Bay conforming an interesting interaction between rhodoliths and marine sponges. These function primarily as refuges, where the rhodolith beds (red coralline algae) grow in association with a high biodiversity of marine sponges and form functional complexes to counteract the hydrodynamic effect generated by the current flow.The main objective in this chapter was to highlight the community structure of cryptic, epibenthic, and sessile fauna with a multifactorial analysis including: the ecological analysis via alpha diversity, the oceanographic dynamicsof the Bay, and the intrinsic complexity of the habitat, including the chemical-biological interaction between sponges and rhodoliths comparing those variables between two different localities (CFE and PED) subjected to different dynamics at the Lagoon System of Magdalena Bay.

The sample area covered five hundred square metres at each site, covering a total area of $250m^2$, reviewing a total of 70 rhodoliths with 7,223 associated invertebrates; these belonged to 52 orders and 105 families, with a total of 110 species (42 determined to species and 68 to genera level) analysed. The morphological arrangement of the rhodolith varies as a response to the physical regimens presented by the author's results revealed that higher number of individuals is supported in PED (52.48± 29.86) organisms cm^2 when rhodoliths present 7 to 4 branches in a volume of 250 to 200 ml.The cryptic and epibenthic mobile groups include: Crustaceans that contributed 67% of the total abundance in both localities, Molluscsat levels of 5.3% in CFE and 8.3% in PED, representing the higher richness taxa, Polychaetes atlevels of 19.6% in CFE and 14% in PED and Ophiurida at levels of 8.1% in CFE and 9.9% in PED. In terms of the richness, molluscs, Crustaceans and polychaetes are the principal components of the 42 species in CFE and 54 species in PED (this last with a high Simpson ds= 6.24). The epibenthic sessile fauna was integrated by: Cnidaria that was monospecific in both localities and represent more than 87% of the sessile abundance, Ascidacea provided 3% of the total abundance and the Porifera taxa representing only 8% of abundance but with a higher coverage and richness.This dovetailing provides a complex habitat represented bya high biodiversity. It is important to highlight that this chapter presents the results of an extensive study of the invertebrate community associated with rhodoliths in the Mexican region of the eastern Pacific.

Chapter 5 - Rhodolith beds are aggregations of free-living, non-geniculated coralline algae with a worldwide occurrence. Their distribution and survivorship is regulated by many factors, from which light, temperature, and sedimentation are considered as main restrainers of growth and structure. Rhodoliths are bioengineers and provide a three-dimensional habitat for associated species. Distinctive faunal and floral assemblages of high diversity are harbored by rhodolith thalli by virtue of their branching and interlocking nature. Rhodolith communities are of ecological and economical importance around the world and subject to periodical events disturbance. Periodic rotation is necessary, but does not need to be frequent, to prevent sedimentation and to allow light reaching all surfaces of the thalli. This rotation is a natural disturbance, stochastic phenomena recognized as an important determinant of presence and abundance of associated species. After a disturbance event the associated community with any substratum starts a recovery process that might eventually return it to what it used to be. This process can take from days up to years. In this chapter, the role of disturbance in structuring the invertebrate assemblages associated with two rhodolith beds was studied. The authors artificially stabilized rhodolith thalli for periods of 6 months to analyze the response of the associated invertebrate community to stability and disturbance. Samples from untied thalli (natural) were compared to artificially stabilized rhodoliths. Results demonstrated that the influence of disturbance on invertebrate assemblages is not simple and will strongly depend on the characteristics of the disturbance as well as the physiological, morphological and behavioral capabilities of the species present at any given time. Factors such as the time of the year when disturbance events happen can modify the recovery patterns of the assemblage diversity. The patterns of the species associated with natural and artificially stabilized rhodoliths emerge at three scales: a) between stability periods (two, four and six months), b) between levels of stability (natural and stabilized thalli), and c) among assemblages in experiments started on different dates. The value of rhodoliths as a unique biotope is under threat by different kinds of human activities around the world. The lack of regulation on small and large scale trawling, reduction of water quality, modification of water circulation, ocean acidification, and any other activities that affect rhodolith stability and disturbance are of special concern. Invertebrate assemblages on rhodolith beds provide recruits for many commercial fisheries around the world. It is necessary to take into consideration the response of these assemblages to both natural and artificial disturbance in the development and implementation of coastal management plans around the world.

Chapter 6 - The Gulf of California, also known as Sea of Cortez, and the Mediterranean Sea are ecosystems considered as spots of high marine biodiversity. Both regions are comparable from various physiographical and climatic perspectives. Both are enclosed seas characterized by a negative hydrological balance derived from high evaporation rates that are partially offset by the inflow of surface oceanic currents and, to a lesser extent, by precipitation and freshwater discharges from rivers; as a result, these seas are saltier than oceans. The Gulf of California, however, has a broad communication with the Pacific Ocean and is bordered by the coasts of a single country (Mexico), in contrast with the Mediterranean Sea, which connects with the Atlantic Ocean through the Gibraltar Strait and involves littorals of European, Asian and African countries, many of which share a complex centuries-long history of coastal settlements and exploitation of marine resources. These two basins are regarded as having a high biological diversity and share a similar number of marine mollusk species (2,113 in the Mediterranean Sea and 2,193 in the Gulf of California). It is worth noting that, when the surfaces covered by these basins are compared, the Gulf concentrates this high biodiversity within a rather small area. Furthermore, considering that studies focused on the Gulf of California are still scarce, it is estimated that the number of faunal records in the area will double as investigation works progress. Of the 293 families of marine mollusks reported in the literature reviewed for both basins, at least 120 share between one and four genera. This chapter aims to compare the Mediterranean Sea and the Gulf of California in terms of physiographic features, the diversity of mollusks inhabiting both basins by family, genera and number of species represented, and the role of these organisms as objects of commercial exploitation in fisheries.

Chapter 7 - The Tropical Eastern Pacific islands ranging from Revillagigedos to the Galapagos Islands are very poorly studied. Throughout this region, there is a lack of studies about the biodiversity of sea squirts. During adulthood, ascidians are sessile organisms and have a variety of shapes and colours, but have unique characteristics to identify them in the field, such as a tunic that covers the body and two siphons, one oral and one atrial. Research in the southern Gulf of California, Southern Mexico, Costa Rica and Panama project the presence of three orders, 11 families, 16 genera and 35 species. Of these, four are the most represented families and eight are the most abundant species. Among the species recorded are five invasive species.

Chapter 8 - Meiofaunal assemblages in seagrasses from Northwest Mexico were studied using NaGISA protocol. The authors evaluated spatiotemporal changes by sampling in the subtropical lagoon Laguna San Ignacio in December 2008 and April 2009. The authors took triplicate sediment core samples (10x15 cm dimensions) to quantify the abundance and taxa composition of invertebrates. The authors found 8 invertebrates taxa in the seagrass meadows of Zostera marina (ascidians, anomurans, amphipods, isopods, diplostraca, mollusks, ophiuroids, and polychaetes) with high spatiotemporal variability in their abundances. Mollusks, amphipods and polychaetes were the most frequent taxa observed. There were significant differences ($p<0.05$) between the abundances of amphipods found in the Mouth and Pitahaya zones. One of the ecological services provided by meiofaunal assemblages is that gray whales (Eschrichtius robustus) that visit this lagoon in the winter season, feed on the amphipods that live within the rhizomes of Zostera marina. Seagrass meadows in Northwest Mexico provide essential ecosystem functions; they form a community on which many species rely on whether as a habitat, food source or as a fishery ground for men. The ongoing monitoring of these meadows will serve as a baseline time series from which the effect of climate change on this community will be assessed. Natural Geography in Shore Areas (NaGISA) protocol represented a good method for the author's biological estimations in relation to other seagrass monitoring protocols reviewed during this study.

Chapter 9 - The *Sargassum* forests are one of the most characteristic and critical habitats for conservation in the rocky shores at the Gulf of California. Therefore, it is necessary to know more about the associated populations. The study was carried out in 2008 in five localities around La Paz Bay, México. The biodiversity associated with sub-tidal *Sargassum* forest was monitored by applying a modification of the NaGISA rocky shore protocol. The richness of fish species was estimated for long of transects of 50 m. The cover percent of the biotic and abiotic component of substrate, the composition of epibenthic invertebrates and the richness of macroalgae species were estimated inside of 1 m^2 quadrants, and the density (thalli number) and size of *Sargassum* were estimated inside 625 cm^2 quadrants. The substrate was composed principally of rock, *Sargassum* and other seaweeds. In total, 27 species of fishes, 23 species of macroalgae, 38 species of macro epibenthic and 17 micro epibenthic invertebrates were registered. Differences in species composition of macroalgae associated with *Sargassum* forests were observed between the locality El Saladito and The Sauzoso. During the study, the size of *Sargassum* showed temporal variation suggesting growth differences. The total number of epibenthic invertebrates found was 633 (mean = 3.7 individuals/m^2), which were classified into six taxa; the Cnidaria was the most abundant. The cover percent of substrate components, the community structure of fauna and morphometric variables of the algae showed a spatial-temporal variation. The results show the baseline that can be obtained, which in turn will provide the information necessary to understand the ecological role of these brown algae for the maintenance of biodiversity and the community-associated changes. *Sargassum* forest is a critical habitat for conservation in the Gulf of California because it is sensitive to environmental and human-induced impacts. Their role in coastal areas is related to species recruitment, the fixation of CO_2 and drift material, which are extremely important to sandy beach stability and food sources for insects from terrestrial environments. *Sargassum* forest comprise a relevant near shore ecosystem where species related to fisheries, protected species and those with high biodiversity treat it as a refuge There are several fisheries related who are close to this ecosystem and legal protection does not exist to protect this critical habitat for conservation in the Gulf of California.

Chapter 10 - This chapter presents a description and analysis of metazoan morphogenesis during evolution and development using topological and fractal concepts and methodology. The topological approach makes it possible to consider the succession of spherical surgeries, which change the topological genus of the surface. Topological singularities inevitably arising and transforming during early development destroy the preexisting pattern of symmetry. Symmetry breaking determined both genetically and epigenetically plays a key role in biological morphogenesis. Some developmental events are interpreted in terms of symmetry breakdown and related to well-known mathematical theorems. In addition to such classical symmetry of biological forms as rotating (radial), mirror (bilateral) and translational ones, the scale symmetry is widespread in fractal morphogenesis. In some animal taxa an increase of genus of the body surface up to high values is evident; these patterns evolved on the basis of fractal systems. Colonial animals with repeating modular elements are fractal organisms. For example, colonial rhizocephalan barnacles (Crustacea: Rhizocephala) have irregular fractal organization at the parasitic stage of their life cycle due to asexual reproduction by budding. Biological fractal structures as irregular forms exhibit fluctuating asymmetry. To evaluate the relationship between order and chaos in the structure of branching fractal channels of the gastro-vascular system in the jellyfish *Aurelia aurita* and the larval tracheal gills in the mayflies *Siphlonurus immanis* and *Parameletus chelifer* these patterns in symmetrical parts of the organisms were compared with deterministic fractal trees. During ontogenesis, the spatial and temporal variability of the branching systems increases and results in the breaking of the radial symmetry in each individual medusa and bilateral symmetry in mayfly larvae. The epithelial surface is an interface between the internal medium of an organism and the outside environmental medium; topological and fractal transformations during metazoan evolution and development increase this interface so ensuring better adaptation of organism to the environment. Fractal structures as well as toroid forms can be considered as a functionally optimized design in Metazoa. Topological methodology reveals a certain set of topological rules constraining and directing biological morphogenesis during evolution and development. The detailed analysis of topological and fractal form dynamics in phylo- and ontogenesis may provide additional possibilities for biological description and classification.

In: Invertebrates: Classification, Evolution and Biodiversity ISBN: 978-1-62618-789-4
Editor: Rafael Riosmena-Rodriguez

Chapter 1

Arthropod Community Structure and Diversity from Galician Upland Peatlands

Noela Carrera* and María J. I. Briones
Departamento de Ecología y Biología Animal, Facultad de Biología,
Universidad de Vigo, Vigo, Spain

Abstract

Peatlands are typically reported to sustain low diverse soil fauna communities due to cold temperatures and waterlogging conditions which restrict biological activities. However, very few studies have undertaken a comprehensive survey of soil invertebrate communities in these systems and in most cases the information available is only based on very few records taken on few sampling occasions. In the case of arthropods, difficulties in obtaining accurate identifications and quantifications due to the fact that some groups only stay in the soil for short periods of time, add complications to their study. Their ecological role as litter transformers and predators is determinant in the decomposition and turnover of soil organic matter. Therefore, in this chapter we provide a detailed study of the Arthropod community from two contrasting climatic and vegetational peatland sites to evaluate their habitat selection and their responses to changes in environmental conditions. The Arthropoda community at the two investigated upland peatlands was represented by three subphyla (Chelicerata, Myriapoda and Hexapoda), but macroarthropods only accounted for 1% of the total number of invertebrates collected. Our results indicate that habitat characteristics can have a strong influence on the structure of the Arthropoda community within a particular peatland area, both in terms of population numbers and diversity. Overall, significant higher abundances of most groups (exceptions were Aranea, Pseudoescorpionida, Opilionida and Insecta) were observed at the drier locality (Chan do Eume) than at the most oceanic and wettest site (Chan do Lamoso). Although at both localities Oribatid mites dominated the arthropod community and with Entomobryomorpha collembolans and Prostigmata mites being the two following more abundant groups, their percentages differed slightly

* E-mail address: eunoela@uvigo.es.

between sites and thus, at Chan do Eume oribatids represented 58% of the total numbers, Entomobryomorpha 17% and Prostigmata 11%, whereas at Chan do Lamoso they followed the order Oribatida (37%) > Entomobryomorpha (29%) > Prostigmata (7%). Interestingly, 15 out of the 48 Acari families and half of the insect orders identified were only recorded at one site; in addition, proturans and symphylans were only found at the most continental locality but pseudoscorpions and opilions at the most oceanic one. These differences in the Arthropod community could be partly related to the particular microclimatic conditions at each site, with a strong effect on soil moisture, but also to the different vegetation. Our results suggest that more frequent waterlogged conditions and marine proximity have negative effects on the two dominant groups in these systems (mites and collembolans) and therefore, alterations in the nutrient cycling dynamics associated to climatic changes are anticipated.

Introduction

The importance of invertebrate fauna in peatlands started in the 70s when increasing concerns of these habitats being progressively deteriorated and rapidly disappearing as a result of peat extraction, agriculture conversion and land transformation became an important conservation issue [1-3]. Arthropods form a large proportion of the epigeic fauna of bogs [4], with insects and arachnids being exceptionally well represented: for example, 96% of all aerial invertebrates captured at the Moor House Nature Reserve using sticky traps were Diptera [5], whereas Acarina and Collembola are usually the most abundant groups living in the soil [1]. In addition, their populations are typically very diverse: for example, Higgins listed around 40 species of spiders [6] and highlighted the common presence of centipedes (Chilopoda) in Irish peatlands, whereas up to 169 species of spiders were recorded in Quebec [7] and at Moor House, Hale recorded 56 species of collembolans [8], Block identified 78 species of mites [9] and in their review work, Coulson and Whittaker [1] indicated that the most diverse groups were Diptera (453 species), followed by Coleoptera (115), Acarina (107), Hymenoptera (85), Araneida (71), Hemiptera (71) and Lepidoptera (70). Furthermore, habitat heterogeneity in blanket bogs seems to determine the community structure of the soil invertebrates and certain studies have suggested that soil arthropods could be related to certain plant species [10, 11]; however, very few species have been associated with a specific area [1] and for example, in this review study, only six of the 16 most abundant species of Collembolans recorded at Moor House were reported on *Juncus squarrosus*.

Spiders, opilions, beetles and mites are the main predators of the mesofauna [1, 4]; therefore, they play an important role in controlling the abundance and diversity of their prey [12] and hence exerting a top-down regulation on the soil food web [13]. In contrast, herbivorous and saprophagous mites, plant feeding groups and collembolans grazing on fungi have a greater impact on soil organic matter decomposition and nutrient mineralisation [14-16] and as a result, increases in decomposition rates up to 20 to 50% have been reported in several ecosystems [17-19]. However, their influence on soil processes become largely determined by the seasonal abundance of food resources (plant productivity, soil fertility and prey emergence/abundance) and thus, their populations are more influenced by bottom-up forces [13].Therefore, in this chapter, we will examine the structure of the soil arthropod community from two different blanket bogs located in Serra do Xistral (NW Spain) with different vegetation and climatic properties to evaluate their habitat seiection and their responses to changes in environmental conditions.

Arthropoda Phylogeny and Systematic

The name of the phylum is related to the presence of jointed appendages which are often modified to perform different functions, such as feeding or locomotion [20]. In addition, arthropods have heteronomous segmentation, i.e. including different specialised regions or "tagmas" and a chitinous exoskeleton. The respiratory system of most terrestrial arthropods consists of "stigmas" or "spiracles" which allow the air to enter in the trachea and/or book lungs. Because the exoskeleton restricts growth, they must replace their exoskeletons by moulting which has led to their inclusion in the clade Ecdysozoa [21]. This phylum was traditionally divided into five subphyla depending on the number of tagmas and the morphology of their appendages: Trilobites (all extinct), Crustacea, Chelicerata, Myriapoda and Hexapoda. However, the phylogeny of this phylum is being continuously modified [22, 23] and the relationships between the four major extant groups are still a matter of debate [24]. For example, some authors [25, 26] suggest the existence of two clades, Pancrustacea (Crustacea plus Hexapoda) and Myriochelata or Paradoxopoda (Chelicerata plus Myriapoda), whereas morphological studies seem to support the traditional classification in which Crustacea is grouped with Insecta (forming Mandibulata) and becoming the sister group of Chelicerata [27-30]. Similarly, the sub-divisions within each subphylum are also being regularly updated in an attempt to reflect their phylogenetic affinities. For this reason, in this chapter we have adopted the more widely accepted hierarchical classification.

The *subphylum Chelicerata* includes three classes (Merostomata, Pycnogonida and Arachnida) but only the arachnids include terrestrial forms. The relationships between the different arachnid groups are not yet fully resolved, partly due to the inclusion or not of certain taxa in the genetic or morphological analyses which led to different results [31-36]. The arachnids have been traditionally divided into two lineages based on their morphology [31]: Micrura (Acari plus Aranea) and Dromopoda (Solifuga+Opiliones+ Scorpionida+Pseudoscorpionida). The inclusion of additional morphological characters and ribosomic sequences, together with a greater number of taxa, resulted in a similar cladogram [37]. However, more recent studies have questioned Aranea as a valid sister group of Scorpionida+Pseudoscorpionida and consequently, the clade Acaromorpha has been proposed as the sister group of the remaining groups of arachnids [32], although most authors still follow the classification given by Giribert et al. [38]. Since the monophyly of Dromopoda seems to be well supported [39], the arachnids collected in Serra do Xistral could be sub-divided into three subclasses: Micrura (spiders), Dromopoda (opilions plus pseudoscorpions) and Acari (mites). However, it should be noted that the phylogenetic relationships between these groups are also under debate depending on whether fossil specimens are included in the analyses [38]. In the case of the Acari, it is usually accepted that they consist of two superorders, Parasitiformes and Acariformes, although their monophyly is still controversial [40, 41]. The Parasitiformes collected in the Galician peatlands included members of the order Mesostigmata and two orders of Acariformes, Trombidiformes (with the suborder Prostigmata) and Sarcoptiformes (with the suborders Oribatida, Astigmata and

Endeostigmata); all of them have been confirmed as separated taxa by rDNA analyses [42]. Molecular and morphological studies have also validated the monophyly of the *subphylum Myriapoda*, which is divided into four classes: Chilopoda, Diplopoda, Symphyla and Pauropoda [28]. The *subphylum Hexapoda* is also considered to be a monophyletic group [28, 43, 44] and divided into two classes: Entognatha (Collembola+Protura+Diplura) and Ectognatha (Hexapoda=Insecta). The monophyly of the Entognatha is still questioned (reviewed by Carapelli et al. [45]). In addition, Carapelli et al. [46] suggested that Collembola could be considered the sister group of Diplura+Insecta; however, the addition of new gene sequences led to the conclusion that Entognatha is monophyletic (although with weak support) and the sister group of Insecta [28, 44]. The relationships between the different groups of collembolans are also pending on clarification [47]; they have been traditionally divided into two suborders, Arthropleona (Poduromorpha+Entomobriomorpha) and Symphypleona [48, 49], but the latter group was later subdivided into two groups [50, 51]. Accordingly, in modern classifications collembolans are divided into four groups [52]: Poduromorpha, Entomobryomorpha, Symphypleona and Neelipleona, at the same taxonomical level (order or suborder) depending on whether Collembola is considered to be a class [53, 54] or an order [49, 55, 56]. However, the suggestion of Tomoceroidea as a new taxon by Xiong et al. [57] clearly reflects that the classification is still far from being complete. Therefore, in this chapter we have adopted the initial division of the collembolans into three groups [48, 49]. In contrast, Insecta is widely accepted as monophyletic group [28, 44] and usually divided into two subclasses: Apterygota (with the orders Thysanura and Archaeognata) and Pterygota (including 28 orders) [57]. Although Hemiptera has been historically placed in two orders or suborders (Homoptera+Heteroptera or Homoptera+Hemiptera [58]) and that some recent identification keys still follow this classification [59], molecular analyses indicate that they should be combined into the single monophyletic group, Hemiptera [60, 61].

Soil Arthropod Community in Galician Blanket Bogs

The site under study was the upland active blanket bog in the Serra do Xistral (Galicia, NW Spain). Two peatland microhabitats with different dominant vegetation and located at a similar altitude (about 970 m a.s.l.) were investigated: (i) Chan do Lamoso (43°30'N, 7°32'W) with a more oceanic microclimate, wetter soils (81.6% water content) and dominated by Cyperaceae (*Carex durieui* Steud. and *Eriophorum angustifolium* Honck) and (ii) Chan do Eume (43°47'N, 7°57'W), with a more continental climate and slightly drier soils (on average, 70.4%) and dominated by *Molinia caerulea* L. (for a more detailed description of the sites see Carrera and Briones this book).

Fifteen soil cores (PVC cylinders, 10 cm diameter x 20 cm deep) were taken in February and June during three consecutive years (2003 to 2005) at Chan do Lamoso and on the same months in 2006 at Chan do Eume. The intact soil samples were transported back to the laboratory, where the animals were extracted using the Tüllgren method [62], fixed in 70% ethanol and finally identified to family or order level by means of a stereomicroscope (Nikon SMZ-U) and an optical microscope (Nikon Eclipse 80i). Results showed that the Arthropoda community of the blanket bog was represented by three subphyla (Chelicerata, Myriapoda and Hexapoda) and within each one Acari, Chilopoda and Collembola, respectively, were the dominant groups (Table 1).

Table 1. Abundance (average numbers per m^2 ± standard error, S.E.) of Arthropoda taxa at both localities. Different letters indicate significant differences between sites ($p < 0.05$)

Arthropoda Taxa	Chan do Lamoso	Chan do Eume
Chelicerata		
Aranea	138.7 ± 25.4 a	168.0 ± 47.7 a
Pseudoscorpionida	5.3 ± 3.8 a	0.0 ± 0.0 a
Opilionida	8.0 ± 4.6 a	0.0 ± 0.0 a
Acari	15,354.5 ± 1,649.5 a	131,813.1 ± 62,670.1 b
Myriapoda		
Chilopoda	2.7 ± 2.7 a	136.0 ± 37.6 b
Symphyla	0.0 ± 0.0 a	128.0 ± 44.2 b
Hexapoda		
Collembola	14,989.1 ± 1,554.2 a	37,966.1 ± 5,488.8 b
Protura	0.0 ± 0.0 a	184.0 ± 121.1 b
Insecta	565.4 ± 68.2 a	760.1 ± 161.6 a

Subphylum CHELICERATA

Their body is divided into two regions, "prosoma" (=cephalothorax) and "opisthosoma" (=abdomen). The cephalothorax is specialised in feeding and locomotion and bears the cephalic (one pair of chelicerae and one pair of pedipalps) and the thoracic appendages (four pairs of walking legs). The abdomen lacks appendages and the only visible structures are the openings of the respiratory, excretory, digestive and reproductive systems [34].

Subclass Micrura

Order ARANEAE

There are about 34,000 species of spiders distributed throughout the world [34, 63]. They are usually more than 5 mm in length and are characterised by the presence of a more or less narrow pedicel that connects the prosoma with the opisthosoma. The cephalotorax is covered by a sclerotised and convex shield overlied by sensory setae (air vibrations senses) and bearing several ocelli o simple eyes, although their vision is poor [20, 34, 58]. The ventral "epigastric furrow" of the opisthosoma separates the respiratory region from the epigyne. One to four pairs of movable telescopic organs called "spinnerets" open in the vicinity of the anus and are connected to silk glands and their shape and size are important taxonomical characters [20, 58].

Spiders are predators with extra-intestinal digestion, sucking the solubilised tissues of their preys once the released gastric enzymes have liquefied them. Because of this feeding strategy and their habitat being mainly the litter and soil surface, spiders do not usually play a very important role in soil processes and only when their densities are very high, they exert an indirect effect on microartrophod populations [63, 64]. In the case of Serra do Xistral, the numbers of spiders were low and similar between the two studied localities (Table 1).

Subclass Dromopoda

Order OPILIONIDA

Opilionida, commonly known as "harvestmen", usually range from 5 to 10 mm in length, although some edaphic species only reach 2 mm. The prosoma and opisthosoma are fused so the body is oval in shape. The external segmentation is usually apparent and there is one pair of ocelli on the top of a protuberance ("tuber oculorum") on the anterior region of the prosoma. The coxae of the four pairs of walking legs are well developed and a special type of coxal glands, the "repugnatorial glands" (producing a defensive secretion), open near the base of the second pair of walking legs [65, 66]. One pair of tracheal stigmata, the anal opening (at the end of the body) and the genital operculum (between legs IV) can be observed in the ventral side of the abdomen [65].

Harvestmen are mostly omnivorous, ingesting particulate food. They usually move on the surface and their densities are often low [63, 66]. Accordingly, only three individuals were collected at Chan do Lamoso during the investigated years (Table 1).

Subclass Dromopoda

Order PSEUDOSCORPIONIDA

Their body size usually range between 1 and 5 mm in length, although some species can reach bigger magnitudes (> 10 mm). Their external appearance is similar to scorpions but they lack post-abdomen and the terminal venomous stinging apparatus. The two body tagmas are fused, with the prosoma being covered by a large shield and bearing 0-2 pairs of ocelli and the opisthosoma made of 12 segments. The fixed digits of the chelicerae consist of a "flagellum" (a group of rigid setae) plus two "serrulas" (one internal and one external), whereas the mobile digit has an external serrula and a protuberance (spinneret or "galea") which contain the silk glands. The pedipalps, which are covered by numerous sensory setae, are also chelated and connected to poisonous glands for feeding and defence. The coxae occupy almost all the ventral side of the prosoma and the tracheal spiracles, the genital operculum and a small terminal anal shield can be observed in the opisthosoma [63, 67, 68]. Pseudoscorpions have cryptic colouration to match the background: rocks and tree barks where they hunt other small invertebrates (collembolans, mites, nematodes,...) by immobilizing them with the poison released by the pedipalps [63, 64, 68]. Their ecological role could be important as they are very active predators, but their densities are usually very low to be considered as relevant players in soil processes. Only two specimens were collected at Chan do Lamoso on one sampling occasion (Table 1).

Subclass Acari

Mites lack dorsal sclerites in the "podosoma" (region including the four pairs of walking legs), but the lateral and ventral areas are covered by anterior and posterior sclerites; as a result, the body can be divided into two regions, the "gnathosoma" and the "idiosoma" (Figure 1). The gnathosoma or "capitulum" is the anterior region and is made of three fused segments; the second one bears one pair of chelicerae, whereas the pedipalps are located on the third segment and the mouth opens between them (Figure 1).

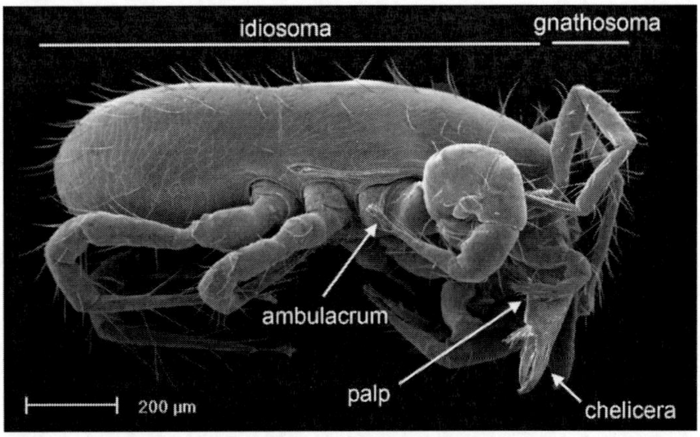

Figure 1. Acari: microphotograph (SEM) of one individual in lateral position showing its main regions and appendages.

Table 2. Abundance (average numbers per m^2 ± standard error, S.E.) of Acari taxa at both localities. Different letters indicate significant differences ($p < 0.05$) between sites per each taxa (uppercase) and between taxa at each locality (lowercase)

Acari taxa	Chan do Lamoso	Chan do Eume
Mesostigmata	1,264.2 ± 130.3 A c	7,313.2 ± 1,184.8 B bc
Prostigmata	2,117.7 ± 309.0 A b	19,363.1 ± 6,046.1 B b
Endeostigmata	112.0 ± 38.3 A e	4,896.8 ± 1,813.1 B cd
Astigmata	346.7 ± 45.5 A d	1,184.2 ± 392.5 B d
Oribatida	11,513.8 ± 1,462.5 A a	99,055.8 ± 62,522.2 B abcd

Most mites have one pair of hypertrophied setae on the "subcapitulum" (ventral region of the gnathosoma), with different shape and function depending on a particular group. The idiosoma results from the fusion of the podosoma with the opisthosoma and is usually covered by sclerotised shields whose shape and position are important for family and species identification.

The walking legs are divided into six segments: coxa, trochanter, femur, genu, tibia and tarsus. The tarsus ends in the pretarsus or "ambulacrum" (Figure 1), a valuable taxonomical character which usually consists of one pair of lateral claws and one highly modified projection between them called "empodium" [69].

In most ecological studies, mites are divided into five taxa: Mesostigmata, Prostigmata, Endeostigmata, Astigmata and Oribatida; all of them present in Serra do Xistral with oribatids dominating at both localities (Table 2).

Superorder Parasitiformes

Parasitiformes usually have the idiosoma covered by a sclerotised dorsal shield, making its separation from the gnathosoma very distinct. The modified setae of the subcapitulum are called "corniculli" and, with the exception of one order with Mediterranean and tropical distribution, they are typically associated with salivary stylets. Unlike the superorder Acariformes, this lineage of mites has free coxae [69-71].

Order MESOSTIGMATA

These mites have a variable size (200-2500 µm) and their body is usually protected both dorsal and ventrally by coloured (from orange to dark brown) shields, although those smaller species are softer and paler. The number, size and ornamentation of these shields, together with the number and arrangement of their setae are useful characters to identify sexes, life stages (larvae, nymphs or adults) and species [72-75].

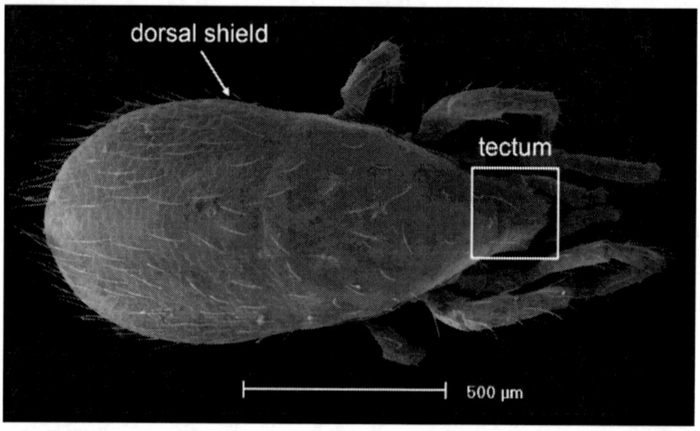

Figure 2a. Mesostigmata: microphotograph (SEM) of the dorsal view of one individual with indication of the dorsal shield and the tectum.

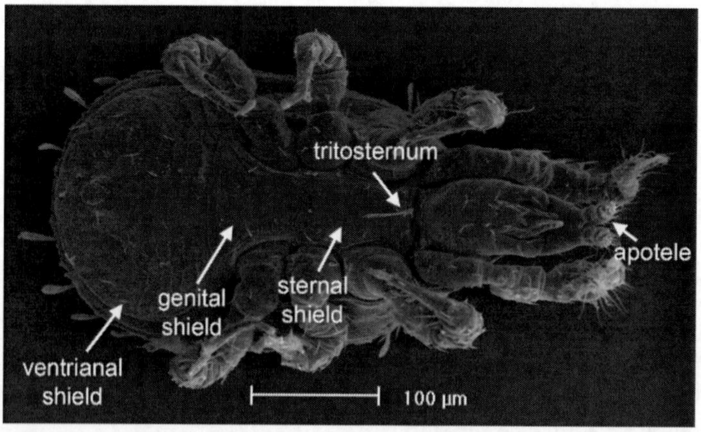

Figure 2b. Mesostigmata: microphotograph (SEM) of the ventral view of one adult female with the arrows showing the apotele, tritosternum and the sternal, genital and ventrianal shields.

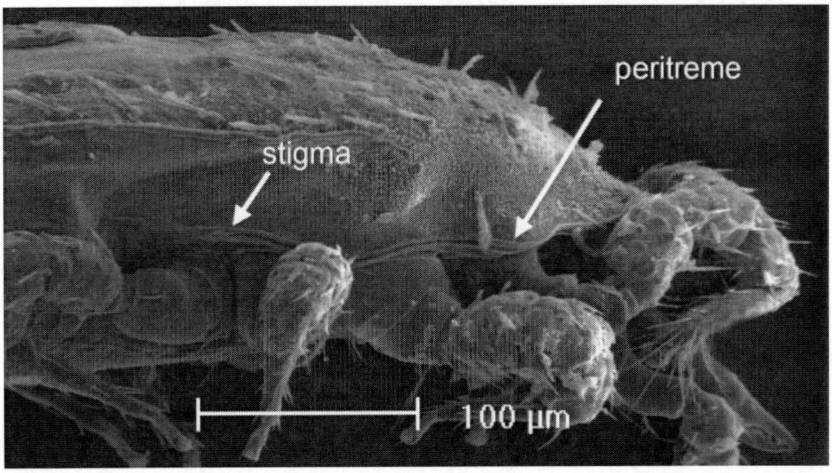

Figure 2c. Mesostigmata: microphotograph (SEM) of the lateral view of one individual with the arrows pointing to one stigma and peritreme.

Table 3. Abundances (average numbers per m^2 ± standard error, S.E.) of Mesostigmata families at both localities. Different letters indicate significant differences ($p < 0.05$) between sites per each family (uppercase) and between families at each locality (lowercase)

Mesostigmata families	Chan do Lamoso	Chan do Eume
Uropodidae	2.7 ± 2.7 A d	2,616.4 ± 526.9 B a
Parasitidae	277.4 ± 58.4 A b	1,208.2 ± 191.0 B b
Veigaiidae	88.0 ± 40.4 A bc	296.0 ± 87.4 B c
Macrochelidae	50.7 ± 15.4 A c	56.0 ± 24.9 A d
Rhodacaridae	845.5 ±112.5 A a	2,784.4 ±760.7 B a
Blattisociidae	0.0 ± 0.0 A d	352.1 ± 109.6 B c

In general, the gnathosoma is covered by the "tectum", whereas the idiosoma is protected by one or two large shields (Figure 2a). The subcapitulum (Figure 2b) bears four pairs of setae, one pair of corniculi and, in most cases, one "tritosternum" located between the insertions of coxae I (Figure 2b). The "deutosternal groove", which transports the alimentary fluid to the mouth, has a series of small teeth and its anterior margin ends in a pair of internal "malae" (membranous processes). In addition, the tined structure of the palptarsal "apotele" (a remnant of the claws at the base of the palptarsus) (Figure 2b) is an important taxonomical character [72-75]. One of the main characteristics of Mesostigmata is the presence of one pair of lateral or dorso-lateral stigmas in the coxal region of walking legs II-IV and well developed "peritremes" (chitinous grooves on the body surface connecting to stigmata openings for filtering the incoming air; Figure 2c) [69].

Recognising whether the collected individual is male or female is very important because the diagnostic characters used in most identification keys refer to adult females.

In Mesostigmata sexual determination is based on the observation of the ventral region: the female has one sternal shield adjacent to the genital one which extends from the tritosternum to the region between coxae of legs III-IV and usually bears four pairs of setae (Figures 2b,c); however, in males the sternal and the genital shields are fused. In contrast, both sexes could present the genital, ventral and anal shields separated or fused, resulting in different shield structures (ventrianal, holoventral, etc.) [72, 74, 75].

Regarding their biology, most edaphic Mesostigmata predate on nematodes, enchytraeids and small microarthropods. Importantly, there seems to be a correlation between the structure of the chelicerae and the type of food captured and thus, if these appendages are dentate, it means that they would feed mainly on nematodes, whereas short and curved chelicerae indicate that collembolans and other mites are their preferred diet [74]. Six Mesostigmata families were identified in Serra do Xistral: Uropodidae, Parasitidae, Veigaiidae, Macrochelidae, Rhodacaridae and Blattisociidae. All of them, except the last one, were present at both localities (Table 3).

Family UROPODIDAE Kramer 1881

The members of this family are commonly known as "turtle mites" because of their dorsal shield covering the gnathosoma almost completely and as a result, only the end of the legs is visible (Figure 3). These mites are morphologically conformed to protect their softer parts and together with the ample dorsal shield, the first pair of coxae is enlarged to hide the gnathosoma and the tritosternum; in addition, they possess a series of concavities or depressions ("pedofosseae") in which legs II-IV could be withdrawn. The two main characteristics to identify this group of mites are the lineal arrangement of the setae on the subcapitulum (triangular in the rest of the families) and the presence of three dorsal setae on tibia I [69, 72].

Uropodidae are very common in organic substrates where they prey on nematodes, insect larvae and weak organisms, although some species also feed on fungi or plant debris [75-77].

In Serra do Xistral, only one individual was found at Chan do Lamoso whereas more than 2,500 individuals per m^2 were collected at the drier locality of Chan do Eume (Table 3).

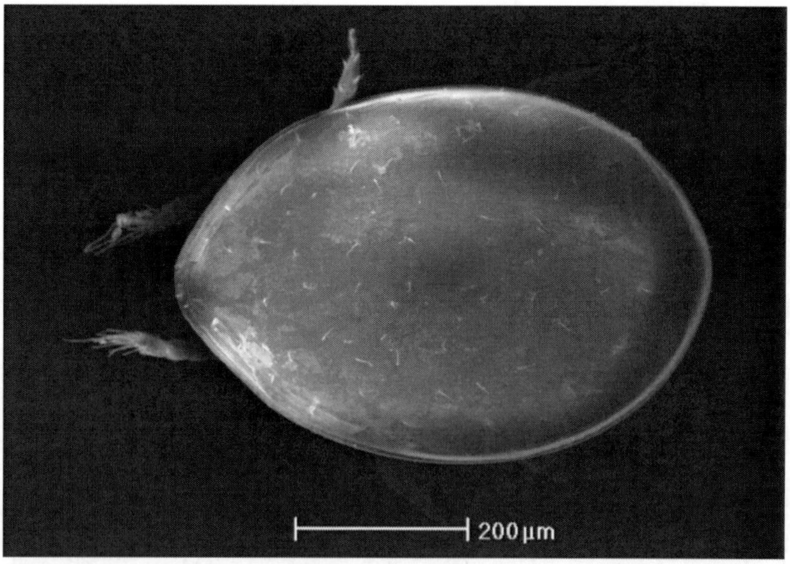

Figure 3. Uropodidae: microphotograph (SEM) of one individual in dorsal position.

Family PARASITIDAE Oudemans 1901

These mites can be large in size (length about 1 mm) and are usually well sclerotised, with one entire dorsal shield (Figure 2a) that, in the case of the females, can be divided into two [75]. Females can be recognised because their large sternal shield bears three pairs of setae and the genital shield is triangular (Figure 4) and surrounded by the former. Males can be identified by the shape of the "spermatotreme", which is the result of the distal fusion of the "spermatodactyl" of the chelicerae with the movable digit involved in sperm transference and by the usual presence of atrophied setae ("spurs") on the second pair of walking legs (Figure 1); in addition, it is common the presence of 14 setae in the tibia I: 4 ventral, 6 dorsal and 4 lateral [69, 74, 75].

Unlike the previous family, Parasitidae move fast on the soil surface and are the dominant predaceous invertebrates in temperate regions, feeding on eggs and immature stages of other microarthropods, nematodes and insect larvae living in organic debris [72, 73, 77]. These mites were well represented in Serra do Xistral, although their densities were notably higher at the most continental locality (Table 3).

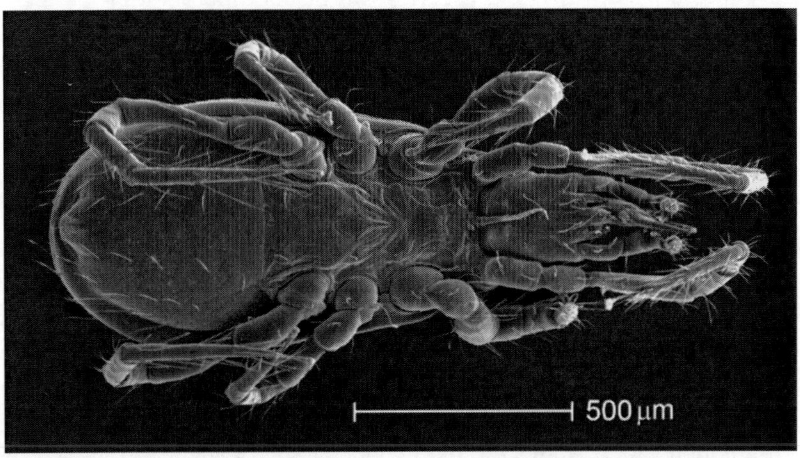

Figure 4. Parasitidae: microphotograph (SEM) of one adult female in ventral position.

Family VEIGAIIDAE Oudemans 1939

This family can be easily identified by the apotele of the pedipalp with three or four tines and a scale-like process. It is also characteristic the internal malae, which in most members of Veigaiidae are divided resembling a moustache. Regarding the sclerotisation, the female dorsal shield can show one pair of deep lateral incisions ("schizodorsal shield") or be totally divided into two smaller shields, whereas the dorsal shield of the male is always entire. The female is also characterized for having a broad sternal shield bearing three pairs of setae, whereas the fourth pair is located on two small sternal shields at each side of the genital one ("metasternal shields"). Finally, the ventral and anal shields are separated in both sexes [69, 72, 74]. Like Parasitidae, Veigaiidae are ferocious predators moving through the litter layer and on the soil surface and feeding on eggs, insect larvae and other microarthropods [76, 77]. However, they are not typical inhabitants of peatland soils and at the two Galician sampled sites their abundances were always less than 300 individuals per m^2 (Table 3).

Family MACROCHELIDAE Vitzthum 1930

Macrochelidae are usually big mites (> 1 mm) that use their first pair of legs not in locomotion but as tactile organs and thus, these appendages are more slender than the rest, lack tarsal claws and are usually raised and directed forward. The most significant character of the family is that the peritreme is deeply curved near the stigmas towards the anterior region of the body. This character and the presence of the fourth pair of sternal setae on metasternal shields are important identifying characters of this family. In addition, the genital shield has well developed lateral sclerites and only bears one pair of setae [69, 72-74]. These mites are rapid colonisers of decaying organic matter such as compost, manure, algae accumulations and nests of mammals or birds and thus, they are not often found in soils. Their diet is mainly based on small microarthropods, immature stages of insects and enchytraeids [72, 77]. This low presence in soils was also reflected in Serra do Xistral, where only about 50 individuals per m^2 were collected at each of the studied sites (Table 3).

Family RHODACARIDAE Oudemans 1902

This family includes very different species with distinct morphology and ecology: there are small (0.25-0.3 mm) and soft animals, pale in colour and inhabiting deep soil layers, but also bigger species (2-3 mm) with brownish well sclerotised shields and commonly found on the soil surface [73, 78]. The animals collected in Serra do Xistral fall into the first morphological group, as they were all small whitish mites (Figure 5a).

There is a number of characters which are found to be exclusive of this family: most specimens have a bifurcated tectum with a long median tine (Figure 5a); the adults (except one genus) have often three "scleronoduli" on the podonotal region, which are usually comma-shaped structures that refract light and appear to be hypertrophied muscle attachments. In addition, the ventrianal shield is large in most species and the female sternal shield is entire and bears four pairs of setae (Figure 5b) [72, 78].

The small size of these mites suggests that they feed on nematodes, collembolans and insect larvae living in more or less deep layers [77]. This family was fairly abundant in Serra do Xistral, with more than 850 and 2,750 individuals per m^2 at Chan do Lamoso and Chan do Eume, respectively (Table 3).

Figure 5a. Rhodacaridae: microphotograph (SEM) of the dorsal view of one individual.

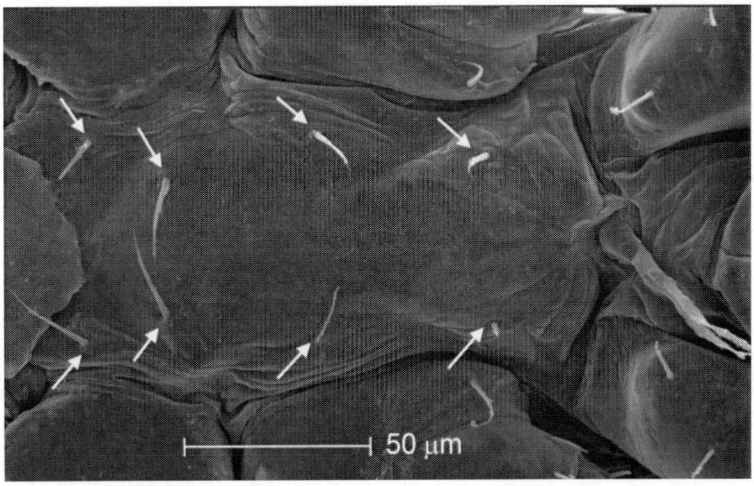

Figure 5b. Rhodacaridae: microphotograph (SEM) of the female sternal shield with indication of the four pairs of sternal setae.

Family BLATTISOCIIDAE Garman 1948

The diagnosis of this group has changed over time and thus, their members (more than 50 species) were initially grouped in one single family [79] and later assigned to different familes, such as Ascidae or Podocinidae [80-82]. In this chapter, we have considered Blattisociidae as a separate family, following the description provided by Lindquist and Moraza [83]. In general, they are intensely sclerotised, with a very ornamented surface and the dorsal setae showing different shapes: tricarinate, forked, etc. (Figures 2b,c). The majority of their members share a series of characters: the pedipalp apotele has only two tines, the tectum is trifurcate and the coxa IV bears just one seta. In addition, the female genital shield ends perpendicular (with respect to its body longitudinal axis) or is slightly convex; it bears one pair of setae and is not fused to the broad ventrianal or to the sternal shields (Figure 2b); three pairs of setae are always present in the sternal shield, being the fourth pair located on the metasternal plates, excepting one genus in which this pair of setae are inserted directly on the cuticle. The sternogenital shield of the male is usually separated but adjacent to the broad ventrianal shield and its anterior and posterior margins are also separated from the dorsal shield [83]. Blattisociidae mites live in the soil, vegetation and stored food; they prey on collembolans, other mites and nematodes, although they can also feed on fungal hyphae, perhaps as a complement to their diet [84]. Their presence in the studied peatland was restricted to the most continental bog, but with very low numbers being recorded (Table 3).

Superorder Acariformes

Acariformes usually have a pronounced sejugal furrow in the idiosoma and as a result, the body can be divided into an anterior region called "proterosoma" ("prodorsus" when referring to the dorsal surface) which includes the gnathosoma plus the "propodosoma" (the region containing legs I and II) and a posterior region called "hysterosoma" ("notogaster" when referring to the dorsal surface) containing legs III and IV. The coxae of the legs are fused to the body and the subcapitulum can bear the "rutella" (instead of corniculli) which is used for food fragmentation and crushing. Their setae are very diverse both in shape and size and most species have one or more pairs of "trichobothria" consisting of sensory setae called "sensillus", with a different morphology to the rest of setae and inserted in a sensory pit called "bothridium" [69, 85].

Order TROMBIDIFORMES

This order includes mites that feed only on fluids and because of this feeding behavior, the rutella, if present, are simple and similar to normal setae. Some Trombidiform groups have peritremes and stigmata in the anterior region of the body [86].

Suborder Prostigmata

This suborder is probable the most heterogeneous one within the Acari, both in their ecology and biology. This diversity is reflected, for example, in the species size, which can range from 100 μm to 16 mm. However, most Prostigmata share in common that they are weakly sclerotised and the fact that the peritremes and stigmata, if present, are located at the cheliceral bases or at the anterior region of the prodorsus (Figure 6a) [69, 75, 87].

Prostigmata lack rutella but their chelicerae and pedipalps are more diverse in shape and size than in the rest of mites. The idiosoma is usually divided by the sejugal furrow although this separation is not always evident.

Several structures of taxonomical importance can be seen in the prodorsum: one or two pairs of ocelli, one or two pairs of trichobothria and one "naso" (a central nose-like lobule that projects over the chelicerae) which can bear one pair of setae (Figure 6b). Regarding the legs, the tarsal claws and the empodium are often highly modified or even absent in some species [73, 75, 87].

The diet of the edaphic Prostigmata is diverse and it is common that different species within the same family are included in two or more trophic groups: although most of these mites predate on nematodes and small arthropods, there are also micro and macrophytophagous species feeding on fungi, bacteria, algae and plant fluids, whereas others are parasites of vertebrates or invertebrates [87].

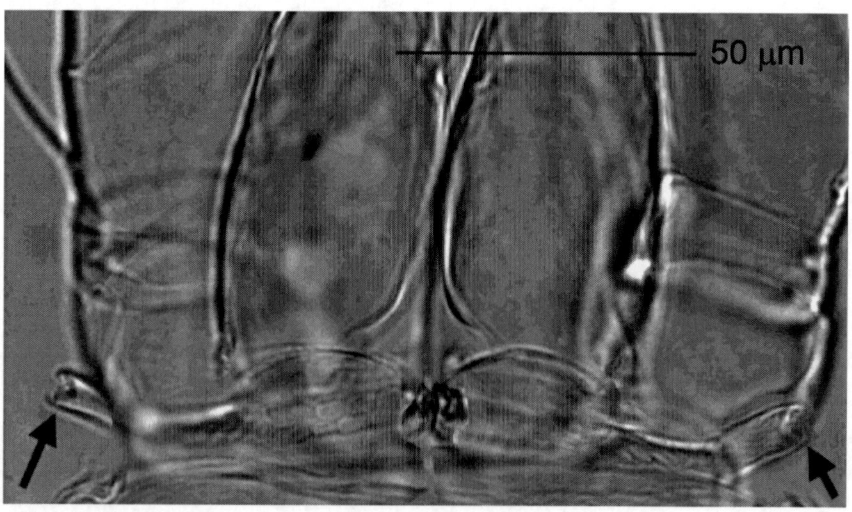

Figure 6a. Prostigmata: anterior region of the prodorsum (optical microscope) in which the stigmata of the emergent peritreme are indicated.

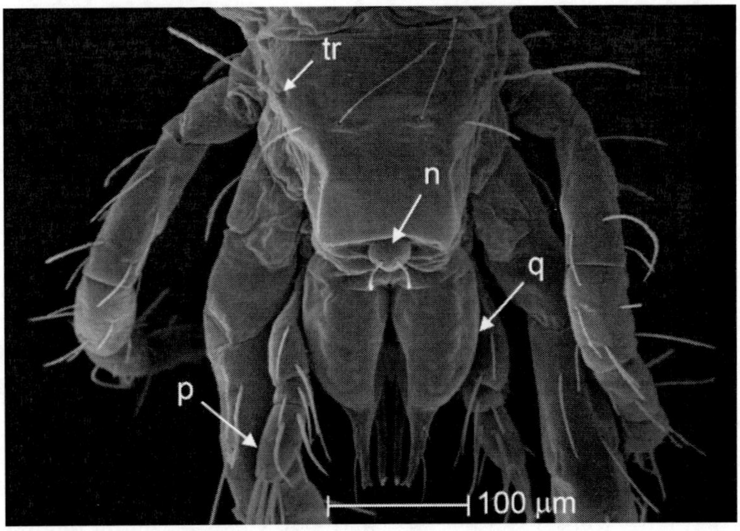

Figure 6b. Prostigmata: microphotograph (SEM) of the prodorsum of one individual with indication of one trichobothrium (tr), chelicera (q), pedipalp (p) and the naso (n).

A total of 14 prostigmatid families were recorded in Serra do Xistral: Rhagidiidae, Eupodidae, Bdellidae, Cunaxidae, Halacaridae, Ereynetidae, Tydeidae, Labidostommatidae, Stigmaeidae, Pygmephoridae, Scutacaridae, Tarsonemidae, Anystidae and Smarididae; in addition, some members of the superfamily Trombidoidea were also collected but not identified to family level. Most families were present at both localities, although populations were significantly more abundant at the driest site (Table 4).

Table 4. Abundances (average numbers per m^2 ± standard error, S.E.) of Prostigmata families at both localities. Different letters indicate significant differences ($p < 0.05$) between sites for each family (uppercase) and between families at each locality (lowercase)

Prostigmata families	Chan do Lamoso	Chan do Eume
Rhagidiidae	13.3 ± 8.8 A ef	3,464.6 ± 479.5 B a
Eupodidae	826.8 ± 158.1 A a	4,744.8 ± 1,377.6 B a
Bdellidae	314.7 ± 51.3 A b	168.0 ± 55.4 A bc
Cunaxidae	21.3 ± 9.8 A de	24.0 ± 17.6 A de
Halacaridae	2.7 ± 2.7 A f	0.0 ± 0.0 A e
Ereynetidae	8.0 ± 8.0 A f	0.0 ± 0.0 A e
Tydeidae	10.7 ± 10.7 A ef	5,504.9 ± 4,296.8 B abcde
Labidostommatidae	0.0 ± 0.0 A f	288.0 ± 77.5 B b
Stigmaeidae	50.7 ± 14.9 A cd	80.0 ± 31.2 A cd
Pygmephoridae	96.0 ± 27.8 A c	3,160.5 ± 1,021.5 B a
Scutacaridae	765.5 ± 156.0 A a	1,816.3 ± 686.6 B a
Tarsonemidae	0.0 ± 0.0 A f	64.0 ± 30.3 B cd
Anystidae	2.7 ± 2.7 A f	0.0 ± 0.0 A e
Trombidoidea	5.3 ± 3.8 A ef	32.0 ± 15.2 B d
Smarididae	0.0 ± 0.0 A f	16.0 ± 16.0 A de

Family RHAGIDIIDAE Oudemans 1922

These soft-bodied mites have pale colours (white, pink or light yellow) and a variable size depending on the species (length ranging from 350 μm to 1.8 mm) (Figure 7). Their chelicerae are extremely large but the pedipalps are simple and made of four segments. The body, which is covered with simple and long setae, has a marked sejugal furrow, lacks peritremes and the stigmata open at the cheliceral bases. These Prostigmata mites have a well developed naso and one or two pairs of trichobothria in the prodorsum (Figure 6b); on their ventral surface they bear four pairs of setae. One nearly exclusive characteristic of this family is the presence of at least two recumbent "solenidia" on the tarsus and tibia of the first pair of legs; these are widened chemoreceptive setae located near the normal setae and arranged more or less horizontally beneath the cuticle. Finally, the tarsi of the legs consist of one pair of claws and one empodium with hairy setae [69, 73, 75, 87].

Rhagidiidae are active predators living in the litter and humus layers but also in deeper horizons [76, 87]. Their densities in Serra do Xistral greatly differed between localities: 15 individuals per m^2 at the wettest site but ca. 3,500 individuals per m^2 at the most continental one and thus becoming one of the dominant prostigmatid families at this site (Table 4).

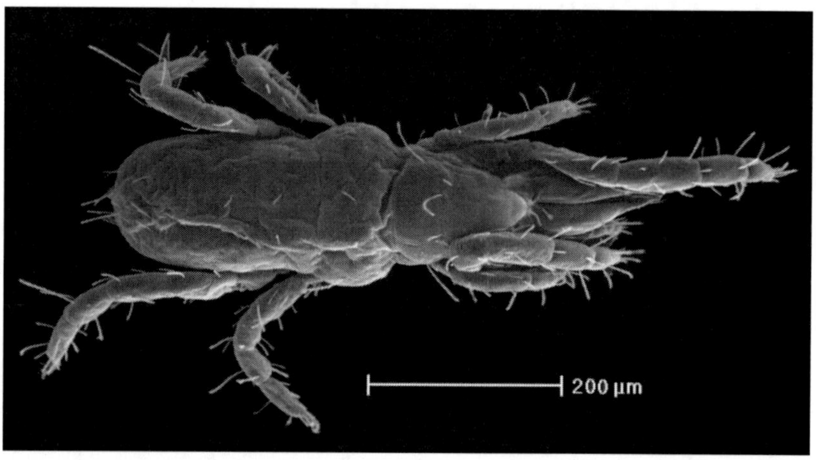

Figure 7. Rhagidiidae: microphotograph (SEM) of one individual in dorsal position.

Family EUPODIDAE Koch 1842

Eupodidae are small to medium-sized mites (150 μm-1 mm) with soft bodies, pale colours and a more or less pronounced sejugal furrow (Figure 8). This family shares several characters with Rhagidiidae: the presence of one naso with one pair of setae, one or two prodorsal trichobothria, linear pedipalps with four segments, tarsi with one pair of claws, a hairy empodium and recumbent solenidia; however, Eupodidae have a maximum number of two solenidia in the first tarsus but none in the tibiae and the subcapitulum only bears two pairs of setae instead of four. Many species of this family often have enlarged femurs in legs IV (Figure 8) which are specialised for jumping [69, 71, 87]. Eupodidae are rapid mites that hunt arthropods and nematodes or graze on algae, bacteria, fungi, etc. [87]. They were well represented at the two studied peatlands, being collected on all sampling occasions, although their abundances at the driest site were nearly five times those recorded at the most oceanic one (Table 4).

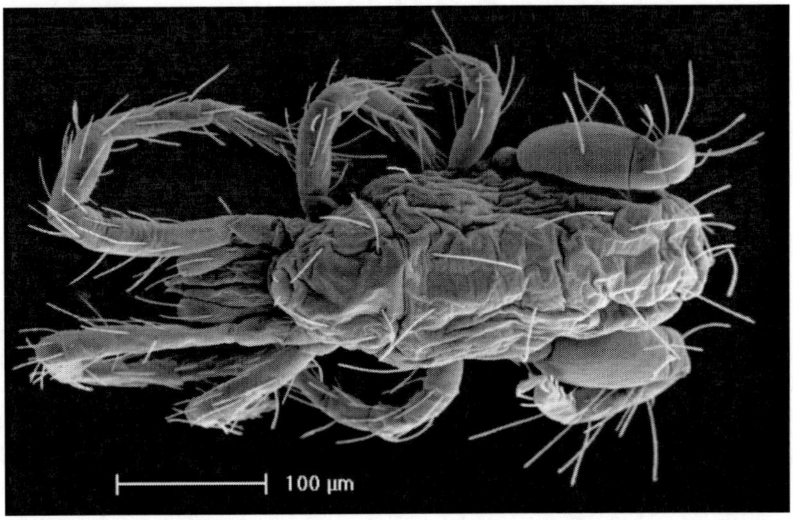

Figure 8. Eupodidae: microphotograph (SEM) of one individual in dorsal position.

Family BDELLIDAE Dugés, 1834

Bdellidae show variable size, from 450 µm to 3.5 mm in length, with soft and colourful body (red, green, brown,…) and are commonly called "nose-mites" because of their long gnathosoma. They lack peritremes and their stigmata open at the cheliceral bases; these chelicerae are separated and usually very elongated. One of the most valuable morphological characteristics for identifying the members of this family is the disposition of the genu and tibia of the pedipalp forming a right angle so that it resembles an elbow with one or two distal setae. Besides bearing one or two prodorsal trichobothria, Bdellidae also have trichobothria on their walking legs (usually on tibiae I and IV and tarsi III and IV), which is a unique character of this and the following family (Cunaxidae). One pair of lateral ocelli or one medium anterior ocellus can also be seen on their body surface. Finally, every pretarsus ends in one pair of claws and an empodium with central setules [69, 71, 87]. Bdellidae prey on other arthropods and on their eggs. Most species are able to produce silk for hunting and for protecting their eggs or themselves when moulting [71]. Despite being collected on all sampling dates, their densities in the studied area were very low (less than 500 individuals per m^2) (Table 4).

Family CUNAXIDAE Thor, 1902

The members of this family are small delicate mites (length 350-500 µm), reddish in colour, with either a smooth, striated or punctuated tegument and, in most species, with one or two dorsal shields. Their pedipalps are used for grasping and therefore, they are large, chelated and provided with spines on its inner margin and claw-like setae on the tarsus. Like Bdellidae, Cunaxidae lack peritremes and the stigmata open at the cheliceral bases; however, unlike them, each chelicera consists of only one mobile digit resembling a small hook. In addition, they have two pairs of filiform trichobothria on the prodorsum and in some cases, one or two pairs of ocelli. An additional trichobothrium is inserted on the dorsolateral side of each tibia IV and the empodium resembles a fork [69, 71, 87].

Like the majority of Prostigmata, Cunaxidae are predators which prey on other mites and insects of similar and smaller size [71]. They are common in the litter and the immediate subjacent soil layer, but also on aerial plant parts and in stored food products [88]. In contrast, peatlands do not seem to be suitable environments for the members of this family since their abundances were very low at both sampled sites, with average numbers being always less than 25 individuals per m^2 (Table 4).

Family HALACARIDAE Oudemans, 1922

These mites are closely related to Cunaxidae, but they form part of Hydracarina, an aquatic group including marine and fresh water mites and thus, they are not typical inhabitants of soils. They are big mites (usually > 500 μm) which lack peritremes and stigmata. The chelicerae have their bases separated and each one consists of a reduced (or even absent) fixed digit and a large, well developed, mobile one. The body is flattened and covered by four dorsal and four ventral shields arranged like two diamonds. The legs are laterally inserted and the first two pairs, joined to the anterior shield, are projected forward, whereas legs III and IV are directed backward; all of them have one pair of claws, usually with serrated margins and some species also bear a claw-like empodium [69, 71, 75, 76].

Despite their aquatic condition, Halacaridae are not able to swim. They usually walk to hunt their prey (crustaceans, insect larvae and other small invertebrates) or to forage on plants and algae [69]. Perhaps their aquatic life style prevented them from colonizing upland peatlands since only one individual was collected at the most oceanic and moist locality (Table 4).

Family EREYNETIDAE Oudemans, 1931

These small (length ranging from 220 to 500 μm) and soft-bodied mites have a striated tegument which varies from colourless to red or yellow. The sejugal furrow is visible and one pair of filiform trichobothria is usually located on the prodorsum, whereas the other one sits at the end of the notogaster; the observation of this second pair is a key taxonomical character for the identification of this family. The peritremes and stigmata are absent and the pedipalps are simple. Legs end in one pair of claws and one hairy empodium, but only the tibia I bear one solenidion, which is recessed into a pouch-like structure called "ereynetal organ" that opens near its distal margin [69, 71, 87].

Ereynetidae are commonly found in mosses, lichens and litter layers and, although some species are parasites of invertebrates and vertebrates, most species are predaceous using their needle-like digit of the chelicerae to hunt [69, 71]. Their densities were very low at the studied peatlands and only 3 individuals were collected at Chan do Lamoso (Table 4).

Family TYDEIDAE Kramer, 1877

These mites are morphologically very similar to Ereynetidae with regard their striated (but also punctuated or reticulated) tegument and their pale colours. Other shared features are the cheliceral bases totally or partially fused, the mobile and needle-like digit (Figure 9a), the simple linear pedipalps (Figure 9b) and the absence of peritremes and stigmata. However, Tydeidae are smaller in size (usually < 300 μm), have a well developed sejugal furrow, bear only one pair of either filiform or pyriform trichobothria on the prodorsum (Figures 9a, b) and the solenidion of tibia I is erect or totally absent [71, 87].

Some species exhibit predaceous feeding preferences by capturing small insects and other mites living in the litter, humus or soil; however, other species are fungivorous or facultative phytophagous [69, 71, 73]. This family was well represented in the studied area, but the population densities were extremely different at the two localities, with more than 5,000 individuals per m^2 (on average) at the most continental site and only 10 individuals per m^2 at the wettest locality (Table 4).

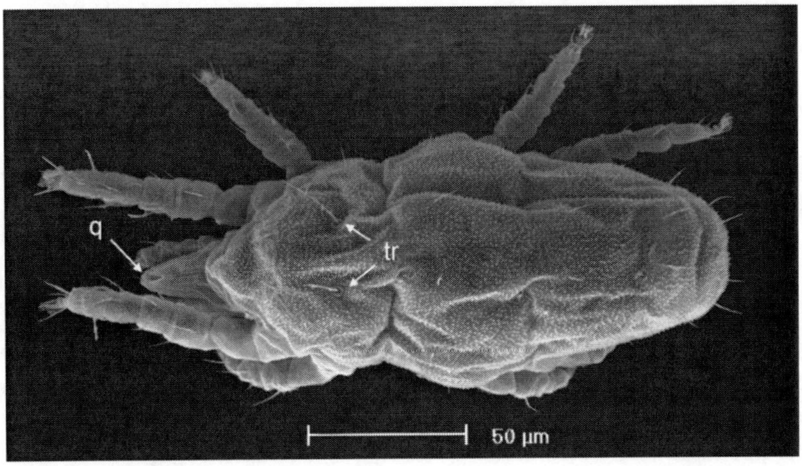

Figure 9a. Tydeidae: microphotograph (SEM) of the dorsal view of one individual showing the fused chelicerae (q) and the trichobothria (tr) with arrows.

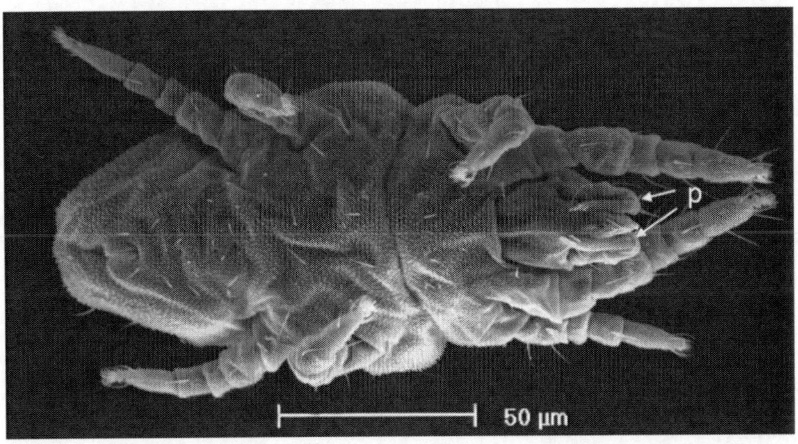

Figure 9b. Tydeidae: ventral view (SEM) indicating the position of the pedipalps (p).

Family LABIDOSTOMMATIDAE Kramer, 1877

These large mites (length 0.5 to 2 mm) are mainly characterised by their well sclerotised and vivid yellow coloured tegument together with a very ornamented dorsal shield. They have two pairs of prodorsal trichobothria, one pair of lateral ocelli and sometimes a medium anterior ocellus; some species also have one pair of lateral organs like lenses located beneath the ocelli. The pedipalps are simple and linear and the stigmata are located at the base of the elongated chelicerae, whose bases are separated. The walking legs are arranged ventrally and their coxae are inserted along rectangular and contiguous body expansions. In addition, legs I end in one pair of claws and are basically tactile, being used as antennae, whereas the rest of the walking legs have two claws and one claw-like empodium [69, 71, 73, 75].

Labidostommatidae are inhabitants of litter and humus layers where they prey on collembolans, insects and small mites [69]. At the Galician peatlands, they were collected only at the continental site and in high densities (Table 4).

Family STIGMAEIDAE Oudemans, 1931

This family includes a wide variety of morphologically different mites (in terms of shape, length, number of shields and ornamentation) but all have in common the lack of trichobothria, naso, peritremes and stigmata. The prodorsum can bear one or two ocelli and the tarsi end in one pair of claws and one empodium with "tenent hairs" (setae with either bifurcated or flattened tips). The separation between coxae II and III is very large and represents a distinct taxonomical character which results in the first two pairs of legs being directed forward and the remaining two backward. The palpal tibia has a modified large seta resembling a claw or nail and the short tarsus is, therefore, displaced to a more ventral position facing the tibial claw. Due to this arrangement, the pedipalp functions like a hand, collecting and holding the food and also cleaning the chelicerae. These appendages consist of a needle-like mobile digit to perforate and their bases are fused or tightly connected [69, 75, 87].

Most Stigmaeidae are predaceous and pierce the preys with their needle-like chelicerae, although some species are phytophagous and perforate plant tissues [75]. Their populations in Serra do Xistral were low and no significant differences were detected between the two sampled sites (Table 4).

Family PYGMEPHORIDAE Cross, 1965

The members of this family measure less than 250 μm and have a soft and flattened body covered by several sclerites (Figure 10).

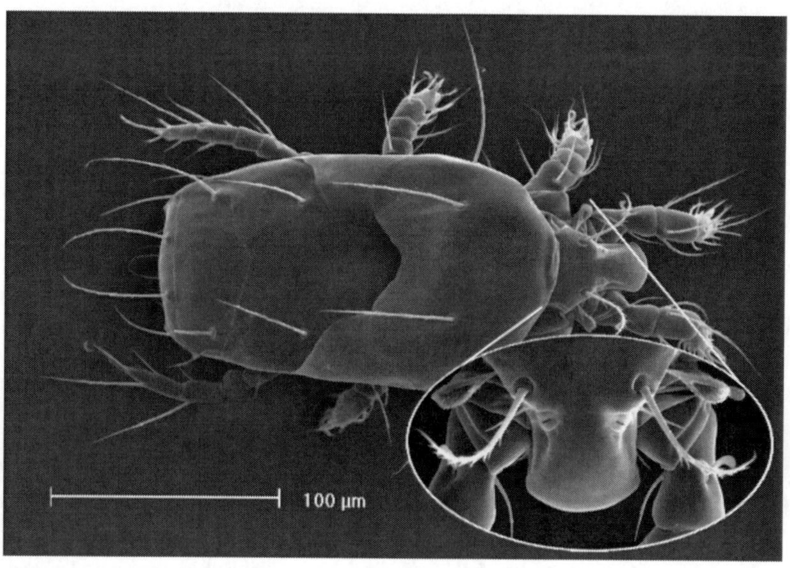

Figure 10. Pygmephoridae: microphotograph (SEM) of one female in dorsal position.

The cheliceral bases are fused forming a capsule ("stylophore") which is integrated into the subcapitulum; the mobile and needle-like digit is retractable but the fixed one is absent. The main distinctive characteristics of these mites are the presence of two or three pairs of prodorsal setae together with three to five setae on femur I (if only three, the dorsal one is not setiform) and the great separation between legs II and III [75].

The prodorsum of the females have pyriform or ampuliform trichobothria and one pair of stigmata (Figure 10), which are absent in males. In both sexes legs IV are longer than the rest, but in males the end of these long walking legs is usually modified for copulation [75].

This family includes free-living species which are predators or micro-phytofagous, others are associated with insects or nests and a great number of them are considered to be pests acting as vectors of pathogenic fungi [71, 87]. In Serra do Xistral they were more abundant at Chan do Eume (on average more than 3,000 individuals per m^2 were collected) than at the most oceanic locality (less than 100 individuals per m^2) (Table 4).

Family SCUTACARIDAE Oudemans, 1916

These small-sized (generally < 200 μm) mites are morphologically similar to those members of the previous family (Figure 11a), with their gnathosoma also resembling a capsule as a result of their chelicerae being fused and forming a stylophore in the subcapitulum. However, Scutacaridae have a rounder body, with most of the prodorsum being covered by an anterior large sclerite which extends over the gnathosoma (Figure 11a). The distance between the insertions of legs I and II is similar to that between legs III and IV (Figure 11b) and, in the case of females, legs IV are very long and end in extremely long setae [71, 75, 87].

Some species of Scutacaridae are parasites of insects, whereas those free-living ones live on mosses, litter and nests of mammals and birds where they feed mainly on fungi [69]. This family was one of the most abundant Prostigmata in Serra do Xistral with their populations being more numerous at the drier locality of Chan do Eume (Table 4).

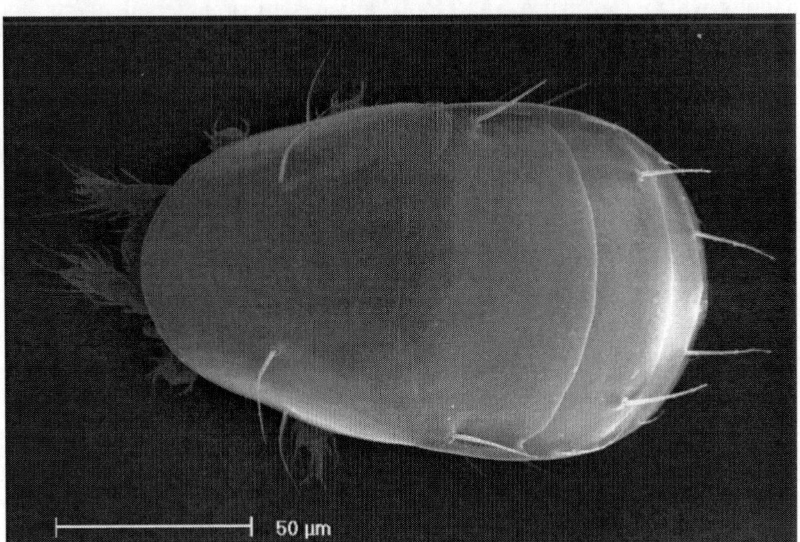

Figure 11a. Scutacaridae: microphotograph (SEM) of dorsal view of one individual.

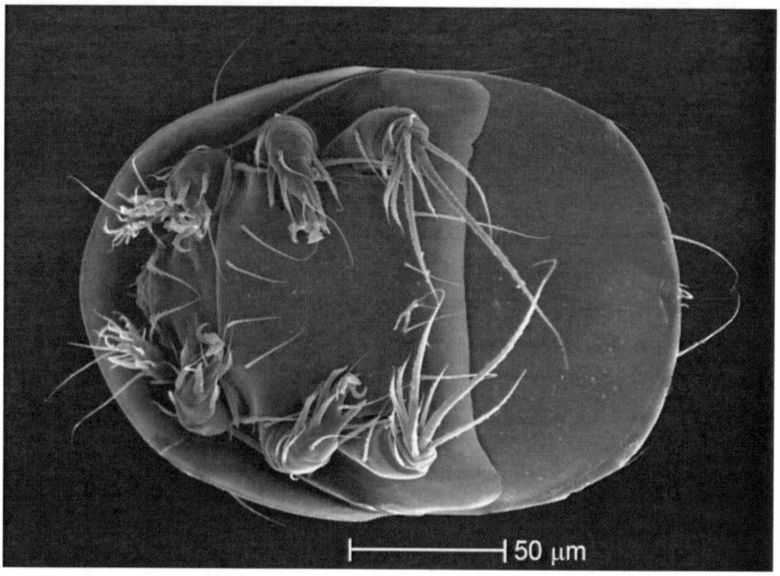

Figure 11b. Scutacaridae: microphotograph (SEM) of ventral view of one individual.

Family TARSONEMIDAE Kramer, 1877

This family shares some morphological features with Pygmephoridae and Scutacaridae: it includes small mites (usually < 250 µm) with a flattened body covered by one to three sclerites. Their needle-like chelicerae are also fused at their bases forming a stylophore. Additionally, there is a marked sexual dimorphism: the female is bigger than the male, has trichobothria and stigmata and its fourth pair of legs is reduced, with the genu and the femur being fused and ending in one or two whip-like setae; in contrast, legs IV of the male are ventrally inserted and modified (usually enlarged) for copulation [69, 71, 73]. Although some species appear to be saprophagous, fungivorous, predators or insect parasites, most species are phytophagous which can cause important damages to ornamental plants and orchards [69]. This family was not well represented in the studied area and only a few individuals were collected (Table 4).

Family ANYSTIDAE Oudemans, 1902

They are medium sized (length 550 µm -1.35 mm) and reddish mites with very long legs whose coxae are radially inserted. The pedipalps are linear with one to three spines at the inner side of the tibia and the tarsus is in ventral position. The bases of the chelicerae are separated; their mobile digit has a distal hook, whereas the fixed digit is lacking. In addition, the linear or slightly curved peritremes, which are located at the basis of the chelicerae, are emergent and the stigmata can be easily observed at both sides of the gnathosoma (Figure 6a). All Anystidae have a well developed prodorsal nose, either without or with (one pair) setae. Finally, the tarsi of the legs show a great variability: the claws could be comb-like, dentate or hairy and the empodium similar to a claw, brush, bell, etc. [69, 73, 87]. Their long and radial legs allow these mites to move very fast in all directions through the litter layer to hunt small mites and collembolans [69, 73]. This family was rare in the peatland area under investigation and only one specimen was collected at Chan do Lamoso (Table 4).

Family SMARIDIDAE Kramer, 1878

These mites are usually long (> 1 mm in length) with very thin and extremely elongated legs that end in just two claws, but without empodium. The presence of more than two solenidia and at least one trichobothrium in tibia I are important taxonomical characters for their identification [69]. One peculiar feature of Smarididae is that their evertable gnathosoma can also be withdrawn into the body, but when extended it can be as long as the idiosoma. Like the previous family, the pedipals have a well developed tibial claw, but in this family the chelicerae are elongated and needle-like. The dorsal surface of the body is slightly convex, sometimes striated due to the existence of longitudinal grooves and is covered by setae which are usually leaf-shaped with pointed hairy tips. The prodorsum of the adults bears two pairs of trichobothria and one or two lateral ocelli; some species also have "crista metopica" (a long dorsal sclerite). The dorsal shield of the larvae bears trichobothria and two pairs of normal setae [69, 73]. The feeding habits of the Smarididae larvae have only been described for one species, an insect parasite. The ecology of the adults is not well studied, but they are considered to be predators inhabiting moss and litter layers and tree barks [69, 71]. The presence of this family in the studied area was very scarce and only two specimens were collected at Chan do Eume (Table 4).

Superfamily TROMBIDOIDEA Leach, 1815

This superfamily includes large mites (> 1 mm in length), red in colour and because most species are "hypertrichous" (with an extraordinary high number of setae) they appear to have a velvet cover. The majority of these mites lack peritremes and their stigmata open at the cheliceral bases or on the anterior region of the prodorsum. The pedipalps have a well developed tibial claw which occupies a terminal position and thus, the tarsus is displaced to a more subterminal location [69].

The prodorsum bears two pairs of ocelli which are often pedunculated and a crista metopica usually bearing one or two pairs of trichobothria [71, 75]. These mites are often found in the soil, litter, tree trunks, etc., where they prey on other arthropods; however, some species also ingest pollen and plant tissues and for many others their diets are still unknown [71, 87]. The abundance of this superfamily in Serra do Xistral was low, with the number of individuals collected at Chan do Eume being significantly higher than at Chan do Lamoso (Table 4).

Order SARCOPTIFORMES

Because Sarcoptiformes mainly feed on plant material and organic debris their rutella are well developed to fragment and crush the food. Peritremes and stigmata, if present, are inconspicuous [69].

Suborder Endeostigmata

These mites are very small in size since they never reach 1 mm in length (usually between 100 and 400 μm) and pale in colour (whitish, pink, orange or brownish). They have a soft body whose shape varies depending on family (extremely globular, ovoid, etc.) (Figure 12) and the sejugal furrow is clearly visible in most cases. The cuticle is folded due to the presence of several grooves, a reflection of their primary segmentation [71, 75, 89].

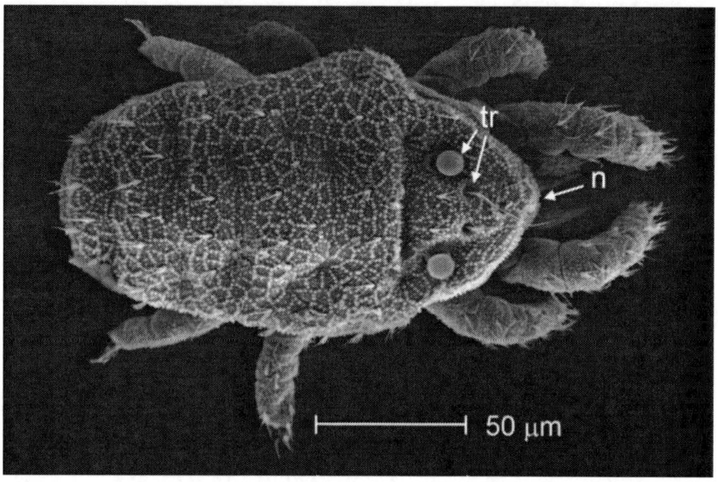

Figure 12. Endeostigmata: microphotograph (SEM) of one individual in dorsal position, with the arrows indicating the trichobothria (tr) and the naso (n).

Table 5. Abundances (average numbers per m^2 ± standard error, S.E.) of Endeostigmata families at both localities. Different letters indicate significant differences ($p < 0.05$) between sites for each family (uppercase) and between families at each locality (lowercase)

Endeostigmata families	Chan do Lamoso	Chan do Eume
Alycidae	64.0 ± 30.5 A a	24.0 ± 13.4A a
Nanorchestidae	10.7 ± 5.28 A a	1,368.2 ± 1,131.5 B ab
Alicorhagiidae	37.3 ± 23.4 A a	3,504.6 ± 1,078.3 B b

Their prodorsum can bear several structures: one or two pairs of trichobothria together with five or, more frequently, four pairs of normal setae, one naso with or without setae and 0, 1 or 3 ocelli (Figure 12). One pair of linear and simple pedipalps made of 4-6 segments and the chelicerae, with separated bases and different morphology depending on the diet, can be observed on the gnathosoma. The stigmata and peritremes are absent and the opisthosoma could be covered by 28-35 pairs of setae or be hypertrichous. The morphology of the pretarsi, which consist of one pair of claws and/or one empodium of varied shape, is very important taxonomical characters for the identification of the different families [71, 75, 89].

Endeostigmata suffer "anamorphosis", i.e. the growth occurs by adding new segments at the end of the body as they progress from larva to adult. Most Endeostigmata seem to base their diet on solid particulate food and therefore, they are considered to be either fungivorous or predators, although the feeding behaviour of many species is still unknown. The exceptions are one family which includes fluid-feeder species and another species that feeds on algae [75, 89, 90].

Three families of Endeostigmata, Alycidae, Nanorchestidae and Alicorhagiidae were recorded in Serra do Xistral (Table 5), with the first one having similar densities between sites but the other two being more abundant (about 100 times more numerous) at the driest locality.

Family ALYCIDAE Canestrini and Fanzago, 1877

The members of this family are very difficult to identify due to the great variability in their morphological characters; for example, the body shape can be globular, rectangular, extremely long, etc. Overall, Alycidae includes very small mites (length 150-400 μm), pale in colour and with striated or reticulated cuticle. In addition, the hysterosoma is often hypertrichous with setae varying in shape. The prodorsum is separated from the hysterosoma by a well developed sejugal furrow and bears two pairs of capitated and/or filiform trichobothria and 2-4 pairs of normal setae (Figure 12). In addition, some species could also have a reduced naso and/or one medium ocellus and one pair of lateral ocelli (Figure 12). The first two pairs of walking legs (I and II) are projected forward while the other two (III and IV) backward and their tarsi are made of two simple claws and one empodium with setules [69, 71, 90, 91].

These mites are commonly found in mosses and forest soils with high content of organic matter, where they feed on nematodes, copepods, tardigrades or fungi [69, 73]. However, their abundances in the investigated area were very low, with no significant differences being observed between the two localities (Table 5).

Family NANORCHESTIDAE Grandjean, 1937

Nanorchestidae are tiny mites (165-300 μm), with a striated cuticle, an elongated or, usually, globular body and white or reddish in colour; the majority are also hypertrichous with pectinate or arborescent setae (Figure 13a). The gnathosoma bears the pedipalps made of four or five segments, the rutella and the finely dentate chelicerae with widened bases. However, the most important characteristic of this region is the presence of a highly sclerotised long "labrum" (equivalent to the upper lip) which often extends over the chelicerae. Although the prodorsal naso lacks setae, there are four pairs of these structures present on the prodorsum accompanied by two pairs of trichobothria and 0, 1 or 3 ocelli (Figure 13b).

Finally, the last pair of walking legs is adapted to jump and their pretarsi lack lateral claws and consist of one empodium with its expanded basal region and claw being both covered by small setae [69, 71, 90, 91].

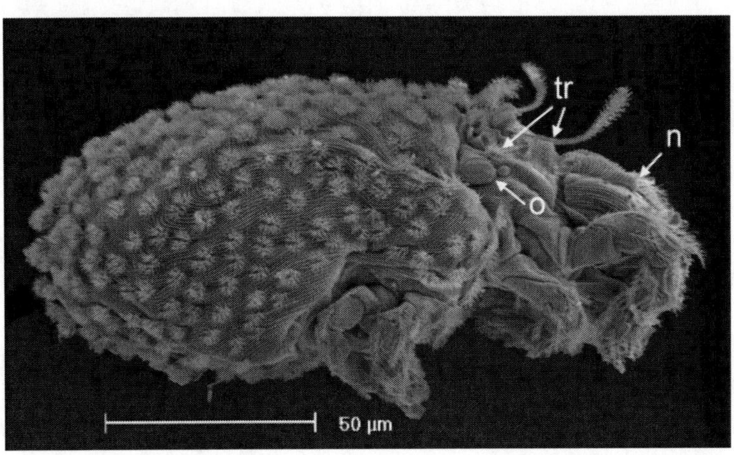

Figure 13a. Nanorchestidae: microphotograph (SEM) of the lateral view of one individual.

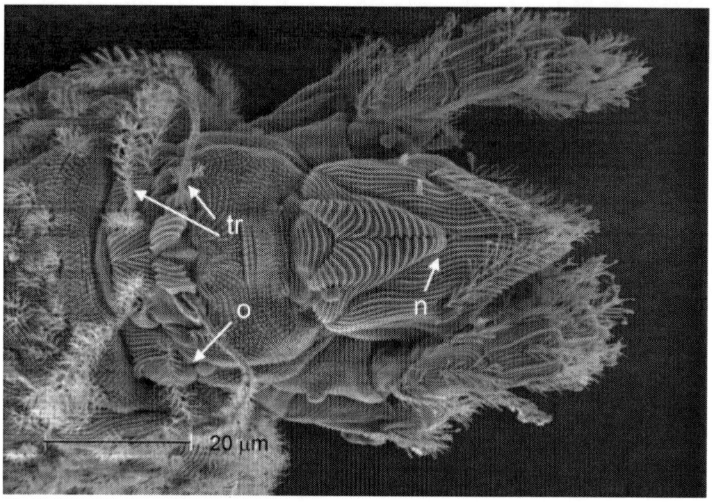

Figure 13b. Nanorchestidae: microphotograph (SEM) of detailed view of the prodorsum with the arrows showing the trichobothria (tr), one ocellus (o) and the naso (n).

These Endeostigmata mites are widely distributed (mosses, grasses, forests,…) and they can withstand severe drought conditions like those found in dry and cold deserts or in sandy beaches. Their feeding preferences cover a wide range of diets as they could be predaceous, fungivorous or phytophagous, although the species living in the proximity of the seashore appear to feed on algae [69, 76, 90].

They were not very abundant at the studied peatlands but their densities were significantly higher at Chan do Eume than at Chan do Lamoso (Table 5). However, their population numbers showed a great variability and up to 150 individuals were counted in just one sample taken from Chan do Eume, whereas almost no individuals were found in the other replicates; this finding points out to an extremely aggregated horizontal distribution exhibited by this family, although the reasons for this are still unclear because the soil moisture content was similar among all the soil samples collected on the same sampling date.

Family ALICORHAGIIDAE Grandjean, 1939

These mites measure between 200 and 300 μm in length and have pale colours (whitish to light brown) but becoming darker towards the end of the body. The overall shape is usually elongated with a visible sejugal furrow (Figure 14). Unlike the rest of Endeostigmata, whose pedipalps are made of four or five segments, in Alicorhagiidae they apparently have six because the femurs are divided. The chelicerae are dentate and each one bears one seta. The prodorsum has a well developed naso provided with one pair of setae, one pair of filiform trichobothria and four pairs of normal setae (Figure 14), but lacks ocelli. In addition, in this area the sclerotised oesophagous can be seen by transparency as one pair of sclerotised linear structures. Like in the previous groups, the pretarsi are important for the identification of this family: in this case, they lack lateral claws and the empodium consist of one simple claw [71, 87, 89, 90]. The observation of darker globules inside their guts and corresponding to remains of undigested food has provided an opportunity for revealing that they feed on soft-bodied invertebrates (mainly nematodes), fungi and algae [69, 89]. The densities of Alicorhagiidae were significantly greater at Chan do Eume than at the wettest site (Table 5).

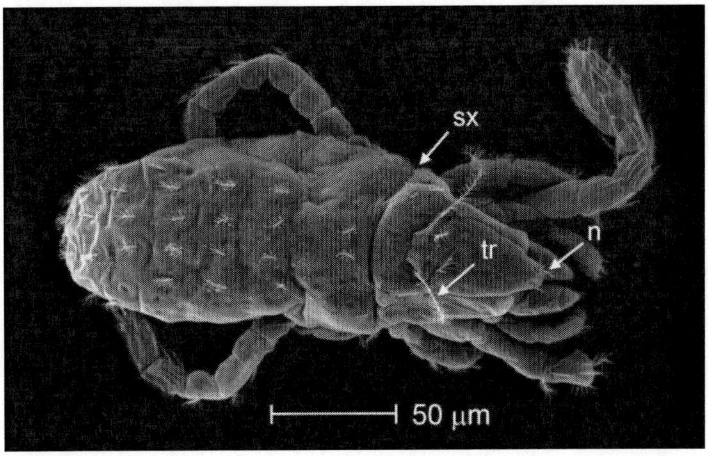

Figure 14. Alicorhagiidae: microphotograph (SEM) of one individual in dorsal position, with the arrows indicating the sejugal furrow (sx), one trichobothrium (tr) and the naso (n).

Suborder Astigmata

Astigmata are a homogeneous group which includes pale mites with slow movements and that lack both stigmata (where the name of the suborder is derived from) and peritremes. The cuticle could be smooth, striated or with some ornamentation; some species even adhere debris and dust to camouflage themselves or to gain protection against predators. The gnathosoma is relatively simple, when compared to the previous suborders: the rutella, with reduced teeth and one pair of ventral setae, cover most of the subcapitulum; the pedipalps are usually made of two segments and like the chelicerae they are reduced in size. This last pair of appendages is often chelate-dentate and functions to break down the solid particles and to scratch debris from the surface [73, 93]. The sejugal furrow which separates the prodorsum from the hysterosoma is clearly visible (Figure 15a). The prodorsum is covered by a more or less rectangular shield bearing at most four pairs of setae, which are important taxonomical characters for identification at species level. In contrast, the hysterosoma is not always sclerotised and bears a maximum number of 18 pairs of setae [92, 93].

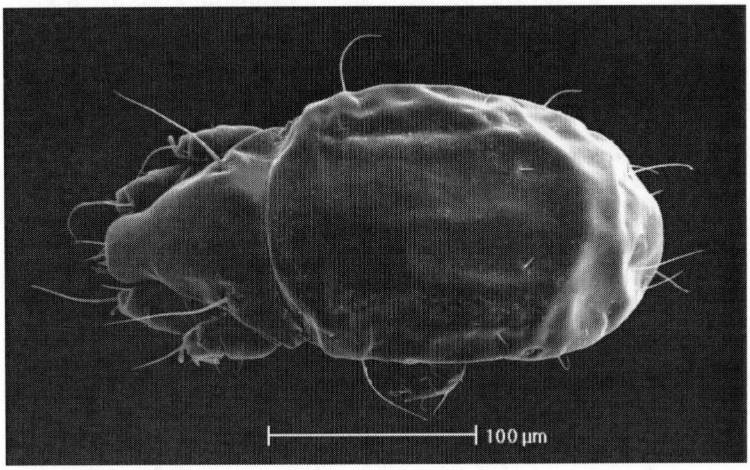

Figure 15a. Astigmata: microphotograph (SEM) of the dorsal view of one adult.

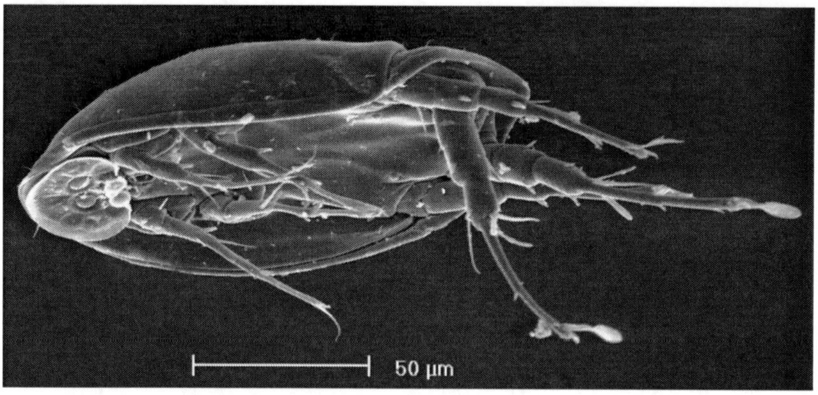

Figure 15b. Astigmata: microphotograph (SEM) of the lateral view of one hypopus.

Table 6. Abundances (average numbers per m2 ± standard error, S.E.) of Astigmata families at both localities. Different letters indicate significant differences (p < 0.05) between sites for each family (uppercase) and between families at each locality (lowercase)

Astigmata families	Chan do Lamoso	Chan do Eume
Acaridae	346.7 ± 45.5 A a	920.1 ± 324.5 A a
Histiostomatidae	0.0 ±0.0 A b	264.0 ± 101.2 B a

Depending on the species, legs can be simple and used for walking or highly modified to allow their adhesion to the host tissues. Typically the leg segments bear sensory setae (solenidia and "famulus") and the pretarsi consist of one scale-like empodium and, in some species, one simple or modified claw [73, 75, 92, 93].

The most peculiar and unique characteristic of the Astigmata is their capability (in most occasions facultative) to adopt an inactive stage of deutonymph called "hypopus", which allows them to withstand adverse conditions and to be dispersed by sticking itself to other animals (mainly other arthropods).

The hypopi are found in soils in greater proportions than the adults so that this nymphal stage is the only known for some families. The body is highly sclerotised and flattened, the gnathosoma is very reduced (the mouth and the chelicerae are absent) and the posterior ventral setae and alveoli are both modified as sucker-like structures used to attach themselves to their potential transporters (Figure 15b); these sucking structures are reduced in those hypopi that are air transported. During this stage the intestine is not functional because the hypopi do not feed [92-94].

Most Astigmata are therefore associated to vertebrates and invertebrates and as a group they should not be regarded as true edaphic taxa; the free-living species are frequently found in stored products and ephemeral habitats such as organic debris, tree holes, caves, rotten logs, etc. [73, 75, 92-95]. Only two families of Astigmata mites were recorded in Serra do Xistral: Acaridae (adults and hypopi) found at both localities and Histiostomatidae (only hypopi) present at the most continental site only (Table 6).

Family ACARIDAE Latreille, 1802

These are small mites, usually 0.5 mm in length or less; pale to light brown in colour, with a broad body and a smooth and shiny tegument overlied by simple or slightly hairy setae. The sejugal furrow is very distinct and in most species, the anterior region of the prodorsum is covered by a shield, whereas in others sclerotisation is very weak (Figure 15a). In order to identify this family, the genital opening located between coxae III and IV is determinant (Figure 16a).

The hypopi of this family can be easily recognised because the coxal fields III and IV are fused together along the middle line and their adhering organ is complete (consisting of modified setae and suckers derived from the cuticle) (Figure 16b) [69, 92, 93].

Acaridae include the greatest number of free-living species within Astigmata and are frequently found in seeds, stored food and in nests of birds and mammals.

Field records include agricultural soils, grasslands and deserts where they live in the litter layer, plant residues, tree barks, fresh or rotting fungi, etc. [69, 96-98]. They were well represented in Serra do Xistral and showed similar densities at both localities (Table 6).

Figure 16a. Acaridae: microphotograph (SEM) of the ventral view of one adult.

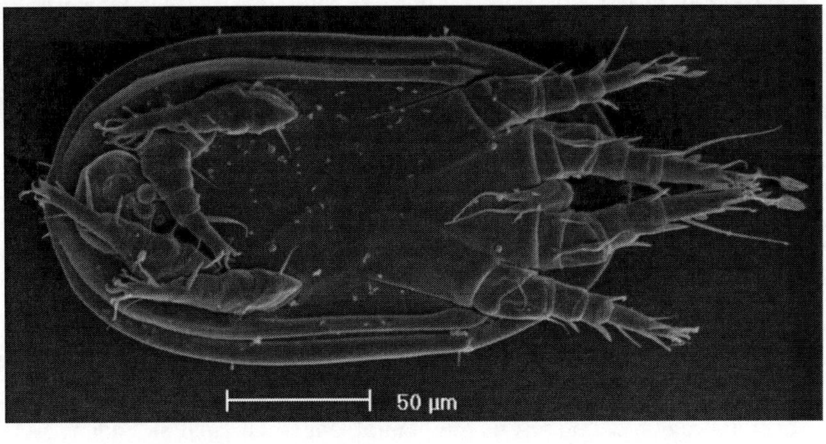

Figure 16b. Acaridae: microphotograph (SEM) of the ventral view of one hypopus.

Family HISTIOSTOMATIDAE Berlese, 1897

Most Histiostomatidae produce hypopi and, in fact, there are several species whose description is solely based on this stage because the adults are unknown. These deutonymphs can be recognised because the prodorsal vertical setae are absent, the femur and the genu of the walking legs are fused and, more easily, because legs III and IV are directed forward (Figure 15b). These mites are commonly found in places with abundant organic matter and/or high moisture such as the sap flowing around tree injuries, rotten fungi or decomposing food, because the majority graze on decomposer microorganisms. The hypopi are usually host specific and some species even establish a mutualistic relationships with their hosts, protecting them from potentially harmful microorganisms: for example, bees and wasps have developed "acarinaria", i.e. specialised anatomical structures (hollowed areas in the thorax or abdominal surface) to facilitate the retention of deutonymphs; it appears that once the hypopi have developed into other active stages they remove microorganisms and possibly other parasites from the pollen and nectar [69, 73, 92, 94]. In the investigated area, this family was only present at the driest locality and in low densities (Table 6).

Suborder Oribatida

Although these mites are now called Oribatida, they were formerly known as "Cryptostigmata" due to the difficulty in distinguishing the stigmata, partly because of their high degree of sclerotisation and which rendered other common names such as "beetle mites" or "armoured mites". To increase protection against predation, the "cero-tegument" (waxy epicuticle) is very thick in some regions, with adhered organic debris and mineral particles and in addition, some sclerotised shields could form a shelf-like projection (tectum) over several articulations [39, 69, 99]. This heavy sclerotisation together with the melanization of the exocuticle are also responsible for their external colouration, which varies from whitish or pale yellow to reddish or almost black. Their body size is also variable, with most species ranging from 300 to 700 μm in length, but others being as little as 150 μm or reaching 1.5 mm. They are slow-moving mites, with low metabolic and reproductive rates and therefore, completing their life cycles can take from several months to two years in temperate forest soils [100].

The sejugal furrow is well developed in oribatid mites and, unlike most Acariformes the hysterosoma is sclerotised not only dorsally but also laterally. Because the rostral tectum or "rostrum" usually extends over the gnathosoma it is difficult to distinguish the pedipalps and chelicerae; these latter appendages are frequently dentate and the length of the mobile digit appears to be related to the type of food ingested [101]. The prodorsum bears 4-6 pairs of setae and sometimes "costulae" (longitudinal ridge or set of ridges), "lamellae" (longitudinal projections as a flat plate or blade, attached along one edge, that protect legs I when retracted), "tutoria" (lateral ridges, ventral and more or less parallel to the lamella which protect the distal segments of legs I when they are retracted) and "pedotecta" (scale-like protuberances arising around the insertion of legs I or II that cover the insertion of the legs and sometimes form a protected space into which the legs can be withdrawn) [39, 102]. The body in Oribatida is ventrally divided into the "coxisternal" and the "anogenital" regions. The first one is composed of the fused coxae (epimera I-IV), each one bearing one pair of legs, while the anogenital region encompasses the genital, adgenital, anal and adanal sclerites [39, 69, 73, 102].

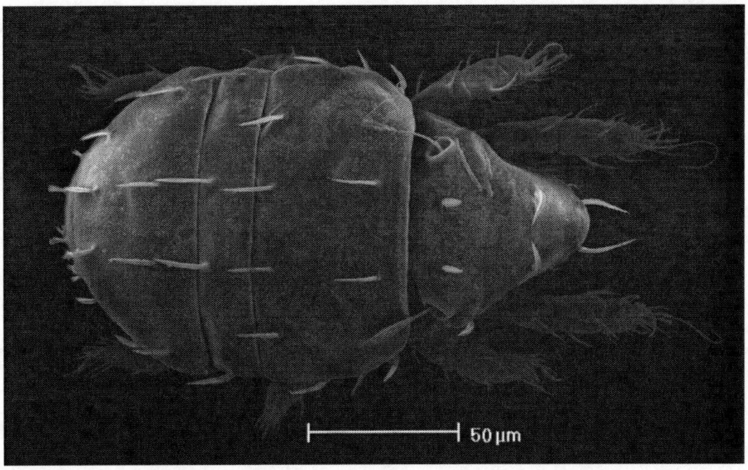

Figure 17a. Oribatida: microphotograph (SEM) of one Macropylina.

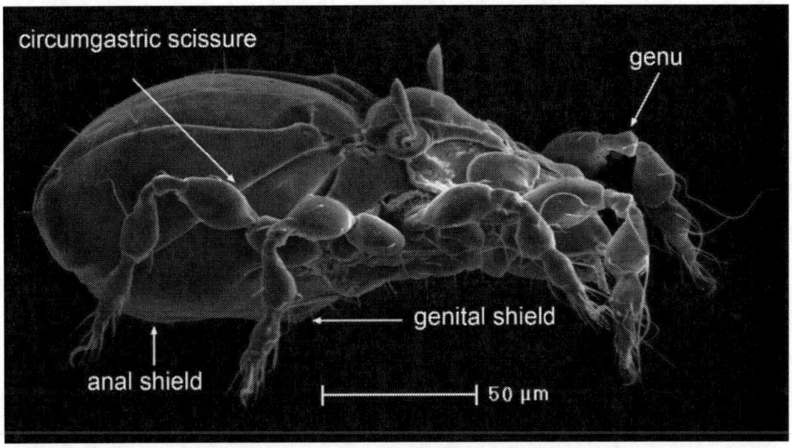

Figure 17b. Oribatida: Brachypylina showing the main diagnostic characters.

Traditionally the suborder has been divided into two large groups, Macropylina (or "lower" oribatids) and Brachypylina ("higher" oribatids), differing in their evolutionary history but also in their morphology. For example, the body shape of Macropylina (Figure 17a) can be "dichoid" (two regions; the sejugal furrow is flexible, allowing the articulation proterosoma-hysterosoma), "trichoid" (three regions; there is a third joint), "ptychoid" (a derived form from dichoid which allows some species to withdraw their legs between the two body regions) or "holoid" (the sejugal furrow is sclerotised and the only flexible articulation is the "circumgastric scissure" which joins the notogaster with the ventral shields); in addition, most species have an entire notogaster, but in others the surface is divided by several transversal fissures (Figure 17a).

In contrast, Brachypilina are characterised by a holoid body form with a distinct circumgastric scissure (Figure 17b) and their dorsal shield could extend over the posterior margin, the circumgastric scissure and one or more pairs of legs forming wing-like or shelf-like lateral extensions called "pteromorphs"; other additional structures include a set of four pairs of porose or sacculate dermal glands which produce oily substances ("octotaxic system") and an unpaired light receptive structure (often with a lens) with circular or oval shape ("lenticulus") [39, 73, 75]. Other important characteristics which help in the distinction between these two suborders are: the ventral region in Macropylina is characterised by the long adgenital sclerite being closely located to the adanal one, whereas in Brachypylina they are fused resulting into a single ventral shield which contains the more or less well separated genital and anal sclerites; in addition, the genu is similar, both in shape and size, to the rest of leg segments in Macropylina, whereas it is very reduced in Brachypylina. Finally, Macropylina appearance changes very little during development, whereas the Brachypylina undergo a drastic metamorphosis and thus, trying to identify their immature stages usually becomes a serious difficult task [39, 102]. Oribatids were represented by 22 families in Serra do Xistral, six of them (Brachychthoniidae, Phtiracaridae, Camisiidae, Malaconothridae, Nothridae and Nanhermanniidae) belonging to Macropylina and the remaining 17 to Brachypylina.

Overall, the dominant families were Nanhermannidae, Mycobatidae, Oppiidae, Camisiidae and Tectocepheidae (Table 7). Furthermore, Malaconothridae showed higher abundances at the most oceanic locality, whereas Brachychthonniidae and Quadroppidae were more abundant at the most continental site (Table 7).

Table 7. Abundances (average numbers per m^2 ± standard error, S.E.) of Oribatida families at both localities. Different letters indicate significant differences ($p < 0.05$) between sites for each family (uppercase) and between families at each locality (lowercase)

Oribatida families	Chan do Lamoso	Chan do Eume
Brachychthoniidae	21.3 ± 8.2 A fg	4,648.7 ± 1,145.3 B bc
Phtiracaridae	53.3 ± 16.0 A ef	320.1 ± 79.2 B ef
Camisiidae	2,195.0 ± 339.4 A b	2,152.3 ± 438.6 A bcd
Malaconothridae	1,965.6 ± 271.0 A b	0.0 ± 0.0 B j
Nothridae	0.0 ± 0.0 A h	600.1 ± 159.8 B ef
Nanhermanniidae	2,509.7 ± 517.5 A b	4,968.8 ± 1,012.1 B ab
Damaeidae	5.3 ± 3.8 A gh	8.0 ± 8.0 A ij
Liacaridae	194.7 ± 78.6 A de	288.0 ± 86.4 A fg
Xenillidae	0.0 ± 0.0 A h	280.0 ± 87.0 B fg
Peloppiidae	213.4 ± 40.2 A cd	16.0 ± 11.1 B hij
Tectocepheidae	445.4 ± 166.8 A c	1,440.2 ± 298.1 B de
Carabodidae	2.7 ± 2.7 A gh	16.0 ± 16.0 A hij
Oppiidae	488.1 ± 230.8 A cd	75,388.1 ± 63,046.3 B a
Quadroppidae	0.0 ± 0.0 A h	3,064.5 ± 954.7 B cde
Thyrisomidae	152.0 ± 33.4 A cd	8.0 ± 8.0 B ij
Passolozetidae	2.7 ± 2.7 A gh	0.0 ± 0.0 A j
Achipteridae	16.0 ± 7.4 A g	888.1 ± 237.0 B ef
Phenopelopidae	194.7 ± 43.9 A cd	808.1 ± 296.7 B ef
Oribatulidae	0.0 ± 0.0 A h	48.0 ± 21.2 A hi
Schlerobatidae	160.0 ± 43.9 A de	88.0 ± 43.8 A gh
Chamobatidae	0.0 ± 0.0 A h	16.0 ± 11.1 A hij
Mycobatidae	2,893.8 ± 418.5 A a	4,008.6 ± 836.5 A b

Family BRACHYCHTHONIIDAE Thor 1934

This family includes species of very small size (less than 250 µm in length) and varying in colours, from almost transparent to dark orange. Their body form is dichoid and thus, the sejugal furrow reaches the ventral region. They can easily be identified because their notogaster is divided by two transversal fissures and as a result, the hysterosoma is separated into three dorsal sclerites, with the first one bearing two pairs of dorsal setae (Figures 17a, 18). The setae on the notogaster are simple, whereas the sensilli are ciliated or barbulated (Figure 17a). Epimera II and III are very close together and epimera III and IV are usually fused (Figure 18) [39, 99, 103]. Brachychthonnidae are common inhabitants of moist soils, mosses, lichens and litter layers, where they fed on fungi, algae and bacteria [76, 104]. In Serra do Xistral their densities were low at the most oceanic peatland, but it was one of the most abundant families at the driest site, reaching ca. 5,000 individuals per m^2 (Table 7).

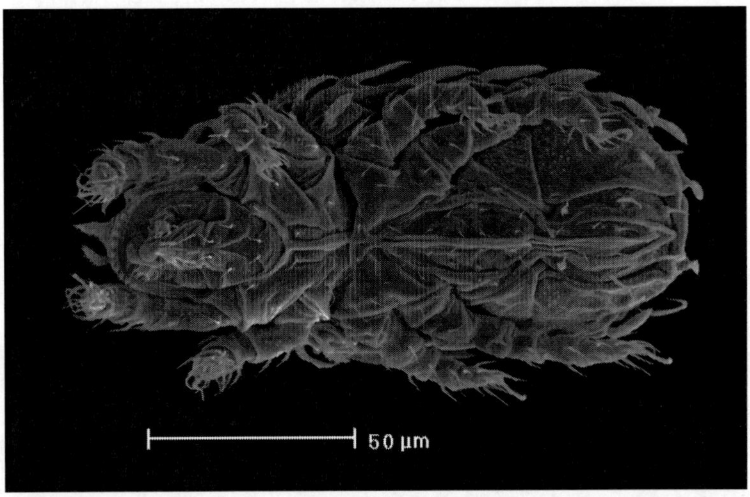

Figure 18. Brachychthoniidae: microphotograph (SEM) of one individual in ventral position.

Family PHTHIRACARIDAE Perty 1841

The main characteristic of this family is their ptychoid body form which allows the retraction of the anterior region of the legs and the prodorsum into the hysterosoma; because of this efficient defensive strategy they are usually called "box mites". Another important taxonomical feature is that the genital and anal sclerites are located in a large and U-shaped ventral shield which is about twice longer than wide. This is the result of the fusion of genital and adgenital sclerites and of the anal and adanal ones. Other important characters are the entire sclerite which covers the hysterosoma and the monodactylous (one claw) legs [39, 69, 103].

The members of this family usually ingest litter and decomposing plant material, although they can also include fungi and other microarthropod remains in their diet. The juveniles are endophages, tunnelling the wood and leaves and drastically changing the physical structure of their environment [76, 105-108]. It has been observed that the density of juveniles of Phthiracaridae in coniferous needles can reach 50,000 individuals per m^2 [109]. In the case of our investigated peatland, their population numbers were very low, with the densities registered at Chan do Eume being significantly higher ($p < 0.05$) than those recorded at Chan do Lamoso (Table 7).

Family CAMISIIDAE Oudemans 1900

These brownish mites are big in size (approximately 1 mm in length) and have a high degree of sclerotisation; in addition, the majority of them are total or partially covered with mineral particles or organic debris. The corporal form is holoid and the notogaster is flattened or concave with generally two or three pairs of longitudinal ridges (Figure 19). Sensilli are usually similar in length or shorter than the interlamellar setae and the posterior setae are often inserted in tubercula (Figure 19). The genital and anal shields are almost fused and surrounded by the adgenital (with two pairs of setae) and adanal ones (which are as wide as the anal shields). The setae on the genital shields are displaced towards its interior margin and separated by a well developed ridge from the posterior region. Epimera II and III are fused and each one bears a maximum number of two pairs of setae, although most species have one pair only. Some Camisiidae are tridactylous, but the majority of the species has only one claw on each pretarsus [39, 99, 103]. These oribatids are frequent in moist soils, living in the mosses and litter layers, but they can also occupy tree canopies and semi-aquatic habitats. Several field and laboratory studies indicate that their diet varies depending on the species but usually includes living or decaying leaves, algae, lichens, fungi, pollen, etc. [76, 107, 108, 110]. This family was well represented in the studied area, with similar densities (more than 2,000 individuals per m^2 on average) being recorded at both localities (Table 7).

Figure 19. Camisiidae: microphotograph (SEM) of one individual in dorsal position with indication of the posterior tubercula with arrows.

Family MALACONOTHRIDAE Berlese 1916

These mites are also holoid in shape but smaller in size (rarely longer than 750 µm) than the members of the previous family; their colouration is yellowish or greyish, with the dorsum slightly convex and covered by a birefringent tegument. Their main identification characteristics are the absence of bothridia and the reduction of the sensilli to normal setae (or even totally absent). In addition, the anterior region of the prodorsum becomes progressively narrower due to the presence of two lateral ridges (Figure 20) [99, 103].

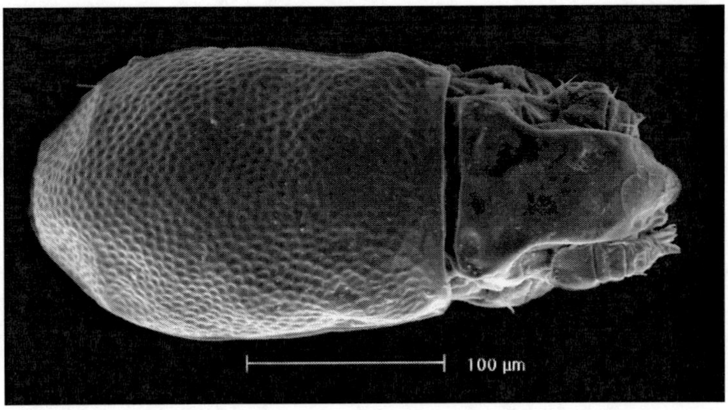

Figure 20. Malaconothridae: microphotograph (SEM) of one individual in dorsal position.

The reduced anal shields are thinner than the adanal ones and, unlike the previous family, the adgenital shields lack setae and the ridge surrounding the genital setae [69, 103].

The preferred diet of Malaconothridae appears to be mosses and litter and the fungi and bacteria living on them [76, 107,108]. This family was completely absent from the more continental site, whereas it was one of the dominant ones at the more oceanic locality (Table 7). These findings are in agreement with previous studies [111, 112], which showed that these mites are particularly abundant in certain peatland microhabitats.

Family NOTHRIDAE Berlese 1896

These holoid and big sized mites (about 1 mm) are moderately sclerotised and their colouration can range from yellow to very dark brown. The prodorsum is characterised by a small incision in the rostrum and by the morphology of their trichobothria: the bothridial bases have numerous short "brachytracheae" (elongated and porose tube-like invaginations) and the sensilli are longer than the interlamellar setae. The tegument is densely covered with alveoli giving a reticulated appearance (Figure 21a).

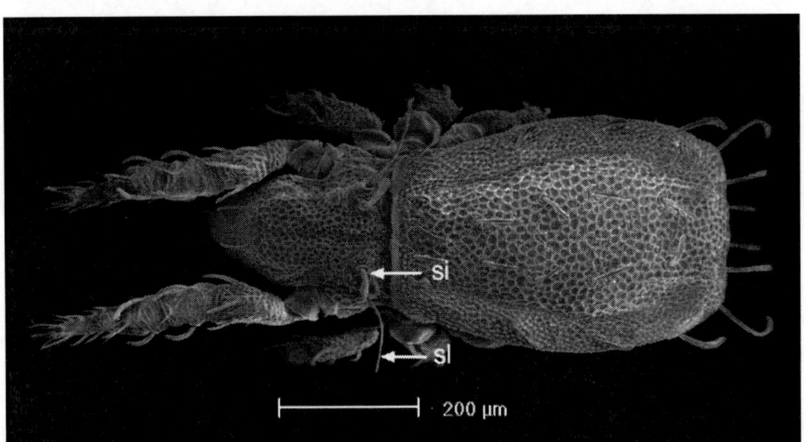

Figure 21a. Nothridae: microphotograph (SEM) of the dorsal view of one individual with the arrows showing the interlamelar seta (si) and the filiform sensillum (sl).

Figure 21b. Nothridae: microphotograph (SEM) of the ventral view of one individual.

Ventrally, epimera I and II are fused along the middle line, whereas epimera III and IV are separated (Figure 21b); in addition, each epimeron of legs II bears three or more pairs of setae. Finally, the adgenital shield lacks setae but the genital one has nine pairs, being the last pair displaced from the medial edge of the plate where the rest of setae are located [39, 99, 103]. Nothridae mites live on the soil surface and in mosses; depending on the species, they preferentially ingest fungal material, litter, nematodes, collembolans or their corpses [76, 103, 106-108]. In Serra do Xistral they were only present, in not very great densities, at Chan do Eume (Table 7).

Family NANHERMANNIIDAE Sellnik 1928

This family comprises holoid mites with an ovoid and long brown body measuring from 400 µm to 1 mm. The prodorsum typically contains several tubercula or lamellae directed backward and over the sejugal furrow (Figure 22a), which could cover the anterior region of the hysterosoma, and the inverted T-shaped area with different reticulation patterns. However, the most important taxonomical character is the absence of circumgastric scissure since the notogaster is fused with the ventral region at its posterior margin and, as a result, the hysterosoma is covered by some kind of armour and only two slits are visible in the genital and the anal regions (whose shields are widely separated; Figure 22b).

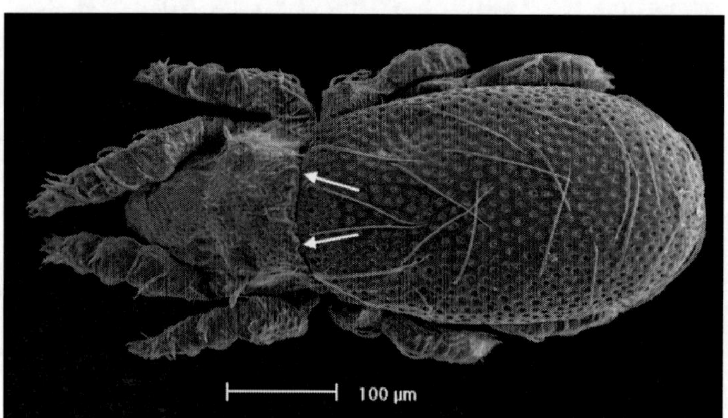

Figure 22a. Nanhermanniidae: microphotograph (SEM) of the dorsal view of one individual with the arrows pointing to the prodorsal tubercula.

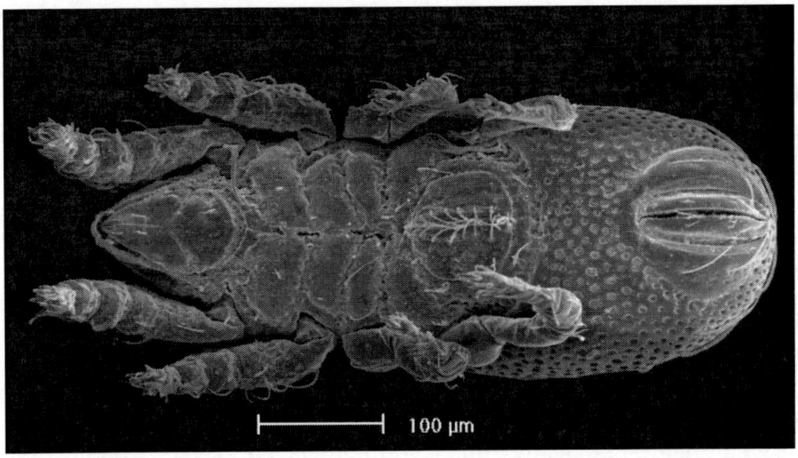

Figure 22b. Nanhermanniidae: microphotograph (SEM) of the ventral view.

Finally, all members of this family are monodactylous [39, 69, 103]. Their diet is very diverse and includes mainly bacteria, fungi, mosses and vascular plants [103, 106, 108, 110], although arthropod fragments have also been found inside the intestine of some species [107]. Nanhermanniidae were one of the most abundant families of oribatid mites at our peatland sites, but the densities recorded at the driest locality (almost 5,000 individuals per m^2) were twice those found at the wettest one (about 2,500 individuals per m^2; $p < 0.05$, Table 7).

Family DAMAEIDAE Berlese 1896

This family includes dark coloured and large mites (up to 1.6 mm in length) which are easily confounded with spiders because their legs are longer than their body, the hysterosoma is almost spherical and the prodorsum is triangular and more or less elongated. The presence of well sclerotised areas ("apophyses"), both on the dorsal and ventral regions, together with one pair of sharply pointed projections towards the anterior margin of the notogaster are distinctive characters. The arrangement of the setae on the notogaster is also important: two longitudinal rows, each one with eight pairs of setae, run from the sejugal furrow to the posterior margin and two marginal rows, each one with three setae, are arranged at the end of the body. The legs are "moniliform", i.e. femur and tibia are wider distally and the tarsus, which bears one claw only, is wider proximally but narrower near the apotele. Another important character of the legs of Damaeidae is that the tibial solenidia are free and separated from the rest of the setae, instead of being joined to them [39, 103, 113]. The habitat of the members of this family is species specific and, although some species can inhabit caves, tree barks and logs, others are saxicolous and the majority prefer forest soils where they feed on fungi [107, 113]. Previous results have suggested that peatland soils might not be adequate habitats for Damaeidae as they have not been previously recorded in organic soils. Accordingly, their very low presence in Serra do Xistral was not surprising (Table 7).

Family LIACARIDAE Sellnick 1928

This family includes ovoid mites, varying in size (length 500 µm-1.2 mm), with a smooth and shiny surface and ranging from yellowish brown to dark brown in colour. The prodorsum bears one pair of lamellae that converge anteriorly and the anterior region of the notogaster has two pairs of humeral setae, whose presence is an important taxonomical character to distinguish this family from other similar ones. The insertions of the walking legs, in particular III and IV, are displaced towards a more ventral position than in the majority of the oribatid families. Finally, the anal (surrounded by three adanal setae) and the genital shields are widely separated [69, 103, 113, 114].

These mites can ingest living or decaying plant material, mosses or lichens. The low presence of juveniles in the litter layers contrasts with the high densities of adults, which have led to the suggestion that many species have endophages juveniles tunnelling decaying wood and leaves [39, 105, 106; 108, 110]. Liacaridae were not very abundant at our investigated peatlands and their densities were similar at both localities (Table 7).

Family XENILLIDAE Wooley and Higgins 1966

These mites are morphologically similar to the members of the previous family but can be distinguished from them by its rougher surface as a result of numerous tubercula and alveoli covering the prodorsum and the hysterosoma. Like Liacaridae, they only have one pair of prodorsal (but wide) lamellae, two pairs of humeral setae and three pairs of adanal setae. The shape of the sensilli varies depending on the species: club-shaped, fusiform, lanceolated or filiform. In addition, legs are inserted far from the lateral margin and always end in three claws [103, 113]. As mentioned earlier, most Xenillidae living in the Iberian Peninsula are typical inhabitants of semiarid soils; others live in the litter layer, but they usually climb up the above-ground vegetation (e.g. shrubs or even trees). The juveniles are considered to be endophages which live and feed inside the wood and decomposing plants. The feeding behaviour of the adults is only known for one species which is believed to ingest living and decaying plant material but unable to digest fungi [103, 110, 113]. Their presence in Serra do Xistral, although in low densities (Table 7), represents the first citation in a peatland soil.

Family PELOPPIIDAE Balogh 1943

The main characteristics of these oribatids are their elongated prodorsum (around one half to two thirds of the total length of the hysterosoma) and the rounded notogaster. Similarly to the two previous families, the walking legs are inserted ventrally, the prodorsal lamellae are convergent and there are three pairs of adanal setae. However, Peloppiidae lack humeral setae and the notogaster bears a maximum number of three pairs of them; in addition, their lamellae are narrow and elongated with a well developed cuspid [99, 103, 113]. Peloppiidae are commonly found in mosses and litter layers. Although they are considered to be mainly fungivorous, the examination of their gut contents showed the presence of pollen and fragments of vascular plants, bacteria and remains of other invertebrates [76, 107]. In the investigated area, this family was more abundant at the wettest locality (Table 7) and consisted of one solely species, *Ceratoppia bipilis* (Herman 1804), which was curiously reported as a xerophilous taxon by Behan-Pelletier and Bisset [115] in Canadian peatlands.

Family TECTOCEPHEIDAE Grandjean 1954

These long and small oribatids (< 500 μm) usually have a dark coloured tegument covered by granules. They lack a clear separation between the notogaster and the prodorsum, which is almost one half of the total length of the hysterosoma (Figure 23). There is one pair of broad lamellae which are transversally connected at the rostrum by the translamella, although this connection could be absent in other species [99, 103].

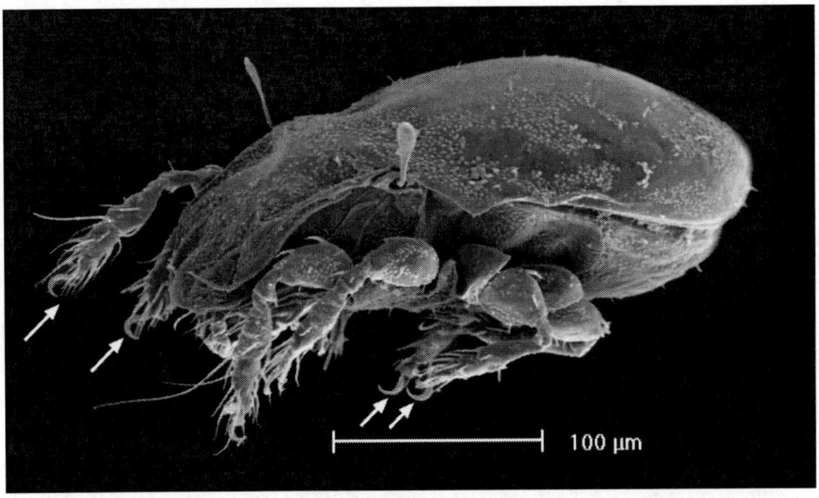

Figure 23. Tectocepheidae: microphotograph (SEM) of one individual in lateral position in which the arrows indicate the single claw on the tarsi.

The prodorsum of Tectocepheidae also bears fusiform or feathery sensilli but the interlamellar setae are reduced or absent. Finally, these mites are always monodactylous (Figure 23) [99, 103]. Plant debris seems to be the main food source of Tectocepheidae, although they can also ingest fungal material and lichens [108, 110]. They were well represented at the studied peatlands, although their population numbers at the most oceanic site were about one third (ca. 500 individuals per m^2) of those collected at Chan do Eume (ca. 1,500 individuals per m^2; $p < 0.05$, Table 7). Curiously, this family was much more abundant in the Galician bog than the English fen studied by MacFadyen [116].

Family CARABODIDAE Koch 1837

The members of this family include elongated and medium-sized mites (length 350-600 µm) with a very sclerotised and ornamented external surface (stairs, circles, reticulations, granules, etc.).

They have one pair of prodorsal lamellae which are fused with the lateral margin of the prodorsum and their apices are orientated towards the rostrum; the prodorsum also bears one pair of interlamellar setae which in this family are inserted near the rostrum and the bothridia look like funnels, with a very visible lateral opening. Ventrally, the genital shield is well separated from the anal one and bears four pairs of setae. In addition, their walking legs are relatively short, the femurs II-IV have very pronounced ventral lamina and the tarsi are functionally fused with the tibiae and end in one claw [39, 69, 103, 113]. The adults of Carabodidae are non-specialised feeders, ingesting plant and fungal material in the litter layer. However, the juveniles are endophages living in decaying wood or fungi [107, 108]. They were very rare at the two Galician peatlands (Table 7) and only few specimens were collected on two sampling occasions.

Family OPPIIDAE Grandjean 1951

Oppidae are one of the most diverse oribatid families comprising about 900 species belonging to 150 different genera based on their great range in body sizes (length less than 100 μm to nearly 500 μm) and the shape of the sensilli (pectinated, globular, filiform, etc.). Because of this great diversity, it is difficult to choose a set of morphological characters which allow an easy identification of the members of this family. However, they all have in common the absence of true lamellae (Figure 24a), although some species show costulae (without free lateral margins) whose length is always shorter than those of the prodorsum. The notogaster is usually smooth, without cuticle thickenings or adhered debris and bears 9-13 pairs of setae. The genital and anal shields are distantly located, at a longer distance than the total length of the anal shield (Figure 24b); the genital shields bear 4-6 pairs of setae and the anal ones have a lateral fissure whose position (anterior or posterior) is species specific. In addition, the number of adanal and adgenital setae is usually three and one pair, respectively.

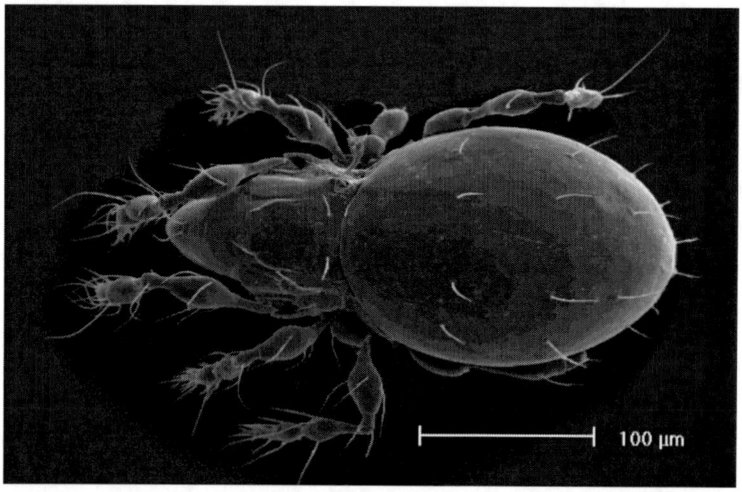

Figure 24a. Oppiidae: microphotograph (SEM) of the dorsal view of one individual.

Figure 24b. Oppiidae: microphotograph (SEM) of the ventral view of one individual.

Finally, epimera III and IV are fused [39, 103, 117]. The members of this family occupy very different habitats, from arid areas to peatland soils, forest soils, trees, etc. [117]. Isotopic analyses [108] have revealed that these mites are mainly mycophagous but they also feed on soil invertebrates (collembolans, nematodes, enchytraeids,...) and thus, they are considered to be omnivorous. Oppiidae was one of the dominant families within the oribatid community of Serra do Xistral, with the densities at Chan do Lamoso being significantly lower ($p < 0.05$) than those found at Chan do Eume (Table 7). These findings are in agreement with previous observations suggesting that this family is well represented in European peatlands [111, 112, 116].

Family QUADROPPIIDAE Balogh 1983

These oribatids are small (rarely longer than 200 μm) and their body is markedly globular. One of the most important morphological characters is a pair of longitudinal and very sclerotised ridges occupying more than one third of the surface of the notogaster; each ridge starts at the external margin of a protuberance ("humeral corniculus") located at the anterior region of the notogaster (Figure 25a). This humeral corniculus bears two kneels (one external and one internal) giving this structure the shape of an inverted "U", whose length is also an important taxonomical character. The prodorsum also has one pair of slightly convergent costulae which are longer than half of the length of the prodorsum and that are joined at their distal end by a translamellar costula [39, 103, 117].

In most species, an ovoid or elliptic-like structure is also located in front of this transversal costula. Furthermore, all members of this family have club-shaped sensilli (Figure 17b). Finally, the genital shields are similar in size to the anal ones and the separation between them is equal to or less than the total length of the shields (Figure 25b) [39, 103, 117]. The majority of Quadroppiidae living in the Iberian Peninsula prefer moist habitats with great accumulations of decomposing organic matter, although some species are able to tolerate drier areas such as agricultural soils [117]. Regarding their feeding preferences, they all seem to base their diet on fungal material [76].

Figure 25a. Quadroppiidae: microphotograph (SEM) of the dorsal view of one individual with the arrows indicating the humeral corniculi.

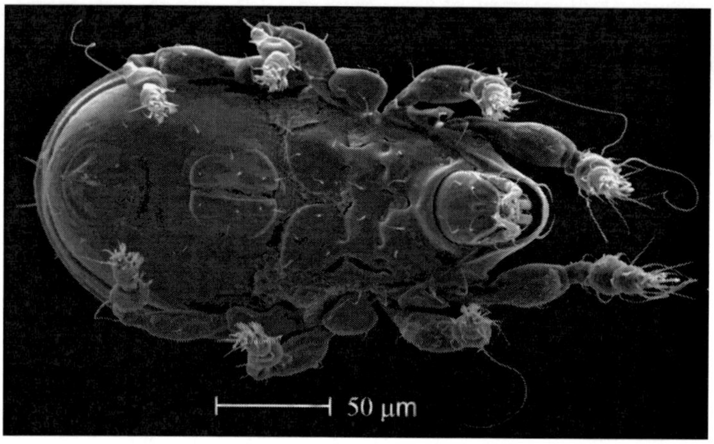

Figure 25b. Quadroppiidae: microphotograph (SEM) of ventral view of one individual.

In Serra do Xistral these mites were only collected at the most continental site and became one of the most abundant families at this locality, with densities reaching around 3,000 individuals per m^2 (Table 7).

Family THYRISOMIDAE Grandjean 1954

These brownish and medium sized mites (length 300-500 µm) have a smooth and shiny surface. The prodorsum usually bears one pair of more or less long costulae which can be distally joined; in addition, the sensilli are long and either filiform, lanceolate or fusiform. Thyrisomidae can be recognised because the genital and anal shields are typically very large and extend nearly over all ventral region; these shields are similar in size and bear 6 and 2 pairs of genital and anal setae, respectively. Epimera III and IV could appear fused and the latter always has a small tectum on its margin [39, 117, 103]. The members of this family seem to prefer moist habitats (soils, litter, mosses) with high organic matter content, although some species have been reported to be nidicolous [117]. They feed on a wide range of food sources, including fungi, plant material and even ants [76, 113]. In Serra do Xistral, Thyrisomidae were more abundant at the most oceanic site ($p < 0.05$), although their overall densities were low.

Family PASSALOZETIDAE Grandjean 1954

This family includes medium sized (250-460 µm) and more or less ovoid (but rounded rostrum) oribatids. Their body surface, both dorsally and ventrally, is covered by a thick and deeply reticulated tegument. In addition, the prodorsum of most species contains a lenticulus, lamellar setae (which are displaced forward and located close to the rostral ones) and one pair of long and filiform sensilli. Despite having porose areas, they are so reduced that their observation is very difficult. Ventrally, it can be seen four pairs of genital and two pairs of anal setae and either bidactylous (most genera) or tridactylous tarsi [102, 103]. Most Passalozetidae species are xerophilous, living in dry poor soils and agricultural systems [102, 103, 118]; however, other studies [76] indicate that some species prefer forest soils, being concentrated at the litter layer, where they feed on fungi. Our study represents the first record of the presence of this family in a peatland soil, although the fact that only one individual was collected suggests that this is not their preferable habitat (Table 7).

Family ACHIPTERIIDAE Thor 1929

These mites are large in size, measuring from 400 μm to 800 μm in length. Their tegument is very dark, black in some occasions and although it is heavily sclerotised, its surface is usually smooth and shiny (Figure 26a). They have one pair of laminar tutoria (Figure 26b) and one pair of broad prodorsal lamellae which could be separated or fused (always along less than half of their total length).

The notogaster has one pair of pteromorphs without articulation, i.e., they are not mobile and in most genera these expansions are prolonged in sharp teeth towards the anterior region exceeding the sejugal furrow (Figure 26b).

In addition, the notogaster has four pairs of porose areas and one inconspicuous lenticulus, although both structures could be totally absent. The genital shields bear six pairs of setae and the genu IV is extremely long and thin (when compared to genu III), sometimes even longer than tibia IV [39, 102, 103].

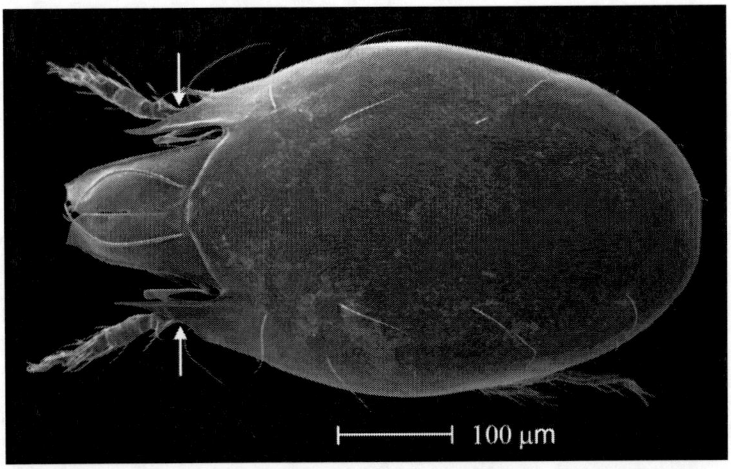

Figure 26a. Achipteriidae: microphotograph (SEM) of the dorsal view of one individual in which the arrows indicate the pteromorphs.

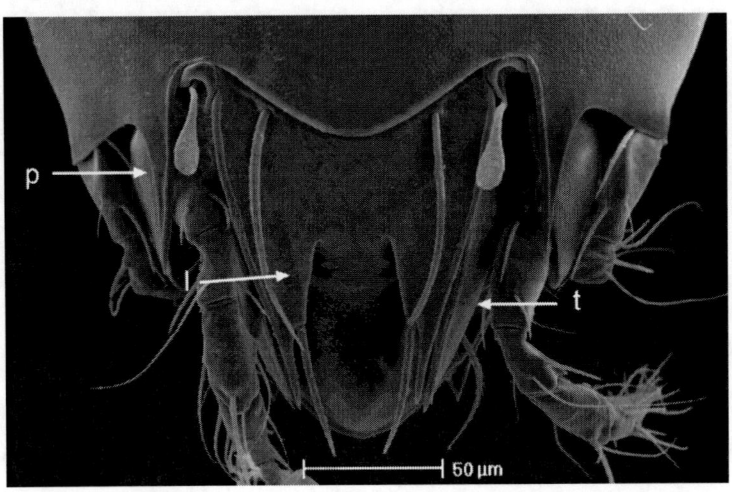

Figure 26b. Achipteriidae: microphotograph (SEM) of detailed view of the prodorsum with arrows indicating one tooth in the pteromorph (p), one lamella (l) and one tutorium (t).

Achipteriidae can be found in forest soils, pastures or grasslands, inhabiting lichens, mosses, stones, trees… where they feed on fungi, living or decaying plant material and pollen [102, 107, 108]. This family was better represented at the most continental site (densities around 900 individuals per m^2) when compared to the most oceanic one (less than 20 individuals per m^2) (Table 7).

Family PHENOPELOPIDAE Petrunkevitch 1955

These oribatids have a globular body of a notable size, with some species reaching 1 mm in length. Their body surface is covered by a thick layer of cero-tegument of varied colouration, ranging from yellow-brown to nearly black. The interlamellar setae extend over the narrow and pointed rostrum and can either be filiform and reduced or flat and broad (Figure 27a). The prodorsum has well developed tutoria at both sides (Figure 27b) and its posterior margin is covered by the notogaster and reinforced by two chitinous transversal bands; a unique character of this family is the presence of other three structures that arise perpendicular from the second band and can be observed by transparency. The anterior margin of the notogaster also contains an oval lenticulus and two large and mobile pteromorphs at either side.

In addition, the central pair of setae in the terminal region is usually longer than the rest of the dorsal ones (Figures 27a,b). Many Phenopelopidae species also have "pelopsiform chelicerae", that is, their cheliceral bases are expanded and sharply narrowed to form long and thin appendages ending in small chela. Finally, the genital shield bears six pairs of setae and the tridactylous legs have enlarged trochanters III and IV and femur IV (Figure 27b) [39, 69, 103, 119]. Phenopelopidae can be found in soils (grasslands, forest soils with moist litter, dry and arid soils, etc.) but also on trees (leaves, branches and trunks) or even on stones covered in lichen [102].

They ingest mainly fungi, but also plant material (litter, grass, dead roots), in particular its parenchyma, which they grasp using their pelopsiform chelicerae [103, 105, 107, 108, 110]. These oribatids were more abundant at Chan do Eume than at Chan do Lamoso ($p < 0.05$), although their densities were always below 1,000 individuals per m^2 (Table 7).

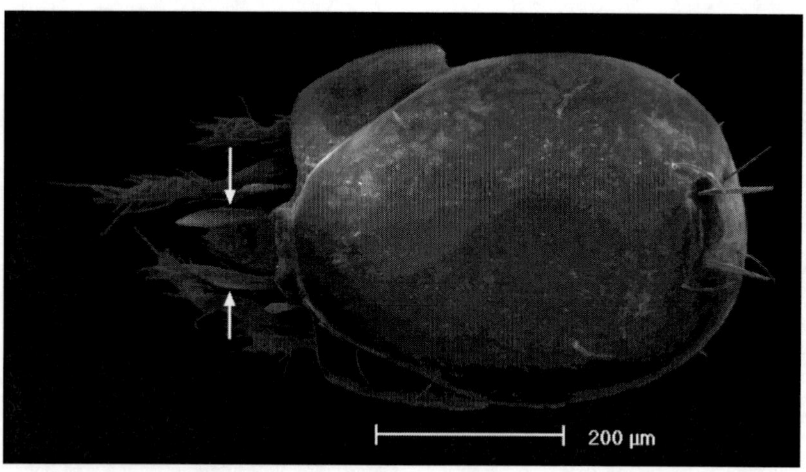

Figure 27a. Phenopelopidae: microphotograph (SEM) of the dorsal view of one individual with the arrows indicating the interlamellar setae.

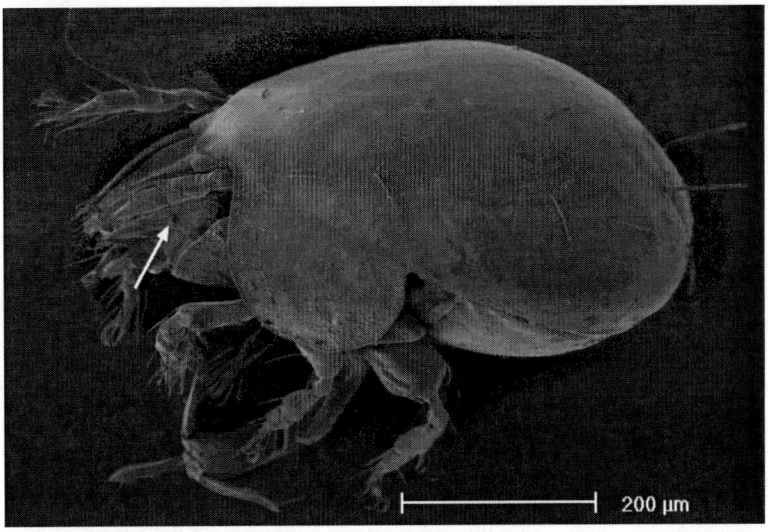

Figure 27b. Phenopelopidae: microphotograph (SEM) of the lateral view with the arrow pointing to the tutorium.

Family ORIBATULIDAE Thor 1929

Oribatulidade are highly diverse mites regarding their ecology and morphology. For example, most species measure between 350 and 550 µm in length but there are species as little as 250 µm and others can reach 900 µm in length.

All members of this family have lamellae and, although their shape varies among genera, they always lack free cuspids and are longitudinally fused with the prodorsal surface. In addition, the sensilli can be fusiform or club-shaped but never filiform and pteromorphs are absent [102, 103].

The notogaster usually has four pairs of porose areas but some species only have two pairs. Most species have ventrally four setae on each of the genital valves and the legs are always tridactylous [102, 103]. These mites feed on fungi and litter that they collect from soils and tree canopies where they live [76, 102, 108, 110]. Despite being well represented in Swiss [112] and Canadian peatlands [115], in our Galician area only a few animals were collected at the driest site (Table 7).

Family SCHELORIBATIDAE Jacot 1935

These small to medium sized oribatids (200-650 µm) have a well sclerotized tegument ranging from yellow to brown in colour. They are morphologically similar to Oribatulidae, not only in their overall aspect but also because their lamellae lack cuspids and there are four pairs of genital setae on the ventral region.

However, Scheloribatidae have immobile and visible pteromorphs which are triangular in shape and relatively small in size. The prodorsum lacks translamella but presents "sublamellae" (under the prodorsal lamellae and joined to the lamellae at their distal end) and, in most species, also "prolamellae" (ridge extending from the anterior end of the lamellae to the rostrum margin).

The prodorsum is slightly covered by the notogaster (without porose areas but with sacculi) because the sejugal furrow is slightly arcuated. Most species have tridactylous legs, but some genera are monodactylous [102, 103].

Members of this family can be found in soils and litter layers, but they also represent one of the mite groups more frequently found on tree canopies [103, 120]. They are considered to be secondary decomposers due to their preferential feeding on fungi, but some species are predators on other small invertebrates [76, 108]. Although Scheloribatidae were present in the investigated area, their densities were low at both localities (Table 7).

Family CHAMOBATIDAE Grandjean 1954

The mites belonging to this family are small to medium sized oribatids (length 290-700 μm) with a very convex, short and broad body. In general, they are heavily sclerotised and their tegument varies in colour from light brown (more or less yellowish) to nearly black. The lamellae have very short or insconspicous cuspids whereas the tutoria are as well developed as the laminae and with its distal end being free. The notogaster bears one pair of large pteromorphs bent towards the ventral side and in most species the four pairs of porose areas can be easily observed. The setae covering the dorsal surface of the body are reduced or absent; ventrally, the genital opening is displaced forward and epimera I are approximately located at the anterior margin of these genital shields [102, 103]. These mites can be found in semi-aquatic areas, on mosses, tree-barks, stones, lichens and in forest litter layers [76, 103]. Many studies indicate that they feed both on plant material (litter) and fungi [106, 108, 110]. Its presence at the Galician peatlands was scarce since only two specimens were collected at Chan do Eume (Table 7).

Family MYCOBATIDAE Grandjean 1954

These small to medium sized oribatids (length 240-730 μm) typically have a very convex body, nearly spherical. The rostrum is smooth, without teeth or hollows and the lamellae have visible cuspids and are connected by a translamella (Figure 28).

In addition, Mycobatidae have tutoria with long free projections, whereas the pteromorphs are small and mobile, connected by a bridge-like structure which is an expansion of the notogaster (Figure 28).

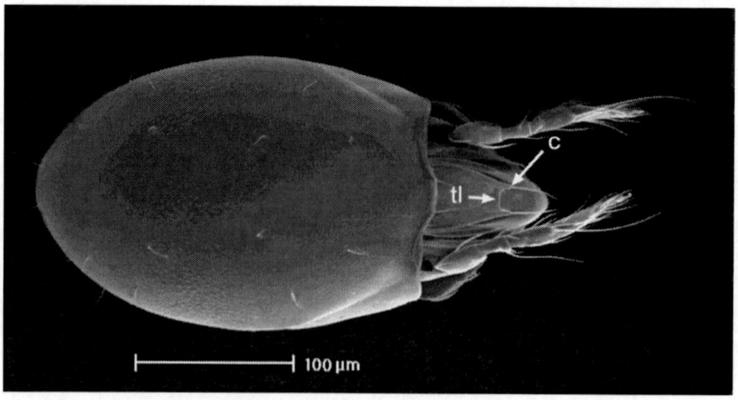

Figure 28. Mycobatidae: microphotograph (SEM) of one individual in dorsal position with the arrows indicating one lamellar cuspid (c) and the translamella (tl).

Many species also have a protuberance at the distal end of their tibiae [102, 103]. Mycobatidae colonise the litter layers of cultivated soils, inundated areas, etc., where they feed on fresh or rotting fungi, leaves, decaying plant material and remains of other arthropods [76, 102, 107, 110]. In Serra do Xistral they represented a dominant family at the most oceanic site and one of the most abundant one at the continental one, with densities higher than 2,500 individuals per m^2 at both localities (Table 7).

Subphylum MYRIAPODA

Myriapods are terrestrial arthropods with numerous legs whose body is divided into two regions: head and thorax. The head bears three or four pairs of appendages: one pair of antennae with sensorial function, one pair of mandibles, one or two pairs of maxillae and depending on the groups, a variable number of ocelli. The thorax is elongated and consists of a series of similar segments, each one usually bearing one pair of walking legs [20, 34]. The cuticle is often very permeable and therefore, these invertebrates are more commonly found in moist habitats than in drier areas. Their ecological role in the decomposition of soil organic matter is not very important because their densities are typically low in most soils [20, 34]. All Myriapoda collected from Serra do Xistral belonged to two classes, Chilopoda (centipedes) and Symphyla, with the later only present at the most continental site (Table 1).

Class CHILOPODA

This class includes very active myriapods with a total number of legs varying between 15 and 181. Their more or less flattened body measures from 0.5 to 6 cm, although some species can reach 30 cm [63, 64, 121]. These invertebrates can be easily recognised by the modification of the first thoracic appendages forming a pincer-like structure denominated "forcipules", which can be used for capturing preys and injecting venom. The rest of the body segments are similar in size or alternate between wide and narrow ones. Each pair of legs is slightly longer than the previous pair and the last one could be two times longer than the first one and have sensorial or defensive function. The final segment bears the telson and also includes the genital openings, the "gonopodes" and the anus [63, 122]. All the specimens collected in Serra do Xistral belonged to the order Lithobiomorpha, which is characterised by having 16 segments which alternate in width. The stigmata are paired and open laterally and the tarsi of the legs consist of several joints or tarsomeres [121, 122]. Centipedes are predaceous although in some occasions they can ingest plant debris. They move rapidly to hunt their prey (such as earthworms, enchytraeids, collembolans, insects…) which they immobilise with their poison. The bigger species usually hide during the day to avoid desiccation and become active at night, whereas the smaller specimens can found protection inside soil pores, burrows or litter and thus, their movements are not so restricted by moisture changes [63, 64]. The number of Lithobiomorpha collected at the most continental site of Serra do Xistral was significantly higher ($p < 0.05$) than at Chan do Lamoso (Table 1).

Class SYMPHYLA

Symphylans are small myriapods (2-10 mm), either white in colour or without pigment and resembling small centipedes. However, they only have 12 pairs of legs and their antennae show deep constrictions between segments which give them a rosary-like appearance. In addition, the head lacks ocelli but a sensory postantennal organ is located at the base of each antenna; the tracheal stigmata also open at each side of the head [123, 124]. Their dorsal body surface is protected by 15 to 22 shields, whereas their lateral and ventral sides are covered by a layer of tiny chaetae which are probably sensorial. The first pair of legs is always smaller in size than the rest and do not present the associated sensory styli. The genital openings are located on the anterior region of the body, on the ventral side of the third segment of the thorax. The final body segment bears two sensory trichocothria and two cerci which contain spinnerets at their distal extreme and as a consequence, silk filaments are often observed at the end of their body; the function of the produced silk is still unknown [63, 123, 124]. All the individuals collected in Serra do Xistal belong to the family Scutigerellidae, which are characterized for measuring more than 4 mm and having 15 large shields with their posterior margins being rounded. The first pair of legs is shorter than the rest but longer than half the length of the second pair and the basal styli are well developed [123, 124]. Most symphylans are voracious animals with an omnivorous diet feeding on algae, bacteria, fungi, litter and dead matter. Those species that base their diet on living plant material (roots) are usually considered to be pests and can produce serious damages on crops. They prefer moist habitats such as those under litter, mosses or stones or use the fissures and tunnels created by the roots to get to the deeper layers, up to 60 cm deep [63, 123, 124]. In the case of Serra do Xistral, symphylans were only collected at Chan do Eume with average densities of 130 individuals per m^2 (Table 1).

Subphylum HEXAPODA

Most hexapods are terrestrial invertebrates with their body divided into three regions: head, thorax and abdomen. The head bears the cephalic appendages (usually a pair of antennae, mandibles, the first pair of maxillae and the fused secondary maxillae forming the "labium") and one pair of compound eyes (some families also have a variable number of ocelli). The thorax has three segments (prothorax, mesothorax and metathorax), each one bearing one pair of legs made of five segments (coxa, trochanter, femur, tibia and tarsus) and, depending on the groups, the meso- and metathorax also support the wings [20]. Hexapods can be divided into two classes based on the arrangement of the mouthparts: Entognatha and Ectognatha.

Class ENTOGNATHA

This class includes hexapods with their mouthparts retracted within the head capsule and articulated mandibles [20]. The Entognatha collected in Serra do Xistral belonged to two orders: Collembola and Protura. However, whereas the first group was well represented at the two studied peatlands, the densities of Protura were very low (Table 1).

Order COLLEMBOLA

Collembolans are small animals, usually ranging from 1 to 5 mm long, although some specimens could be as tiny as 0.5 mm or reach 15 mm in length [48, 49]. Their head bears two antennae, the mouthparts, two optional postantennal organs and the compound eyes.

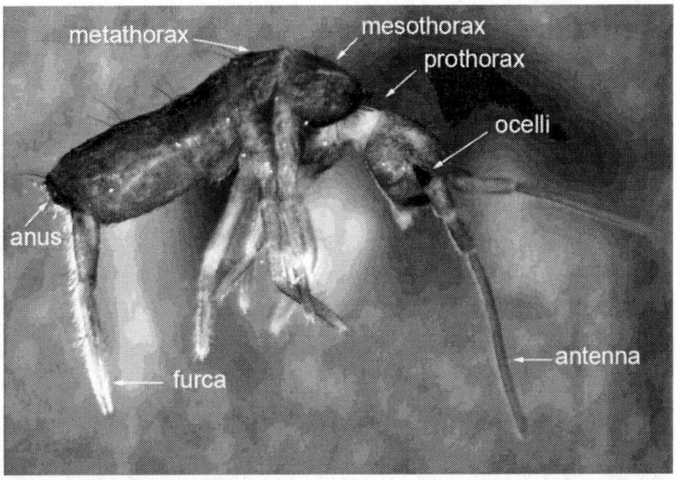

Figure 29. Collembola (stereomicroscope): lateral view of one individual with arrows indicating the main body regions and appendages.

Table 8. Abundances (average numbers per m^2 ± standard error, S.E.) of Collembola suborders at both localities. Different letters indicate significant differences ($p < 0.05$) between sites per each suborder (uppercase) and between suborders at each locality (lowercase)

Collembola suborders	Chan do Lamoso	Chan do Eume
Entomobryomorpha	8,966.8 ± 1,163.1 A a	29,764.8 ± 4,536.3 B a
Poduromorpha	4,283.4 ± 536.2 A b	7,249.2 ± 1,057.0 B b
Symphypleona	1,738.9 ± 226.6 A c	952.2 ± 191.0 A c

The abdomen is divided into five segments and a terminal periproct which, depending on the group, may be indistinct. The first abdominal segment has a unique structure within the Hexapoda, the “collophore”, a ventral tube used for fluid balance and for adhering to slippery surfaces; the genital opening is located on the fifth segment and the anus opens on the sixth (Figure 29). Their most characteristic feature is the “furca” (Figure 29) on the fourth abdominal segment; it is a jumping organ which enables them to be propelled several times their body length when escaping from predators. The furca is folded under the abdomen and clasped by another collembolan unique structure, a bifurcated organ called “retinaculum” on the third abdominal segment [31, 51, 125].

Collembolans are very selective feeders [64], having a strong preference for hyphae and fungal spores [126-128]. However, compound specific isotopic studies have revealed that some species prefer nematodes over fungi [129]. Three main suborders of Collembola are usually considered: Entomobryomorpha, Poduromorpha and Symphypleona. Members of the Entomobryomorpha are usually bigger in size, with a more or less elongated body shape and comprising highly active surface-dwelling species. Poduromorpha have a more oval shape and short legs, whereas Symphypleona have a much more rounded body shape and are mostly attractively patterned, surface-living species [51].

These three taxa were very well represented at both localities in Serra do Xistral, although Entomobryomorpha was clearly the dominant group, followed by Poduromorpha

(Table 8). In contrast, Symphypleona showed the lowest densities ($p < 0.05$) and was the only group being equally abundant at both localities (Table 8).

Suborder Entomobryomorpha

This suborder contains the slimmest species, usually with long antennae and more than three abdominal segments can be easily distinguished in its abdomen (Figures 29, 30).

The prothorax is poorly developed and the dorsal shield is difficult to recognise; in addition, the dorsal setae can be present or absent [49, 50]. They were the most abundant collembolans in Serra do Xistral, in particular at the drier site where their population numbers were nearly three times those found at Chan do Lamoso ($p < 0.05$; Table 8).

Figure 30. Entomobryomorpha: microphotograph (SEM) of one individual in dorsal position.

Suborder Poduromorpha

A minimum number of four out of the total six segments can be distinguished in the abdomen of these collembolans (Figure 31). In addition, the first thoracic segment is well developed and bears setae all over its surface [49, 50]. Poduromorpha were fairly abundant in Serra do Xistral, although the densities recorded at the most continental site were almost double than those found at the wettest locality (Table 8).

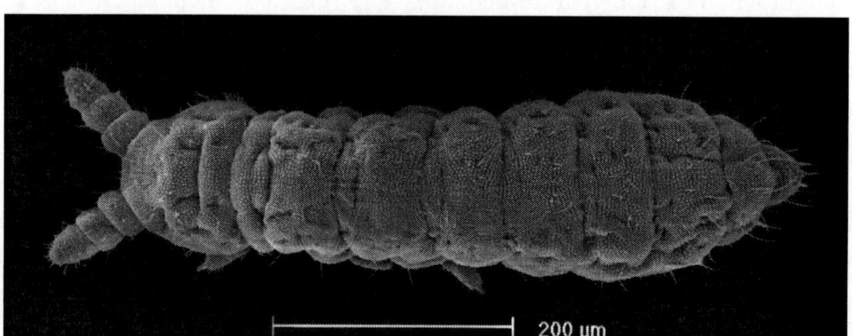

Figure 31. Poduromorpha: microphotograph (SEM) of one individual in dorsal position.

Suborder Symphypleona

These collembolans are almost spherical in shape and thus, it is very difficult to differentiate between the thoracic and abdominal segments. The suborder includes two separate lineages which are considered as separated orders/suborders by many authors: i) Symphypleona, comprising animals whose antennae are equal or slightly longer than the length of their head (Figure 32a) and ii) Neelipleona, which includes very small collembolans with short antennae (Figure 32b) without eyes [49, 50].

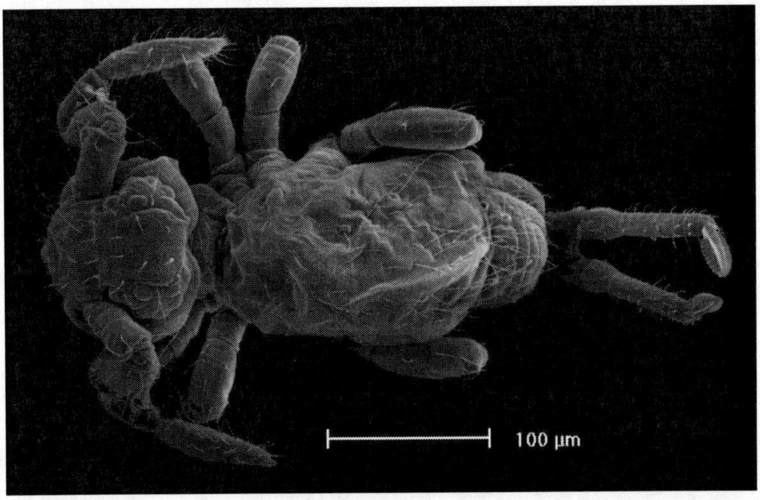

Figure 32a. Symphypleona *s.l.*: microphotograph (SEM) of the dorsal view of one individual with long antennae.

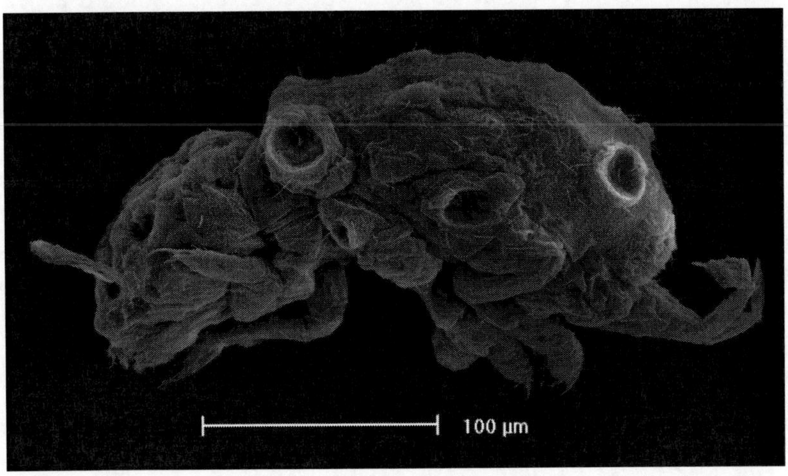

Figure 32b. Symphypleona *s.l.*: microphotograph (SEM) of the lateral view of one individual with short antennae.

However, the taxonomical status of Neelipleona remains disputed between being ranked as order/suborder or settled as family Neelidae, whereas recent molecular analyses have suggested that they could represent the most ancient lineage of springtails [130]. For this reason, in this study, we have grouped both lineages under a single group, Symphypleona *sensu lato* [53, 54]. In Serra do Xistral, they were the least represented group of collembolans at both sampled sites and, unlike the other two suborders, they were less abundant at Chan do Eume (though not significant; Table 8), which could be explained as a result of their strong sensitivity to decreasing soil moisture [131].

Order PROTURA

Proturans are one of the smallest hexapods (< 2 mm), hardly pigmented and with an elongated body covered by a slightly sclerotised cuticle and bearing short legs. They lack antennae and thus, the first pair of thoracic legs acts as sensory appendages and are directed forward. They also lack eyes and ocelli but have one pair of "pseudoculi" at each side of the head which is of taxonomical importance.

The mandibles and maxillae are elongated and their distal end is usually modified to cut, chew or suck. The first three abdominal segments bear a series of small ventral appendages, with or without terminal vesicle, which are probably involved in hydrological regulation. Finally, the "striate band" (a band of setae on the anterior margin of the eighth abdominal segment) is another important feature for species identification; in some genera the posterior margin of this segment also bears an additional group of setae resembling a "comb" [48, 63, 64, 132].

They occupy either the surface or the deeper soil layers where they preferentially feed on mycorrhizal fungi. Although their actual diet is unknown, some studies indicate that their abundances are higher in those soils with a dense root system when compared to acid soils where the radical systems are less well developed [63, 132].

Despite their numerical importance in certain habitats, they hardly play an important role in the decomposition processes due to their small contribution to the total soil fauna biomass [63]. Furthermore, although they prefer wet soils, waterlogging conditions appear to have a negative effect on their populations [63, 132]. Accordingly, in our investigated area, they were totally absent from the wettest locality whereas their densities reached 184 individuals per m^2 ($p < 0.05$) at the most continental site (Table 1).

Class ECTOGNATHA = INSECTA

This class includes hexapods whose mouthparts are not protected by the cephalic capsule [20]. The number of described species surpasses one million. Their average size is about 2-3 cm, but they also include smaller species (about 1 mm) and others reaching 20 cm of length. The internal skeleton of the head ("tentorium") offers extra protection to the cephalic internal organs and reinforces the head capsule [59].

Finally, they lack abdominal appendages, although some species have one pair of sensory cerci at the end of the body [59, 125]. Some insects are soil inhabitants during their entire life cycle, but the majority only spend their immature phases in the soil (Figure 33) [63]. All the specimens collected in Serra do Xistral belonged to six orders: Thysanoptera, Hemiptera, Psocoptera, Coleoptera, Lepidoptera and Diptera.

Overall, their densities were low and only Coleoptera and Diptera populations presented, on average, more than 100 individuals per m^2 at both sites (Table 9). Coleoptera were more abundant at the drier site (Table 9) and almost one half of their populations were in adult stage (40%), whereas 75% and 90% of Diptera were larvae collected at Chan do Lamoso and Chan do Eume, respectively.

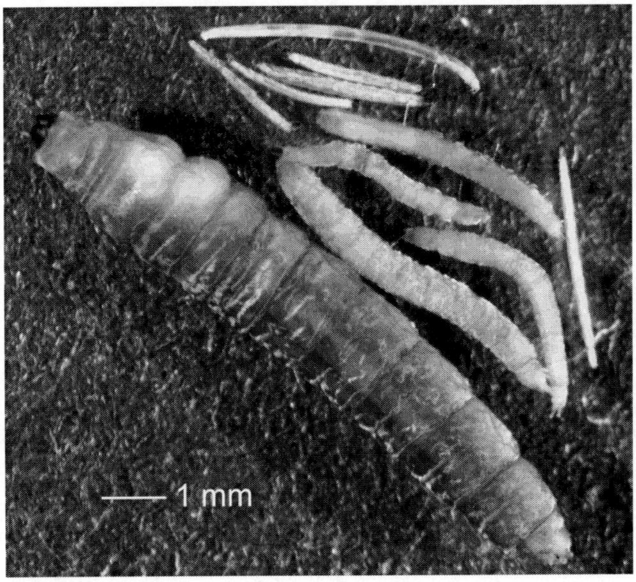

Figure 33. Diptera (stereomicroscope): larvae of different species.

Order Thysanoptera

Thysanoptera are small and elongated insects (usually less than 5 mm in length). They are considered piercing-sucking hexapods, since their mouthparts are arranged forming a cone-like structure with the left unpaired mandible and maxillae bearing stylets. Most of these insects have two pairs of narrow wings with a ciliated fringe [133]. Five families have been described in Spain [133]; however, the soil samples collected from Serra do Xistral only contained members of the family Thripidae, which are characterised by having pubescent wings, a serrated ovipositor directing downwards and antennae made of 6-9 segments with Y-shaped sense cones (trichomes) on antennal segments 3 and 4 [133]. The diet of these insects is variable, including fungivorous and phytophagous species (sometimes considered as pests since they can transfer diseases among plants) but also predators. Therefore, they can be found in litter, mosses, soils, lichens, tree barks, etc. [133]. Our results suggest that the acidic and climatic conditions at the two Galician upland peatlands were not favourable for Thripidae since their densities were very low at both sites (Table 9).

Table 9. Abundances (average numbers per m^2 ± standard error, S.E.) of Insecta orders at both localities. Different letters indicate significant differences ($p < 0.05$) between sites per each order (uppercase) and between orders at each locality (lowercase)

Insecta orders	Chan do Lamoso	Chan do Eume
Thysanoptera	16.0 ± 7.4 A c	0.0 ± 0.0 A c
Hemiptera	136.0 ± 49.2 A b	40.0 ± 16.6 A b
Psocoptera	16.0 ± 6.3 A c	0.0 ± 0.0 A c
Coleoptera	122.7 ± 22.2 A b	440.1 ± 84.6 B a
Lepidoptera	0.0 ± 0.0 A d	8.0 ± 8.0 A bc
Diptera	274.7 ± 30.0 A a	272.0 ± 109.6 A a

Order Hemiptera

The immature stages of Hemiptera are commonly found in edaphic samples and the unequivocal character that allows their identification is the arrangement of their sucking mouthparts, where the mandibles and the maxillary laciniae form a proboscis, sheathed within a modified labium to create a "rostrum" capable of piercing tissues and sucking fluids [134]. The remaining external characters are highly variable: presence of wings, number and position of the ocelli, number of antennal segments, etc. [125]. These insects are commonly found on the vegetation because most species suck plant sap, with many being species specific; however, some Hemiptera are carnivorous and others are ectoparasites of vertebrates feeding on blood and consequently, they can act as vectors of diseases [125]. At the studied peatlands, these insects were better represented at Chan do Lamoso than at Chan do Eume, although the differences were not significant (Table 9).

Order Psocoptera

Psocoptera are terrestrial insects that can be easily identified by the shape of their head, large and globular, bearing one pair of compound eyes (in some species there are also three ocelli between these eyes), one pair of very long and mobile antennae made of several segments (from 13 to 50), a very well developed "postclypeus" (area between the antennae) and chewing mouthparts. Most adults have wings and as a result, the prothorax is reduced, whereas the mesothorax is expanded. The abdomen lacks cerci and tergites 8-10 are usually hardened forming a dorsal sclerotised semi-ring. Finally, the tarsi are composed by two or three joints [59, 135, 136]. These insects are not truly soil inhabitants and can be found in the foliage, litter, tree trunks, on the surface of stones and in anthropic habitats, causing damages in stored products. Furthermore, some species are associated with vertebrates appearing on the surface of squirrels, dogs or human hair. They feed on algae, lichen, pollen, fungal spores and hyphae, dead insects, organic debris, etc. [59, 136]. Because of the low numbers collected in the studied soils (only a few individuals were recorded at Chan do Lamoso; Table 9), we can conclude that upland peatlands are not their preferential habitats.

Order COLEOPTERA

Coleoptera is an order of insects commonly called "beetles" which includes species very variable in both size and shape. Adults have the first pair of wings sclerotised forming more or less hard "elytra" which usually covers the second pair (if present) when the animals are resting.

They are not good flyers and in fact, many groups are incapable of flying. The meso- and metathorax are small in relation to the prothorax. The head bears one pair of compound eyes, ocelli (in some species) and one pair of antennae made of eleven segments. Although some species absorb liquid substances, the majority have their mouthparts modified to chew with well developed mandibules and the maxillary and labial palps moving the food towards the mouth [137].

The larvae are characterised by their well sclerotised head, ocelli and well developed mandibles because, like the adults, they are predaceous [137]. These insects were well represented at the driest site, being their densities almost four times higher than those recorded at the most oceanic site (Table 9).

Order Lepidoptera

This order of insects is typically found in the soil in its larval stage; they can be recognised by having three pairs of thoracic legs made of three segments ending each one in one claw. The abdomen consists of ten segments and usually bears four pairs of "false legs" from the third to the sixth segment. Lepidoptera larvae have their mouthparts arranged to chew which most species use to ingest leaves, fruits, shoots or seeds, causing serious damages to crops [59, 138]. Again, the organic and acid soils of our Galician upland peatlands appear not to be suitable habitats for these insects since only one specimen was collected at Chan do Eume (Table 9).

Order DIPTERA

The name of the order derives from the fact that the second pair of wings is usually reduced forming two small knobbed structures called "halters". The head bears one pair of compound eyes, ocelli and antennae. The mouthparts are adapted to suck (or for piercing) and therefore, the mandibles are narrowed and the labium is widened. Both prothorax and metathorax are small and fused to a large mesothorax which carries the functional wings. Finally, the tarsi are formed by five joints [139-141]. Most dipterans occupy intermediate habitats between terrestrial and aquatic ones during their life cycle; the immature phases are usually the ones found in the soil, in particular their legless larvae (Figure 33) with a small or reduced hardened head and the mandibles as their main noticeable structures [139-141]. These insects are frequent in organic substrates because most species are saprophagous [140, 141]. In Serra do Xistral they represented the most abundant insects and their population numbers were very similar at the two surveyed localities (Table 9).

Comparison between Sites and with Other Peatlands

Traditionally, research studies in peatland areas have focused on a single faunal group, such as collembolans [10, 142, 143] or mites [111, 112, 144], but only very few investigations have tried to compare the responses of different groups of arthropods to changes in temperature and moisture [145, 146]. Nevertheless, it is now recognized that soil communities show complex interactions with top-down and bottom-up regulations, which could be altered due to changes in the environmental factors [147]. Both localities were dominated by mites and collembolans (Table 1), in agreement with previous studies which have highlighted the importance of these arthropods in terms of abundances [1]. The remaining groups (spiders, harvestmen, pseudoscorpions, myriapods, insects) were found in very low numbers (Table 1) which could be related to the sampling and extraction methods, more adequate to edaphic (soil inhabitants) animals; if sticking traps were used, we would have probably found a greater diversity as well as higher numbers. In addition, only a few groups of these invertebrates were abundant enough to show significant differences between sites (myriapods, proturans and coleopterans) and they seem to prefer lower soil moisture than the one measured in soils from Chan do Lamoso (always above 70%) as suggested in other studies [63, 132].

Regarding the most abundant microarthropods, i.e. mites and collembolans, the majority of studies have indicated that the former animals are commonly more numerous than the latter in peatland soils, principally due to the great numbers of oribatid mites [146, 148-150]. In the Serra do Xistral, the number of Acari was three times higher than those of springtails at Chan do Eume, whereas these two invertebrate groups reached similar densities at the most oceanic locality (Table 1). These differences could be partly related to habitat heterogeneity as Cragg observed in English moorlands [151]: mites are usually the most abundant group in poor organic soils, such as ombrothrophic peatlands, whereas nutrient rich mineral soils tend to promote collembolan populations. Similarly, oribatids dominate in forest soils [152-155], whereas it is in agricultural soils and grasslands where similar densities of mites and collembolans are often recorded [156-160].

In the case of the Serra do Xistral, one of the main differences between sites was the dominant vegetation: sedges in Chan do Lamoso and grasses in Chan do Eume. In this context, some studies have suggested that there is an apparent relationship between invertebrates and plant species in poor soils such as peatlands [10, 11], in contrast to forest soils or grasslands where no association is detected [161]. Similar to our results, Mcfadayen [145] found out that the presence of certain vegetation had a greater effect on population numbers than in community composition in a *Molinia* fen, being the oribatids dominant under all plant types followed by collembolans.

Together with the differences in vegetation, environmental parameters such as soil water content can play an important role in determining the community structure. In fact, the most oceanic locality (Chan do Lamoso) had a greater moisture supply which was reflected in wetter soils (average moisture content of 83.5 and 79.5% in winter and summer, respectively) than those collected from the most continental site (average values of 74.0 and 66.7% of soil water content). It has been shown that microarthropods can respond rapidly to changes in soil water content and thus, population increases have been observed when soil water content decreases, with mites being more resistant than collembolans to the new drier conditions [131]. In accordance, around 8 times more mites and 2.5 times more collembolans were recorded at the drier locality of Chan do Eume when compared to Chan do Lamoso (Table 1).

The community composition of Acari, with Oribatids dominating at both localities and Astigmata showing the lowest values (Table 2), is similar to that observed in other peatlands [9, 146, 148-150] and leads to the conclusion that oribatid mites are the most abundant and diverse invertebrates in acid organic soils [99]. In contrast, Protigmata reach higher densities in certain ecosystems (agricultural areas, dry and cold deserts) [157, 160, 162-164] but Mesostigmata can rapidly become dominant in these areas if the abundances of their preys increase as well [165].

As indicated above, all Mesostigmata families (Uropodidae, Parasitidae, Veigaiidae, Macrochelidae, Rhodacaridae) except Blattisociidae were present at both sites but, with the exception of the scarce Macrochelidae, they were by far more abundant at the most continental and driest site (Table 3). In agreement with our results, Block collected all the above-mentioned families at the Moor House National Nature Reserve [9], with Blattisociidae being only present at one of the four sampling sites. Similarly, Parasitidae, Veigaiidae and Rhodacaridae dominated the Mesostigmata community at a permanent pasture adjacent to an Irish bog [166]. The dominance of Rhodacaridae at both Galician localities suggests that this family is not exclusive of more mineral soils, such as shortgrass steppes [167], uncultivated grasslands [156-158] or beech forests [168], but also well represented in organic soils. The great densities of Uropodidae at Chan do Eume (nearly as high as Rhodacaridae) were somewhat surprising, although it has been observed that their densities vary a lot within one sampled site [9, 146]. Interestingly, the family Zerconidae seems to be fairly abundant in moorland soils [9, 146] although not a single individual was collected at the surveyed Galician peatlands.

Most prostigmatid families recorded in the Serra do Xistral were also present at both localities, but their numbers were higher in Chan do Eume and those ones which were only collected in one site showed very low densities (Table 4). This Prostigmata community was similar to the one found in the moorland soils surveyed by Wood [146], where the greatest densities were reached by Tydeidae, Scutacaridae, Pygmephoridae, Eupodidae and Ereynetidae. Our Galician upland peatlands contained higher densities of these four groups as well, with Scutacaridae and Eupodidae being the dominant families at both sites; in addition, when comparing the two localities, Pygmephoridae and Rhagidiidae reached higher densities at Chan do Eume than at the most oceanic site, whereas Tydeidae were only present at Chan do Lamoso in very few numbers (Table 4). In contrast to the great diversity of Prostigmata recorded at our peatlands and in those sampled by Wood [146], only a few Eupodidae were collected by MacFadyen [145] in a *Molinia* fen and three species composed the Prostigmata community of the moorland site investigated by Block [9]. Curiously, one of these families was Halacaridae [9], also recorded in our study despite being mostly aquatic.

The same three Endeostigmata families collected in the Serra de Xistral were also present in a British moorland [146], although their numbers, and particularly those of Nanorchestidae, were higher than the ones collected at our Galician peatlands; in fact the densities recorded in our study are in the range of those reported in a shortgrass steppe [167]. Alycidae was, on the contrary, the only Endeostigmata recorded in the English moorland studied by Block [9]. The oceanic conditions at Chan do Lamoso appeared to have detrimental effects on Endeostigmata populations since their densities were on average less than 65 individuals per m^2 at this site; although in the case of Alycidae their numbers were also very low at Chan do Eume. In contrast, the other two families, in particular Alicorhagiidae, were very abundant at the more continental locality and dominated the Endeostigmata community at this site (Table 5).

In the case of Astigmata mites, the higher densities Acaridae compared to Histiostomatidae corroborate that the former family is more commonly found in soils than the latter: Acaridae mites are found in several mineral soils [157, 158, 167] but also some moorlands [9, 146], whereas greater abundances of Histiostomatidae were only recorded in a raised bog which had been transformed into an agricultural grassland [166, 169]. Finally, the elevated proportion of hypopi recorded in our studied area is in agreement with the fact that both families produce more nymphs under wet conditions [170].

Oribatids were represented by 22 families in Serra do Xistral, being the dominant ones Nanhermannidae, Oppiidae, Mycobatidae, Camisiidae and Tectocepheidae (Table 7). Overall, the oribatid community structure is similar to that recorded in other peatland areas, but with slightly differences: together with Nanhermannidae and Oppiidae, Mycobatidae were also the most abundant groups in an English fen [116] but not in the Canadian [115] or North and Central Europe peatlands [111, 112, 144]. In the case of Tectocepheidae, this family was not as dominant at the Serra do Xistral as at those bogs investigated by Markkula [111] and Borcard [112] nor as rare as at the fen sampled by MacFadyen [145]. In contrast, most of these studies have also observed an important number of aquatic oribatids (family Limnozetidae) which were not present at our Galician upland bogs. Interestingly, these mites were also absent from drained peatlands [116, 171] and reinforces the idea that waterlogging conditions are essential for their survival. Similarly, Oribatulidade are well represented in Canadian [115] and Swiss peatlands [112], but only a few animals were collected at the driest site in our Galician area (Table 7).

Similar to the other groups of Acari, about half of the oribatid families present in Serra do Xistral were more abundant at the driest site and among them, the family Xenillidae, which includes species typical of arid systems, was recorded in a peatland soil for the first time. In contrast, three families (Malaconothridae, Peloppiidae and Thyrisomidae) had significantly higher densities at the wettest locality or were solely present at this particular site (Table 7), suggesting an apparent need of higher soil moisture. In fact, previous studies had already highlighted that some families such as Malaconothridae were particularly abundant in just certain microhabitats within the whole peatland [111, 112].

Finally, two groups of collembolans, Entomobryomorpha and Poduromorpha, were collected in greatest densities at the driest area, in agreement with previous studies which indicated that their abundances increased in drier and/or forested areas within a peatland [8, 10, 143, 148]. In addition, Vannier [131] demonstrated that Symphypleona are more sensitive to reductions in soil water content whereas Entomobryomorpha are more resistant to drier conditions. The community structure observed at our Galician peatlands is also very similar to that found under the *Molinia* fen investigated by MacFadyen [145] where, with independence of the plant microhabitats, Entomobryomorpha dominated the collembolan community and Symphypleona showed very low presences.

Conclusion

The Arthropoda community at the two investigated upland peatlands of Serra do Xistral was represented by three subphyla: Chelicerata, Myriapoda and Hexapoda. However, the abundances of bigger sized arthropods (i.e. spiders, pseudoscorpions, harvestmen, centipedes, symphylans, proturans and insects) were typically below 150 individuals per m^2, except for Coleoptera whose densities reached up to 750 individuals per m^2; overall, these macroarthropods only accounted for 1% of the total number of invertebrates collected. In contrast, the population numbers of the smaller representatives (mites and collembolans) dominated the community with their densities reaching averaged values of 73,500 and 26,500 individuals per m^2 and constituting around 73 and 26% of the arthropods present in the area, respectively. Overall, significant higher abundances of most groups (exceptions were Aranea, Pseudoescorpionida, Opilionida and Insecta) were observed at the drier locality (Chan do Eume) than at the most oceanic and wettest site (Chan do Lamoso). Although at both localities Oribatid mites dominated the arthropod community and with Entomobryomorpha collembolans and Prostigmata mites being the two following more abundant groups, their percentages differed slightly between sites and thus, at Chan do Eume Oribatids represented 58% of the total numbers, Entomobryomorpha 17% and Prostigmata 11%, whereas at Chan do Lamoso they followed the order Oribatids (37%) > Entomobryomorpha (29%) > Prostigmata (7%). Interestingly, 15 out of the 48 Acari families and half of the insect orders identified were only recorded at one site; in addition, proturans and symphylans were only found at the most continental locality and pseudoscorpions and opilions at the most oceanic one. Our results indicate that habitat characteristics can have a strong influence on the structure of the Arthropoda community within a particular peatland area, both in terms of population numbers and diversity. For example, more frequent waterlogged conditions and marine proximity appears to have negative effects on both mites and collembolan populations, the two dominant groups in these systems and therefore, alterations in the nutrient cycling dynamics associated to climated changes are anticipated.

Acknowledgments

We would like to thank Dr. Esther Barreal, Dr. Pedro Pablo Gallego and Dr. Pablo Ramil-Rego for their assistance in the field. Thanks are also due to Dr. Pablo Martín for their help with oribatid identification.

References

[1] J. C. Coulson and J. B. Whittaker, in *Production Ecology of British Moors and Montane Grasslands*, pp. 52-93, Springer Verlag, Berlin (1978).
[2] J. D. Reynolds, *British Ecological Society Bulletin* 9, 41 (1984).
[3] J. D. Reynolds, in *Ecology and conservation of Irish peatlands*, pp 135-143, Royal Irish Academy, Dublin (1990).
[4] G. Scott, G. S. Oxford and P. A. Selden, *Biological Conservation* 127, 420 (2006).
[5] J. M. Nelson, *Journal of Animal Ecology* 34, 579 (1965).
[6] D. G. Higgins, *Bulletin of the Irish Biogeographical Society* 8, 91 (1984).
[7] S. Koponen, *Memoirs of the Entomological Society of Canada* 126, 41 (1994).
[8] W. G. Hale, *Revue de Ecologie et Biologie du Sol* 3, 97 (1966).
[9] W. Block, *Pedobiologia* 5, 244 (1965).
[10] R. E. Blackith, *Proceedings of the Royal Irish Academy* 74, 203 (1974).
[11] J. P. Curry, *Scientific Proceedings Royal Dublin Society, Series A 6*, 131 (1978).
[12] L. Cole, S. M. Buckland and R. D. Bardgett, *Soil Biology and Biochemistry* 37, 1707 (2005).
[13] P. C. de Ruiter, A. Neutel and J. C. Moore, *Science* 269, 1257 (1995).
[14] L. Heneghan, D. C. Coleman, X. Zou, D. A. Jr. Crossley and B. L. Haines, *Applied Soil Ecology* 9, 33 (1998).
[15] N. Kaneko and E. F. Salamanca, *Ecological Research* 14, 131 (1999).
[16] C. O. Adejuyigbe, G. Tain and G. O. Adeoye, *Nutrient Cycling in Agroecosystems* 75, 47 (2006).
[17] P. F. Santos and W. G. Whitford, *Ecology* 62, 654 (1981).
[18] T. R. Seastedt, *Annual Review of Entomology* 29, 25 (1984).
[19] S. J. Joo, M. H. Yim and K. Nakane, *Forest Ecology and Management* 234, 192 (2006).
[20] C. P. Hickman, L. S. Roberts, S. L. Keen, A. Larson, H. I'Anson and D. J. Eisenhour, *Integrated Principles of Zoology*, McGraw-Hill, New York (2008).
[21] M. Aguinaldo, J. M. Turbeville, L. S. Linford, M. C. Rivera, J. R. Garey, R. A. Raff and J. A. Lake, *Nature* 387, 489 (1997).

[22] R. H. Thomas, *Science* 299, 1854 (2003).

[23] G. E. Budd and M. J. Telford, *Nature* 457, 812 (2009).

[24] J. Caravas and M. Friedrich, *BioEssays* 32, 488 (2010).

[25] D. Pisani, L. Poling, M. Lyons-Weiler and S. Hedges, *BMC Biology* 2, 1 (2004).

[26] J. C. Regier, J. W. Shultz, A. R. D. Ganley, A. Hussey, D. Shi, B. Ball, A. Zwick, J. E. Stajich, M. P. Cummings, J. W. Martin and C. W. Cunningham, *Systematic Biology* 57, 920 (2008).

[27] D. G. Edgecombe, *Arthropod Structure and Development* 39, 74 (2010).

[28] J. C. Regier, J. W. Shultz, A. Zwick, A. Hussey, B. Ball, R. Wetzer, J. W. Martin and C. W. Cunningham, *Nature* 463, 1079 (2010).

[29] O. Rota-Stabelli, L. Campbell, H. Brinkmann, G. D. Edgecombe, S. J. Longhorn, K. J. Peterson, D. Pisani, H. Philippe and M. J. Telford, *Proceedings of the Royal Society B: Biological Sciences* 278, 298 (2011).

[30] O. Rota-Stabelli, E. Kayal, D. Gleeson, J. Daub, J. L. Boore, M. J. Telford, D. Pisani, M. Blaxter and D. V. Lavrov, *Genome Biology and Evolution* 2, 425 (2010).

[31] J. W. Shultz, *Cladistics* 6, 1 (1990).

[32] J. W. Shultz, *Zoological Journal of the Linnean Society* 150, 221 (2007).

[33] P. Weygoldt, *Experimental and Applied Acarology* 22, 63 (1998).

[34] S. A. Miller and J. P. Harley, *Zoology 7th ed.*, McGraw-Hill, New York (2007).

[35] J. A. Dunlop and G. Alberti, *Journal of Zoological Systematic and Evolutionary Research* 46, 1 (2008).

[36] S. E. Masta, S. J. Longhorn and J. L. Boore, *Molecular Phylogenetics and Evolution* 50, 117 (2009).

[37] W. C. Wheeler and C. Y. Hayashi, *Cladistics* 14, 173 (1998).

[38] G. Giribet, G. D.Edgecombe, W. C. Wheeler and C. Babbitt, *Cladistics* 18, 5 (2002).

[39] R. A. Norton and V. M. Behan-Pelletier, in *A Manual of Acarology 3rd ed.*, pp. 430-564, Texas Tech University Press, Lubbock (2009).

[40] H. Klompen, M. Lekveishvili and B. Black, *Molecular Phylogenetics and Evolution* 43, 936 (2007).

[41] A. Repato, C. da Rocha and J. Dunlop, *BMC Evolutionary Biology* 10, 235 (2010).

[42] M. Dabert, W. Witalinski, A. Kazmierski, Z. Olszanowski and J. Dabert, *Molecular Phylogenetics and Evolution* 56, 222 (2010).

[43] M. J. T. N. Timmermans, D. Roelofs, J. Mariën and N. M. van Straalen, *BMC Evolutionary Biology* 8, 83 (2008).

[44] K. Meusemann, B. M. von Reumont, S. Simon, F. Roeding, S. Strauss, P. Kück, I. Ebersberger, M. Walzl, G. Pass, S. Breuers, V. Achter, A. von Haeseler, T. Burmester, H. Hadrys, J. W. Wägele and B. Misof, *Molecular Biology and Evolution* 27, 2451 (2010).

[45] A. Carapelli, F. Nardi, R. Dallai and F. Frati, *Pedobiologia* 50, 191 (2006).

[46] A. Carapelli, P. Lio, F. Nardi, E. van der Wath and F. Frati, *BMC Evolutionary Biology* 7, S8 (2007).

[47] Y. X. Luan, J. M. Mallatt, R. D. Xie, Y. M. Yang and W. Y. Yin, *Molecular Biology and Evolution* 22, 1579 (2005).

[48] M. Gaju, in *Bases para un curso práctico de Entomología*, pp. 323-347, Asociación Española de Entomología, Barcelona (1988).

[49] K. A. Christiansen, in *Soil Biology Guide*, pp. 965-995, John Wiley and Sons, New York (1990).
[50] S. P. Hopkin, *A Key to the Springtails (Insecta: Collembola) of Britain and Ireland, AID GAP (Aids to Identification in Difficult Groups of Animals and Plants) test version*, Field Studies Council, Preston Montford, Shrewsbury (2000).
[51] S. P. Hopkin, in *Encyclopedia of Soil Science*, pp. 207-210, Marcel Dekker Inc., New York (2002).
[52] P. F. Bellinger, K. A. Christiansen and F. Janssens, *Checklist of the Collembola of the Worl, http://www.collembola.org* (2009).
[53] C. A. D'Haese, *European Journal of Entomology* 101, 385 (2003).
[54] Y. Xiong, Y. Gao, W. Y. Yin and Y. X. Luan, *Molecular Phylogenetics and Evolution* 49, 728 (2008).
[55] F. Nardi, A. Carapelli, P. P. Fanciulli, R. Dallai and F. Frati, *Molecular Biology and Evolution* 18, 1293 (2001).
[56] J. Najt, W. M. Weiner and P. Grandcolas, *Zoologica Scripta* 34, 305 (2005).
[57] Q. Xie, X. Tian, Y. Qin and W. Bu, *Molecular Phylogenetics and Evolution* 50, 310 (2009).
[58] J. A. Barrientos, in *Bases para un curso práctico de Entomología*, pp. 115-152, Asociación Española de Entomología, Barcelona (1988).
[59] J. A. Pechenik, *Biology of the Invertebrates*, McGraw-Hill Higher Education, New York (2010).
[60] C. D. von Dohlen and N. A. Moran, *Journal of Molecular Evolution* 41, 211 (1995).
[61] P. J. Gullan, *Entomologica* 33, 101 (2001).
[62] A. Tüllgren, *Zeitschrift für angewandte Entomologie* 4, 149 (1918).
[63] G. Eisenbeis, in *Intestinal Microorganisms of Termites and other Invertebrates*, pp. 3-53, Springerlink, Berlin (2006).
[64] C. Coleman, D. A. Jr. Crossley and P. F. Hendrix, *Fundamentals of soil ecology 2nd. ed.*, Academic Press New York, New York (2004).
[65] J. A. Barrientos and M. Rambla, in *Bases para un curso práctico de Entomología*, pp. 143-152, Asociación Española de Entomología, Barcelona (1988).
[66] L. Edgar, in *Soil Biology Guide*, pp. 529-581, John Wiley and Sons, New York (1990).
[67] J. A. Zaragoza, in *Bases para un curso práctico de Entomología*, pp. 107-114, Asociación Española de Entomología, Barcelona (1988).
[68] W. B. Muchmore, in *Soil Biology Guide*, pp. 503-527, John Wiley and Sons, New York (1990).
[69] W. Baker and G. W. Wharton, *An introduction to Acarology*, MacMillan, New York (1952).
[70] D. E. Johnston, in *Synopsis and Classification of Living Organisms vol. 2*, pp. 111-117, McGraw-Hill, New York (1982).
[71] J. B. Kethley, in *Synopsis and Classification of Living Organisms vol. 2*, pp. 117-145, McGraw-Hill, New York (1982).
[72] O. Evans and W. M. Till, *Transactions of the Zoological Society of London* 35, 145 (1979).
[73] B. McDaniel, *How to know the mites and ticks*, William C. Brown Company Publishers, Dubuque, Iowa (1979).

[74] W. Krantz and B. D. Ainscough, in *Soil Biology Guide*, pp. 583-665, John Wiley and Sons, New York (1990).
[75] O. Evans, *Principles of Acarology*, CAB International, Wallingford, England (1992).
[76] M. Smith, E. Lindquist and V. M. Behan-Pelletier, in *Assessment of species diversity in the mixedwood plains ecozone*, pp. 1-38, Eastern Cereal and Oilseed Research Centre, Otawa (1998).
[77] H. H. Koehler, *Agriculture, Ecosystems and Environment* 74, 395 (1999).
[78] N. G. Bregetova and G. I. Shcherbak, in *Key to the Soil-Inhabiting Mites. Mesostigmata*, pp. 256-307, Nauka, Leningrad (1977).
[79] D. A. Chant, *Canadian Journal of Zoology* 41, 243 (1963).
[80] A. Enkegaard and H. F. Brødsgaard, *BioControl* 45, 285 (2000).
[81] A. Abou-Awad, A. M. Korayem, M. F. Hassan and M. A. Abou-Elela, *Journal of Applied Entomology* 125, 125 (2001).
[82] O. L. Makarova, *Entomological Review* 84, 667 (2004).
[83] E. E. Lindquist and M. L. Moraza, *Zootaxa* 2479, 1 (2010).
[84] D. E. Walter and E. E. Lindquist, *Canadian Journal of Zoology* 67, 2797 (1989).
[85] L. Moraza, *Sociedad Española de Acarología* 26, 281 (1999).
[86] D. E. Walter, E. E. Lindquist, I. M. Smith, D. R. Cook and G. W. Krantz, in *A Manual of Acarology 3rd ed.*, pp. 233-420, Texas Tech University Press, Lubbock (2009).
[87] B. Kethley, in *Soil Biology Guide*, pp. 667-756, John Wiley and Sons, New York (1990).
[88] J. Den Heyer, *Zoosymposia* 6, 34 (2011).
[89] D. E. Walter, *Australian Journal of Entomology* 40, 207 (2001).
[90] D. E. Walter and H. C. Proctor, *Mites in soil: an interactive key to mites and other soil microarthrpods*, CSIRO Publishing, Collingwood (2001).
[91] H. Womersley, *Transactions of the Royal Society of South Australia* 68, 133 (1944).
[92] B. M. O´Connor, in *Synopsis and classification of living organisms*, pp. 146-169, McGraw-Hill, New York (1982a).
[93] J. R. Philips, in *Soil Biology Guide*, pp. 757-778, John Wiley and Sons, New York (1990).
[94] B. M. O´Connor, *Annual Review of Entomology* 27, 385 (1982b).
[95] M. Luxton, *Pedobiologia* 39, 238 (1995).
[96] R. R. Weil and W. Kroontje, *Soil Biology and Biochemistry* 11, 669 (1979).
[97] J. C. Noble, W. G. Whitford and M. Kaliszweski, *Journal of Arid Environments* 32, 329 (1996).
[98] J. Arroyo, J. C. Iturrondobeitia, A. I. Caballero and S. González-Carcedo, *Sociedad Española de Acarología* 32, 73 (2003).
[99] R. A. Norton, in *Soil Biology Guide*, pp. 779-803, John Wiley and Sons, New York (1990).
[100] H. Marjorie, in *Encyclopedia of Entomology*, pp. 3463–3466, Springerlink, Dordrecht (2008).
[101] N. Kaneko, *Revue d'Ecologie et de Biologie du Sol* 25, 353 (1988).
[102] C. Pérez-Íñigo, *Acari. Oribatei, Poronota*, Museo Nacional de Ciencias Naturales, CSIC, Madrid (1993).
[103] S. Gilyarov, *A key to soil-inhabiting mites. Sarcoptiformes*, Izdatel'stvo Nauka, Moscow (1975).

[104] M. Moritz, *Mitteilungen aus dem Zoologischen Museum in Berlin* 52, 227 (1976).
[105] G. Riha, *Yearbooks of Zoology* 80, 407 (1951).
[106] J. M. Anderson, *Journal of Animal Ecology* 44, 475 (1975).
[107] V. M. Behan-Pelletier and S. B. Hill, *Revue d'Ecologie et de Biologie du Sol* 20, 221 (1983).
[108] K. Schneider, S. Migge, R. A. Norton, S. Scheu, R. Langel, A. Reineking and M. Maraun, *Soil Biology and Biochemistry* 36, 1769 (2004).
[109] E. Edsberg and S. Hågvar, *Pedobiologia* 43, 413 (1999).
[110] H. Siepel and E. M. de Ruiter-Dijkman, *Soil Biology and Biochemistry* 25, 1491 (1993).
[111] I. Markkula, *Annales Zoologici Fennici* 23, 33 (1986).
[112] D. Borcard, *Revue Suisse de Zoologie* 98, 303 (1991).
[113] C. Pérez-Íñigo, A. Arillo and L. S. Subías, *Acari. Oribatei, Gimnonota I*, Museo Nacional de Ciencias Naturales, CSIC. Madrid (1997).
[114] T. A. Woolley, *Transactions of the American Microscopical Society* 77, 1 (1958a).
[115] V. M. Behan-Pelletier and B. Bissett, *Memoirs of the Entomological Society of Canada* 169, 73 (1994).
[116] A. Macfadyen, *Journal of Animal Ecology* 21, 87 (1952).
[117] L. S. Subías and A. Arillo, *Acari. Oribatei, Gimnonota II*, Museo Nacional de Ciencias Naturales, CSIC, Madrid (2001).
[118] J. Arroyo and J. C. Iturrondobeitia, *European Journal of Soil Biology* 42, 259 (2006).
[119] T. A. Woolley, *Transactions of the American Microscopical Society* 77, 258 (1958b).
[120] Z. Lindo and N. N. Winchester, *Soil Biology and Biochemistry* 39, 2957 (2007).
[121] A. Serra, in *Bases para un curso práctico de Entomología*, pp. 305-313, Asociación Española de Entomología, Barcelona (1988).
[122] P. Mundel, in *Soil Biology Guide*, pp. 819-833, John Wiley and Sons, New York (1990).
[123] A. Mas, in *Bases para un curso práctico de Entomología*, pp. 315-321, Asociación Española de Entomología, Barcelona (1988).
[124] A. Edwards, in *Soil Biology Guide*, pp. 891-910, John Wiley and Sons, New York (1990).
[125] F. Hales, in *Invertebrate Zoology 2nd Ed.*, pp. 232-274, Oxford Univ Press, Oxford (2001).
[126] K. Newell, *Soil Biology and Biochemistry* 16, 227 (1984).
[127] J. C. Moore, E. R. Ingham and D. C. Coleman, *Biology and Fertility of Soils* 5, 6 (1987).
[128] R. T. Lartey, E. A. Curl and C. M. Peterson, *Soil Biology and Biochemistry* 26, 81 (1994).
[129] M. Chamberlain, N. P. McNamara, J. Chaplow, A. W. Stott and H. I. J. Black, *Soil Biology and Biochemistry* 38, 2655 (2006).
[130] Y. Gao, Y. Bu and Y. X. Luan, *Zoological Science* 25, 1139 (2008).
[131] G. Vannier, in *Écologie du sol*, pp. 23-258, Editions du Centre national de la Recherche Scientifique, Paris (1970).
[132] T. P. Copeland and G. Imadaté, in *Soil Biology Guide*, pp. 911-933, John Wiley and Sons, New York (1990).
[133] J. Berzosa, in *Bases para un curso práctico de Entomología*, pp. 463-472, Asociación Española de Entomología, Barcelona (1988).

[134] D. Forero, *Revista Colombiana de Entomología* 34, 1 (2008).
[135] M. D. Martínez, in *Bases para un curso práctico de Entomología*, pp. 445-449, Asociación Española de Entomología, Barcelona (1988).
[136] N. García Aldrete, in *Soil Biology Guide*, pp. 1033-1052, John Wiley and Sons, New York (1990).
[137] J. M. Salgado, R. Outerello, P. Gamarra, M. Blas, X. Vázquez, E. Vives and J. C. Otero, in *Bases para un curso práctico de Entomología*, pp. 573-639, Asociación Española de Entomología, Barcelona (1988).
[138] E. Galante-Patiño and J. L. Viejo, in *Bases para un curso práctico de Entomología*, pp. 533-562, Asociación Española de Entomología, Barcelona (1988).
[139] M. Baez, in *Bases para un curso práctico de Entomología*, pp. 503-519, Asociación Española de Entomología, Barcelona (1988).
[140] J. F. McAlpine, in *Soil Biology Guide*, pp. 1211-1252, John Wiley and Sons, New York (1990).
[141] H. J. Teskey, in *Soil Biology Guide*, pp. 1253-1276, John Wiley and Sons, New York (1990).
[142] A. Fjellberg, *Norwegian Journal of Entomology* 23, 181 (1976).
[143] M. Slawska, Pedobiologia 44, 413-420 (2000).
[144] N. Tarras-Wahlberg, *Oikos* (Suppl. 4) 1(1961).
[145] A. Macfadyen, *Journal of Animal Ecology* 21, 87 (1952).
[146] T. G. Wood, *Oikos* 18, 102 (1967).
[147] D. A. Heemsbergen, M. P. Berg, M. Loreau, J. R. van Hal, J. H. Faber and H. A. Verhoef, *Science* 306, 1019 (2004).
[148] N. Silvan, R. Laiho and H. Vasander, *Forest Ecology and Management* 133, 127 (2000).
[149] R. Laiho, N. Silvan, H. Cárcamo and H. Vasander, *Applied Soil Ecology* 16, 1 (2001).
[150] B. Reynolds, J. Hamel, J. Isbanioly, L. Klausman and K. K. Moorhead, *Pedobiologia* 51, 273 (2007).
[151] J. B. Cragg, *The Journal of Ecology* 49, 477 (1961).
[152] A. Jr. Crossley and K. K. Bohnsack, *Ecology* 41, 628 (1960).
[153] J. A. Wallwork, *Annual Review of Entomology* 28, 109 (1983).
[154] N. Lindberg and J. Bengtsson, *Applied Soil Ecology* 28, 163 (2005).
[155] M. A. Tsiafouli, A. S. Kallimanis, E. Katana, G. P. Stamou, and S. P. Sgardelis, *Applied Soil Ecology* 29, 17 (2005).
[156] J. G. Sheals, *Journal of Animal* Ecology 26, 125 (1957).
[157] S. Dhillon and N. H. E. Gibson, *Pedobiologia* 1, 189 (1962).
[158] N. K. Davis, *Journal of Animal Ecology* 32, 49 (1963).
[159] N. López, A. M. Vincini, N. L. Clemente, P. L. Manetti, D. M. Carmona, and H. A. Álvarez-Castillo, *Ciencia del Suelo* 23, 115 (2005).
[160] G. H. R. Osler, L. Harrison, D. K. Kanashiro, and M. J. Clapperton, *Applied Soil Ecology* 38, 71 (2008).
[161] J. A. Salamon, M. Schaefer, J. Alphei, B. Schmidt, and S. Scheu, *Oikos* 106, 51 (2004).
[162] W. G. Whitford, K. Stinnet, and Y. Steinberger, *Pedobiologia* 31, 147 (1988).
[163] H. A. O´Lear and J. M. Blair, *Biology and Fertility of Soils* 29, 207 (1999).
[164] S. Bokhorst, A. Huiskes, P. Convey, P. M. Van Bodegom, and R. Aerts, *Soil Biology and Biochemistry* 40, 1547 (2008).

[165] N. Z. Elkins and W. G. Whitford, *Oecologia* 55, 303 (1982).
[166] J. Whelan, *Scientific Proceedings of the Royal Dublin Society Series A* 6, 175 (1978).
[167] W. Leetham and D. G. Michulmas, *Pedobiologia* 28, 311 (1985).
[168] L. Moraza, *Graellsia* 63, 35 (2007).
[169] J. Whelan, *Pedobiologia* 28, 191 (1985).
[170] P. Garman, *Connecticut Agricultural Experiment Station* 402, 885 (1937).
[171] T. Solhøy, *Fauna Norvegica Series B* 26, 91 (1979).

In: Invertebrates: Classification, Evolution and Biodiversity ISBN: 978-1-62618-789-4
Editor: Rafael Riosmena-Rodriguez

Chapter 2

Oligochaeta Communities from Galician Upland Peatlands

Noela Carrera* and María J. I. Briones
Departamento de Ecología y Biología Animal, Facultad de Biología,
Universidad de Vigo, Vigo, Spain

Abstract

Unlike in other European countries, peatland soils have a limited presence in Spain; in the Galicia region (NW Spain) they survive in upland areas where cold temperatures and continuous moisture supply allow their presence. Very little is known about the invertebrate diversity in these areas, but previous research has highlighted the dominance, in terms of biomass, of oligochaete worms (families Lumbricidae and Enchytraeidae). As a result of their burrowing and casting activities, these two groups play an important role in ecosystem functioning by promoting microbial growth, soil structure and porosity as well as soil organic matter incorporation and mixing and nutrient cycling. Therefore, any changes in their presence and/or abundance in response to climatic variations (namely, temperature and soil moisture) could have important consequences on the amount of carbon being stored in these soils. In this chapter, we will describe the Oligochaeta community composition of two peatland sites with different microclimate and dominant vegetation to determine the relative importance of abiotic and biotic factors in the size and diversity of their populations. The Oligochaeta community from the upland peatlands within Serra do Xistral was mainly represented by two families, Lumbricidae and Enchytraeidae. However, only two species of lumbricids were collected in very low densities (about 50 individuals per m^2) at both localities, whereas enchytraeids showed a far greater diversity (up to 11 species were identified) and their population numbers reached, on average, more than 60,000 individuals per m^2. In addition, although at both sites the enchytraeid community was dominated by one species (*Cognettia sphagnetorum*), its structure showed marked differences between localities. Overall, enchytraeid numbers were significantly higher at the oceanic site (Chan do Lamoso) due to the greater abundance of *C. sphagnetorum, Cernosvitoviella atrata, Marionina*

* E-mail address: eunoela@uvigo.es.

filiformis and *Mesenchytraeus sanguineus*, whereas drier conditions at Chan do Eume significantly favoured higher population numbers of five species and in particular those of *C. hibernica* and *Achaeta* spp. which reached similar densities to those of the dominant *C. sphagnetorum* at this site. Finally, *M. flavus* and *M. lusitanicus* showed similar abundances at both sites. Our results highlight the close link between enchytraeids and soil water content in these Galician peatland habitats. Due to the key role these oligochaetes play on soil organic matter decomposition and nutrient cycling, any alterations in the abiotic factors are expected to have important implications in the functioning of these systems.

Introduction

Peatland ecosystems represent about 3% of the total global surface, but they store more than one third of the carbon pool in the biosphere [1]. They are present in more than 130 countries over the world, being majorly concentrated at high latitudes (boreal and subarctic regions), but also in tropical regions in Asia [1]. In Europe, most peatlands are located in Scandinavia, Russia and the United Kingdom [2] and those in Spain probably represent the most southern limit of their distribution, where they subsist in upland regions. In particular, those within Serra do Xistral (Galicia, NW of Spain) are of special interest because they are the highest in Europe and are considered to be ‘Priority Natural Habitats’ by the European Union (Code 7130) [3].

The environmental conditions in these areas (high moisture, cold temperatures and acid soils) allow peat formation and maintenance, but also put several restrictions to their biodiversity, including plants, microorganisms and soil fauna; therefore, their communities are usually represented by few species and thus, any biodiversity losses could have strong impacts on the functioning of these systems.

Despite nematodes and microarthropods usually being the most abundant groups in these organic soils, oligochaete worms (lumbricids and enchytraeids) dominate these systems in terms of biomass, reaching as much as 50 and 100 g m^{-2}, respectively [4]. These animals play an important role in ecosystem functioning as a result of their tunnelling and casting activities [5]. By building horizontal and vertical burrows they improve soil structure and stability and create favourable channels for water drainage and gas transport, whereas their casts constitute hotspots for microbial activity [6-10]. In addition, their selective feeding could facilitate microbial dispersal [11, 12] or even reduce the incidence of pest disease [13].

Although both groups occur worldwide, earthworms are most abundant in forests and grasslands of temperate and tropical regions [14], whereas large enchytraeid populations are usually associated with temperate rainy climates with moderate or cold summers [15]. As a result of this close link between climatic factors and their geographical distribution, it is foreseen that any alterations in temperature and moisture regimes are likely to produce important changes in their population sizes and thus, on soil organic matter turnover. This is of particular importance in high organic soils, such as peatlands, where climate change is expected to produce temperature increases and more frequent drought periods. Therefore, in this chapter, we will describe the Oligochaeta community composition of two peatland sites with different microclimate and dominant vegetation to determine the relative importance of abiotic and biotic factors in the size and diversity of their populations.

Galician Active Blanket Bogs

The most extensive peatland area within the Galicia region is located in Serra do Xistral. It occupies about 22,964 Ha and is included within the European Natura-2000 network since 2004 (Lic Es 1120015). Among the bogs present in this area, the 'upland active blanket bogs' are of particular importance due to their relict nature. Their acidic soils (pH≈4) are formed from quartzite and other metamorphic rocks such as slate and schist and accumulate high amounts of organic matter (around 32% of organic carbon).

Although weather stations are not available for the area, the one located at Fraga Vella at an altitude of 750 m indicates that, for the period 1999-2011, the average annual temperature was 10.5°C, with a temperature difference up to 10ºC between winter and summer temperatures. The average annual rainfall was 1,450 mm, mainly concentrated in autumn and winter; however, the horizontal precipitation also plays an important role in keeping a high relative humidity (> 75%) in the area (Meteogalicia, Xunta de Galicia; http://www.meteogalicia.es/).

Chan do Lamoso (43°30'N, 7°32'W) and Chan do Eume (43°47'N, 7°57'W) are two representative upland active bogs at 976 and 960 m a.s.l., respectively. Despite their rather similar altitude, the former site is more exposed to the prevailing oceanic winds and according to the bioclimatic classification proposed by Rivas-Martínez and Loidi Arregui [16], it is characterised by an ultra-hyperhumid ombrotype and higher supra-temperate thermo-type, whereas Chan do Eume has a more continental influence which results in a higher hyperhumid ombrotype with higher meso-temperate thermotype. This microclimatic heterogeneity is also responsible for the spatial distribution of the peatland habitats and the depth of peat sediments formed during the Holocene [17].

The microclimatic differences are also reflected in the vegetation communities, being the most northerly location (Chan do Lamoso) dominated by the plant association *Carici durieui - Eriophoretum angustifolii* (*Carex durieui* Steud., *Eriophorum angustifolium* Honck. and *Molinia caerulea* L.), which is typical of shaded and foggy slopes, whereas the association *Carici durieui - Molinietum caeruleae* is present at the most continental site, with *Molinia caerulea* L. as the dominant species followed by *Carex durieui* Steud. and *Deschampsia flexuosa* (L.) Trin. [18].

Soil Oligochaeta Community

Soil organisms were collected from each locality using soil cores (PVC cylinders, 10 cm internal diameter x 10 cm deep) which were taken in the field twice every year (winter and summer), from 2003 to 2006. Once back at the laboratory, the animals were extracted using wet extraction [19]. After fixing the specimens in 70% ethanol they were inspected under the microscope (Nikon Eclipse 80i).

The results indicated that, within Annelids, all the specimens collected belonged to two Oligochaeta families, Lumbricidae and Enchytraeidae, with a clear dominance of the latter (Table 1). Previous studies have highlighted that enchytraeid worms are one of the key soil invertebrate groups in peatland soils [20], reaching great abundances in particular in those areas dominated by *Juncus squarrosus* [4, 21, 22].

Table 1. Abundance (average numbers per m^2 ± standard error, S.E.) of each of the two Annelid families collected at both localities. Different letters indicate significant differences between sites ($p < 0.05$)

Taxa	Chan do Lamoso *(Carici durieui - Eriophoretum angustifolii)*	Chan do Eume *(Carici durieui - Molinietum caeruleae)*
Annelida		
Lumbricidae	53.3 ± 21.4 A	48.0 ± 26.7 A
Enchytraeidae	76,705.6 ± 4,715.1 A	63,306.1 ± 10,264.2 B

In addition, when comparing the two sites, enchytraeids were significantly more abundant at Chan do Lamoso, where the vegetation, as indicated above, is dominated by Cyperaceae (*Carex durieui* and *Eriophorium angustifolium*), than at Chan do Eume. This contrasting abundance could also be partly associated to the existing microclimatic differences between the two localities and for example, the soil samples collected at Chan do Lamoso were always wetter (average moisture content of 81.5%) than those taken from Chan do Eume (average values of 70.3% soil water content).

Family LUMBRICIDAE Rafinesque-Schmaltz 1815

This family includes about 670 valid and 55 uncertain names [23] of the commonly called "earthworms". Their size is variable, from species measuring only a few millimetres to those specimens reaching more than one meter long; however, the majority of the individuals fall within the range of a few centimetres [14], which have led to their inclusion within the soil macrofauna and not mesofauna [24].

The shape of the prostomium (preoral segment) in relation to the peristomium (first segment), the position of the clitellum, the tubercula pubertatis (glandular protuberances surrounding the clitellum and involved in sperm transference) and the first dorsal pore together with the setae pattern are external morphological characters commonly used for earthworm identification. Some internal anatomical characters have also an important taxonomical value; for example, the arrangement of the musculature, the number and position of the seminal vesicles (reproductive structures associated to the testes where the sperm matures) and the spermathecae (small capsules for storing the exchanged sperm after copulation) and the shape of the typhlosole (a dorsal internal fold of the intestine) [25].

Table 2. Abundance (average numbers per m^2 ± standard error, S.E.) of each earthworm species at both localities. Different letters indicate significant differences ($p < 0.05$) between sites for each species (uppercase) and between species at each locality (lowercase)

Species	Chan do Lamoso	Chan do Eume
Dendrobaena madeirensis	29.3 ± 13.04 A a	0.0 ± 0.0 A a
D. attemsi	24.0 ± 10.1 A a	48.01 ± 26.7 A a

Regarding their geographical distribution, earthworms seem to prefer moist areas and to avoid extreme temperatures. Accordingly, they are more abundant in forests and grasslands of temperate and tropical regions (up to 2,000 individuals per m^2 in fertilized pastures), whereas the lowest densities are usually found in arid areas [14]. Additional abiotic factors that also affect earthworm populations are soil type and pH [26]. Therefore, the acid conditions in Serra do Xistral could easily explain the low diversity of earthworms observed in this area, with only two species belonging to the genus *Dendrobaena* (*D. madeirensis* and *D. attemsi*) being collected in very low densities at both localities (Table 2). This diversity was much lower than the one recorded at the Moor House National Nature Reserve where up to 15 species were found with an average density of 400 individuals per m^2 [27]. However, these data correspond to mineral soils with more basic and drier conditions than those typically found in most peatlands and consequently, overall abundance usually lies below 1 specimen per m^2 [4]. These two *Dendrobaena* species are considered, from an ecological point of view, as acid-tolerant species [25, 26].

Genus Dendrobaena Eisen 1873

This genus includes small to medium earthworms with reddish colour, epi- or tanilobic prostomium and widely spaced setae. The first dorsal pore is usually located in an anterior position (5/6-7/8) and the male pores on segment 15. They have, at least, two pairs of spermathecae which open in furrows 9/10 and 10/11 and two to four pairs of seminal vesicles located in segments (9, 10) 11 and 12. The calciferous gland (which releases a concentrated suspension of calcium carbonate) opens directly into the oesophagus and lacks oesophageal pouches in 10. The nephridial vesicles are ocarina-shaped and the nephroducts open ventrally [25, 26].

***Dendrobaena madeirensis* (Michaelsen 1891)**

The colour of this worm is grey in the posterior region but reddish anteriorly; it is 85 mm long and 4 mm wide, with about 129-135 segments. The clitellum extends over segments 32 to 36 and the tubercula pubertatis (Figure 1a) occupy segments 33 and 35 [28, 29].

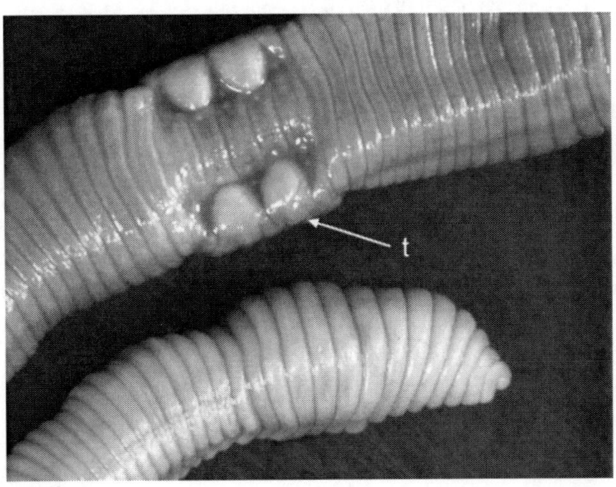

Figure 1a. *Dendrobaena* specimen: ventral and side view of *D. madeirensis* (stereomicroscope; 10X) indicating the tubercula pubertatis (t) with an arrow.

Only 11 individuals of *D. madeirensis* were collected in Serra do Xistral, all of them from the most oceanic site (Chan do Lamoso) (Table 2).

Dendrobaena attemsi Michaelsen 1901

Typically, the segment number of this species is 125-145, its length 40-70 mm and its diameter 0.8-1.4 mm. The body colour is reddish, with a characteristic ventral depigmentation on segments 9-11. The clitellum occupies five segments (from 28, 29-33, 34) and the tubercula pubertatis (Figure 1b) are longitudinal ridges extending from 29, 30-32 [25, 29]. Like the previous species, the densities of *D. attemsi* were very low and no differences in their population numbers were observed between both sampling localities (Table 2).

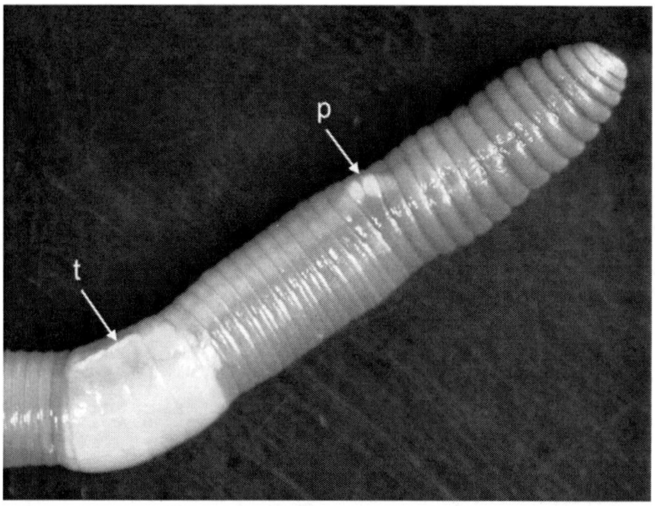

Figure 1b. *Dendrobaena* specimen: lateral side of *D. attemsi* (stereomicroscope; 10X) with the arrows showing the tubercula pubertatis (t) and the male pore (p).

Family ENCHYTRAEIDAE Vejdovský 1879

Enchytraeidae is family of small-sized terrestrial and aquatic worms which includes about 700 species grouped in 31 genera [30] and are commonly found in soils [14].

Most species are hermaphrodites with sexual reproduction, developing a clitellum on segments 12-13 which also surrounds the genital pores. However, there are also species with parthenogenesis or even asexual reproduction by fragmentation [14]. With the exception of the genus *Achaeta*, which includes species lacking setae, the number, shape and size of these structures are important taxonomical characters used in enchytraeid identification. Besides this external character, enchytraeids should also be identified by the morphology of their internal organs and therefore, live specimens are needed. Among them, the shape of the brain and the origin of the dorsal blood vessel, as well as several structures associated with the digestive system: presence and shape of the oesophageal appendages and intestinal diverticula and the number, position and shape of the pharyngeal glands. Adult sexual characteristics are also very informative, such as the shape, size and position of the spermathecae, the length of their conducts and the presence of associated glands, the shape and cellular structure of the clitellum, the presence and size of seminal vesicles, etc.

Table 3. Abundance (average numbers per m^2 ± standard error, S.E.) of each enchytraeid species at both localities. Different letters indicate significant differences ($p < 0.05$) between sites for each species (uppercase) and between species at each locality (lowercase)

Species	Chan do Lamoso	Chan do Eume
Achaeta spp. (including *A. affinis*)	789.5 ± 83.2 A d	11,537.8 ± 2,106.5 B b
Cernosvitoviella atrata	13,079.4 ± 1,378.7 A b	9,761.6 ± 1,782.2 B bc
Marionina filiformis	17,533.5 ± 1,848.2 A b	0.0 ± 0.0 B e
Marionina clavata	0.0 ± 0.0 A f	928.2 ± 169.5 B d
Mesenchytraeus flavus	5.3 ± 0.6 A f	0.0 ± 0.0 A e
Mesenchytraeus sanguineus	138.7 ± 14.6 A e	16.0 ± 2.9 B e
Mesenchytraeus lusitanicus	1,098.8 ± 115.8 A d	1,376.2 ± 251.3 A d
Cognettia sphagnetorum	39,222.3 ± 4,134.4 A a	18,530.9 ± 3,383.3 B a
Cognettia hibernica	4,454.1 ± 469.5 A c	15,490.5 ± 2,828.2 B ab
Cognettia cognettii	117.3 ± 12.4 A e	4,992.8 ± 911.6 B c
Cognettia clarae	0.0 ± 0.0 A f	672.1 ± 122.7 B d

Finally, coelomocytes (size, shape and texture), the position of the first pair of nephridia (each one with a preseptal and a postseptal section located in two consecutive segments with the opening on the latter segment) and their number and shape in the preclitellar segments are also of taxonomical importance [30-32].

The enchytraeid community in the two upland active bogs of Serra do Xistral was represented by 11 species (Table 3) together with some unidentified *Achaeta* spp. This species richness and diversity are similar to that found in different Irish peatland microhabitats [33] where up to 10 different species were identified, with the majority of them (8) also being present in Serra do Xistral. Similarly, seven out of a total of eleven species recorded by Healy [34] in the organic soils of Moor House Reserve (UK) were also collected at the Galician sites. Overall, six enchytraeid species were common to these three Atlantic peatlands: *Achaeta affinis, Marionina filiformis, M. clavata, Mesenchytraeus sanguineus, Cognettia sphagnetorum* and *C. cognettii*. Furthermore, in these three studies the dominant species was *C. sphagnetorum*, which is effectively the most abundant enchytraeid in most acid soils with high organic matter content [22, 35-40].

Genus Achaeta Vejdosky 1877

European species are usually very variable in size, ranging from 2.5 to 9 mm [30]. They are the only terrestrial enchytraeids lacking setae along their body (Figures 2a, b) and their epidermal glands are frequently enlarged and projected into the coelom, the distinct "pyriform glands" [30, 31]. In addition, the number and location of the flat epidermal glands (lens-shaped, with a circular or oval outline) is also an important taxonomical character. The oesophageal appendages, when present, are unpaired and located in the dorsal region of segment 5 or 4 and 5, rarely only in 4.

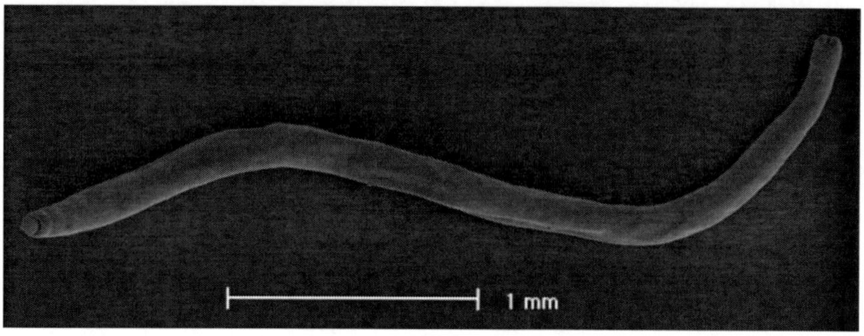

Figure 2a. Microphotograph (SEM) of *Achaeta* sp.: full view of one individual.

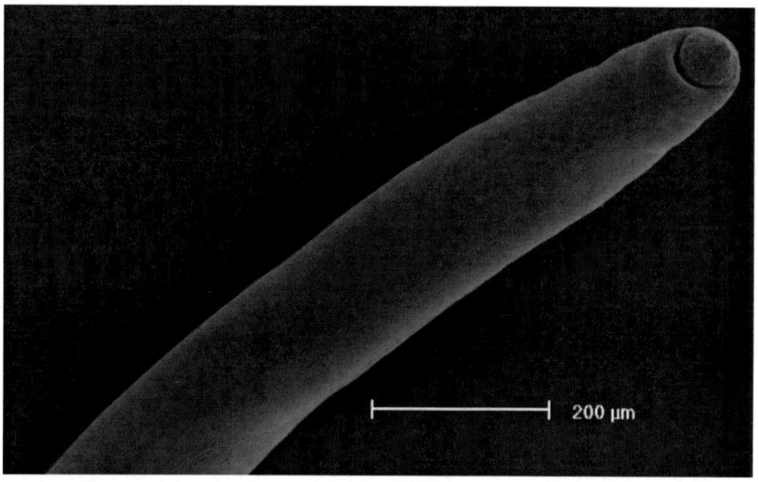

Figure 2b. Microphotograph (SEM) of *Achaeta* sp.: ventral view of anterior region.

These enchytraeids have primary pharyngeal glands located from 4 to 6, usually dorsally connected, but some species also have secondary ventral glands. When the spermathecae are present, they are not connected with the oesophagus. Finally, the brain is elongated (about twice longer than wide) with its posterior margin being rounded or slightly convex [30].

Achaeta affinis Nielsen and Christensen 1959

In the literature, this species is described as having a very variable length: 8-10 mm according to Nielsen and Christensen [31], 4-5.5 mm *sensu* Chalupský [41] and 4.5-7 mm by Rota [42], which has led to the suggestion of *A. affinis* may be consisting of more than one single species [30]. It is characterised by the presence of two pairs of pyriform glands in each segment (with the ventral pair being usually reduced and indistinct). The oesophagus has club-shaped appendages in segments 4 and 5 and three pairs of pharyngeal glands in segments 4, 5 and 6 connected both dorsally and to the ventral lobes, although in segment 6 the ventral lobe is separated from the dorsal one. Two pairs of preclitellar nephridia in 6/7 and 7/8 (anteseptale/postseptale) are present. The dorsal blood vessel originates from segment 7 and the spermathecae and seminal vesicles are usually absent. The posterior extremity of the brain is rounded and the coelomocytes are numerous and contain dark globules [30, 31, 41, 42].

A. affinis is common and widely distributed, although it seems to prefer fairly acid and not excessively moist soils [30, 31, 33]. The rest of the recorded 22 European species have been found in different habitats, such as deciduous forests, sand dunes, marshes, fens or grasslands [33], although many of them seem to prefer moderate to acidic soils [30, 43]. This genus was well represented in our studied peatlands, but their abundances were significantly greater ($p < 0.05$) at Chan do Eume than at Chan do Lamoso (Table 3).

Genus Cernosvitoviella Nielsen and Christensen 1959

The members of this genus are small enchytraeids (2-5 mm) and their sigmoid setae are thin, slender and with conspicuous nodulus (slight swelling in the middle). The total number of setae per bundle is variable, but usually the preclitellar bundles contain a higher number (varying from 5 to 10 setae) than the ventral ones. They lack oesophageal appendages and the brain has a posterior incision [30, 31].

***Cernosvitoviella atrata* (Bretscher 1903)**

The specimens of *C. atrata* collected in Serra do Xistral (Figure 3a) usually have 2-4 setae in the dorsal preclitellar bundles (Figure 3b) and 2-6 in the ventral ones, although higher numbers have also been observed in some specimens. Two pairs of primary pharyngeal glands are located in segments 4 and 5 together with two additional secondary pairs in segments 5 and 6. The spermathecae sit between them in segment 5 [30, 31, 41, 42].

Nephridia are located in 6/7 (one pair), 7/8 (unpaired) and 9/10 (unpaired) [42]. The origin of the blood vessel is located either in furrows 11/12 [42] or in segment 13 [41]. Due to the great variability between specimens in terms of size and shape of the coelomocytes and spermathecae, *C. atrata* is probably a species complex [30, 31, 41]. *C. atrata* has a wide geographical distribution, being often recorded in moist soils, such as peatland areas or lake shores, which suggests that soil water content is a limiting environmental factor for their survival [30, 31, 33, 43, 44]. At our Galician peatlands *C. atrata* was very abundant, but their densities reached the greatest values ($p < 0.05$) at the moister site (Chan do Lamoso) (Table 3).

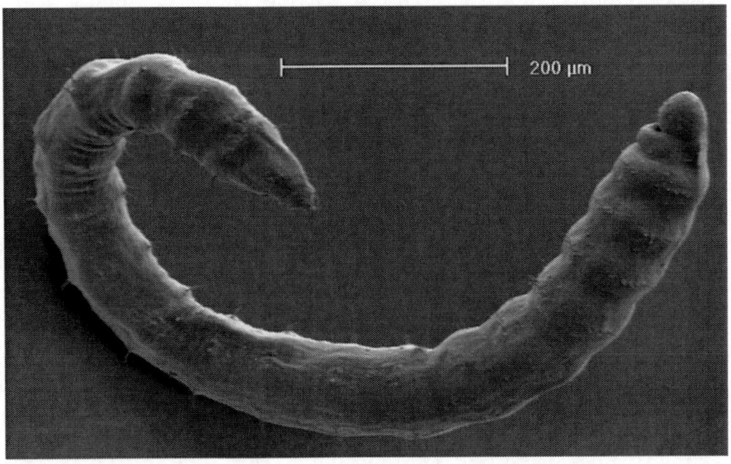

Figure 3a. Microphotograph (SEM) of *Cernosvitoviella atrata*: full view of one individual.

Figure 3b. Microphotograph (SEM) of *Cernosvitoviella atrata*: ventral view of the anterior region.

Genus Marionina Michaelsen 1890

The specimens of *Marionina* are small enchytraeids that usually measure less than 5 mm and rarely reach 7 mm in length. The number of setae is low, rarely more than 3 per bundle and even totally absent from others; these setae are straight although in some species could be sigmoid. There are three pairs of primary pharyngeal glands but lack secondary ones [30, 31]. This genus is very heterogeneous and includes specimens with high morphological variability [31].

Marionina filiformis **Nielsen and Christensen 1959**

This species (Figure 4a) was re-described by Chalupský [41] as a result of some discrepancies with the original description [31]. It is a small enchytraeid (about 5 mm long and 27-30 segments) with two straight, but slightly bent distally, setae per segment (Figure 4b), although some bundles could contain only one setae, a common phenomenon in *Marionina* spp. Its brain is elongated with a narrow and deep incision in its posterior margin and the dark coelomocytes contain brown granules. There are not oesophageal appendages and the primary pharyngeal glands are located in segments 4, 5 and 6: the first pair forms one dorsal and two ventral lobes in segment 5, whereas two dorsal and two ventral lobes are observed in segment 6 (bigger in size than the previous pair). Less than three preclitellar nephridia are present (i.e. one pair followed by a single one) and located in segments 7/8 and/or 9/10. The dorsal blood vessel, with slightly red blood, originates from segment 14 and the spermathecae are tubular, straight, simple and connected to the oesophagus [41, 45]. *M. filiformis* is frequent in organic soils with high water content and consequently, it is considered to be a wet-site indicator. It is usually concentrated at the deeper layers because it can tolerate, to a great extent, anoxic conditions [33, 44].

It was well represented in the studied area, although it was mainly restricted to the wettest site (Chan do Lamoso), where an average number of 17,000 individuals per m^2 were recorded (Table 3).

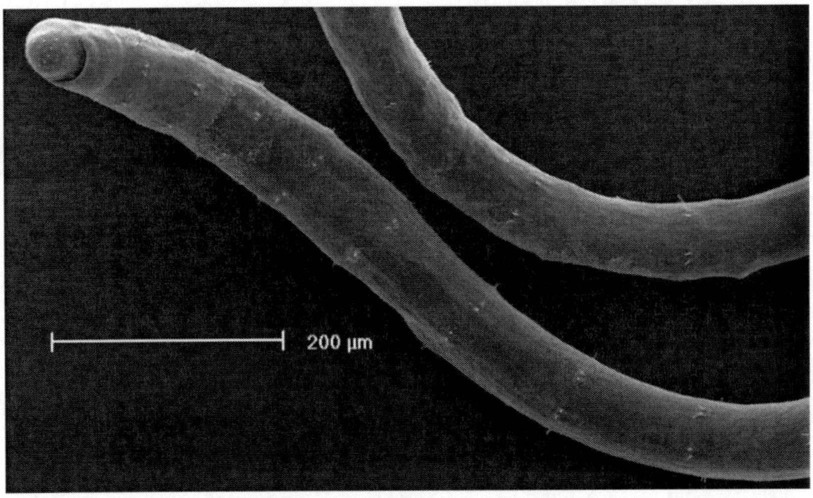

Figure 4a. Microphotograph (SEM) of *Marionina filiformis*: ventral view of the anterior region.

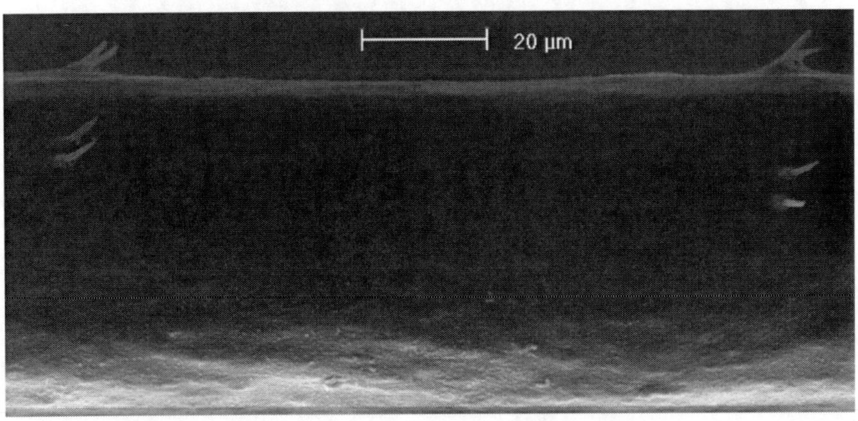

Figure 4b. Microphotograph (SEM) of *Marionina filiformis*: setae.

The presence of *M. filiformis* in Serra do Xistral represents the most southwest record in Europe and was subjected to a more detailed study [45] in order to clarify a few morphological disagreements with the original description [31] and first noticed by Chalupský [41]. From this it was concluded that rather than the existence of two different species, intraspecific variability (e.g. body length, shape of the anterior margin of the brain and the morphology of the clitellar cells) and observational errors (e.g. clitellum extension and the number of cell layers on the spermathecal ectal duct) are the most likely interpretations of the observed differences [45].

Marionina clavata Nielsen and Christensen 1961

This species (Figure 5a) is slightly shorter than *M. filiformis*, with 24-29 segments and length of 4-7 mm. The setae are short and straight, but totally absent from the lateral bundles of segments 9-11 and often just one in those ones of segments 7 and 8 (Figure 5b). The anterior margin of the brain is convex but deeply incised in its posterior one; the coelomocytes are often brownish in colour. One pair of short and tubular oesophageal appendages are present dorso-laterally in segment 4 and the three pairs of pharyngeal glands (4, 5 and 6) are dorsally connected and have ventral lobes in segments 5 and 6. Two pairs of preclitellar nephridia are usually in segments 7/8 and 8/9 although in some cases they can be found in 8/9 and 9/10. The dorsal blood vessel originates from segment 12 with an anterior bifurcation in the first segment and the blood is colourless. The spermathecae are fused proximally, with a long thin ectal duct. Unlike the previously described species, the seminal vesicles are small or even absent [30, 31, 42].

This species is typically found in acid soils [42] and in fact, it could be totally absent if soil pH values reach above 5.5. In addition, it also prefers well aerated but moist soils [43, 44]. The two investigated sites hardly showed any differences in soil acidity; however, *M. clavata* was only found at the most continental site (Chan do Eume), although in low densities (Table 3), possibly associated to a lower water content and thus, with more air spaces at this site.

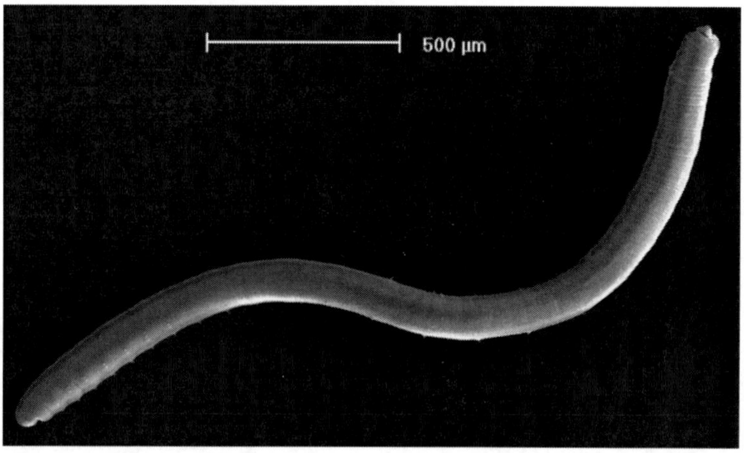

Figure 5a. Microphotograph (SEM) of *Marionina clavata*: full view of one specimen.

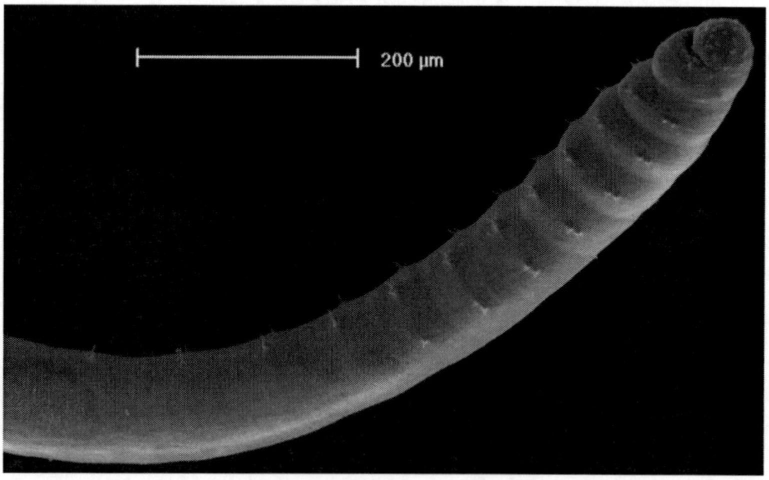

Figure 5b. Microphotograph (SEM) of *Marionina clavata*: ventral view of the anterior region.

Genus Mesenchytraeus Eisen 1878

All the European species belonging to the genus *Mesenchytraeus* are large worms (usually more than 8 mm in length), typically with more than three sigmoid setae in several bundles but, in most species, higher numbers can be counted in the ventral bundles than in the lateral ones. They lack oesophageal appendages and the brain is relatively short with its posterior margin being variable in shape [30, 31].

***Mesenchytraeus flavus* (Levinsen 1884)**

The specimens of *M. flavus* collected at the two Galician sites (Figure 6a) reached 35 mm (59-66 segments) and in some cases up to 13 setae were counted in the preclitellar ventral bundles (Figure 6b). The main characteristic of this species is the intense yellow colour observed in live specimens due to the presence of small coelomocytes containing yellowish refractive granules.

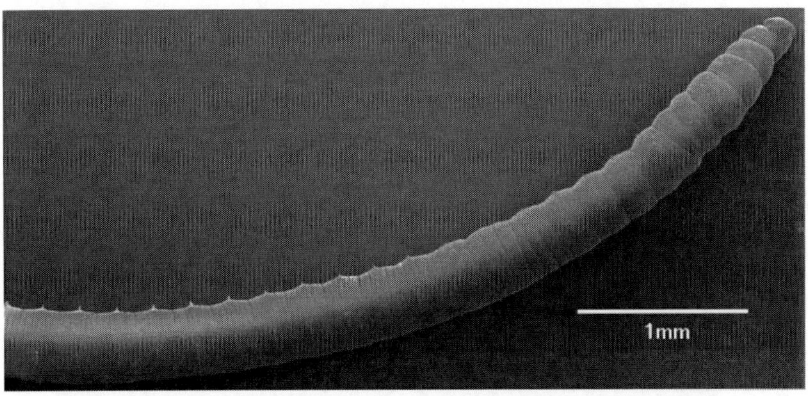

Figure 6a. Microphotograph (SEM) of *Mesenchytraus flavus*: lateral view.

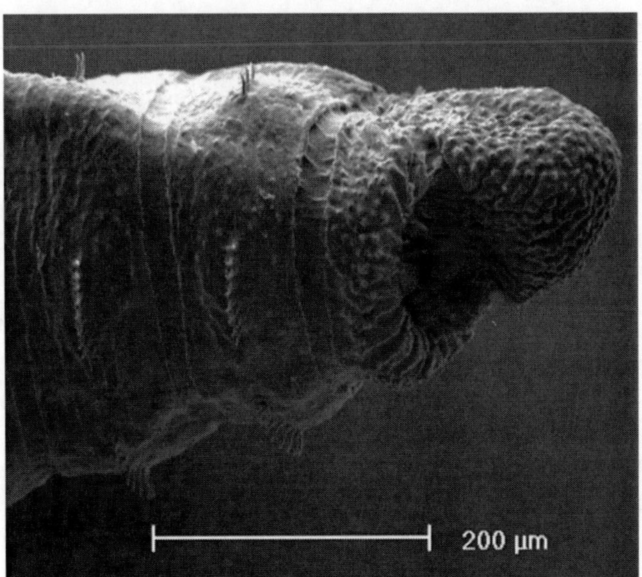

Figure 6b. Microphotograph (SEM) of *Mesenchytraus flavus*: ventro-lateral view of the first segments.

No oesophageal appendages are present and the number of pharyngeal glands is one pair in each one of the segments 4 and 5 plus secondary impair smaller glands (one per segment) in segments 5 and 6. A total number of five pairs of preclitellar nephridia can be found from 6/7 to 10/11 (Schmelz pers. comm.). The origin of the dorsal blood vessel is located in segments 17-18 and the blood is colourless [30, 31].

The spermathecae are connected to the oesophagus and consists of an ectal duct and an asymmetrical diverticulum of similar length, with the ampulla being slightly longer [30, 31].This species is mainly found in fairly wet and cold soils and consequently, it has been often recorded in arctic and subarctic areas [31, 46]. In our Galician peatlands only one specimen of *M. flavus* was found at Chan do Lamoso (Table 3) and thus, its presence in Serra do Xistral should be considered as rare.

Mesenchytraeus sanguineus Nielsen and Christensen 1959

This species (Figure 7a) is slightly shorter than the previous one (length 10-15 mm and about 45 segments) and its specific epithet derives from the colour of its blood (red). Unlike the previous *Mesenchytraeus* species, *M. sanguineus* has a similar number of setae (5 in most cases) in both lateral and ventral bundles. The setae are much thinner at their distal end than at their base (Figure 7b). The coelomocytes contain oval and slightly refractive granules. Two pairs of primary pharyngeal glands dorsally connected are located in segments 4 and 5 and one secondary pair in segment 6 (sometimes also in 7; each one with two lobes). Three pairs of preclitellar nephridia are located in segments 6/7, 8/9 and 9/10 (Schmelz pers. comm.). The dorsal blood vessel originates from segment 13 and the spermathecae, which are not attached to the oesophagus, have a very long ectal duct, twice as long as the diameter of the enchytraeid body [30, 31].

M. sanguineous is a semi-aquatic species and thus, it is commonly found in flooded and not well aerated soils such as peatland areas; furthermore, it prefers acid conditions but occasionally it can be found in more basic soils, with pH values around 5 or slightly higher [30, 31, 33, 43, 44].

Figure 7a. Microphotograph (SEM) of *Mesenchytraeus sanguineus*: lateral view of the anterior region.

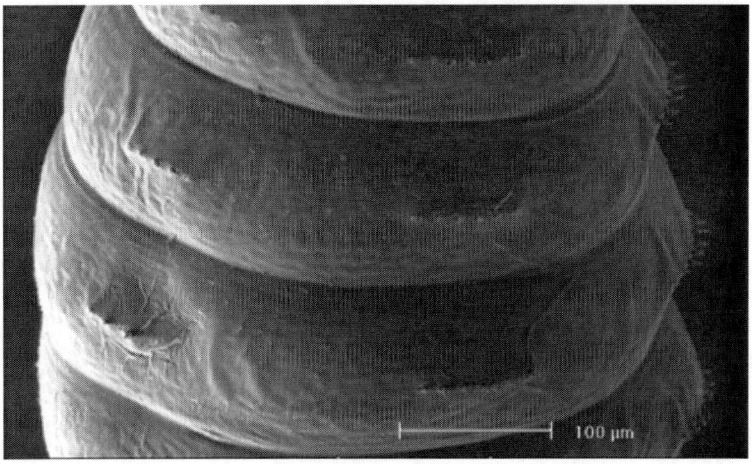

Figure 7b. Microphotograph (SEM) of *Mesenchytraeus sanguineus*: setae.

Although this enchytraeid species was not very abundant in Serra do Xistral, their densities were significantly higher ($p < 0.05$) in the wettest soils of Chan do Lamoso than at Chan do Eume (Table 3).

Mesenchytraeus lusitanicus Collado, Martínez-Ansemil and Giani 1993

The species name derives from being particularly abundant in the NW of the Iberian Peninsula (Galicia and North of Portugal). *M. lusitanicus* is similar in size to the previous species: length about 9-15 mm and 43-52 segments. No more than six setae are present in each bundle, appearing in a similar number in both lateral and ventral ones (Figure 8a), but always less than four setae in those of the postclitellar segments. The shape of the setae is also similar to those of the previous species (very thin at their distal end; Figure 8b). The coelomocytes are abundant, disc-shaped and contain round granules [30, 47].

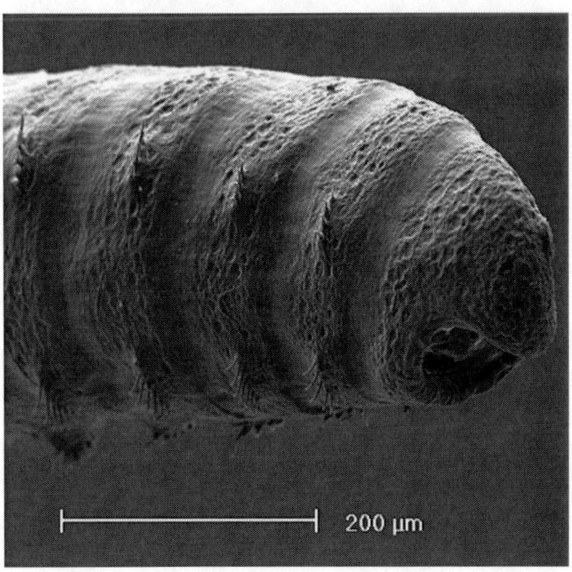

Figure 8a. Microphotograph (SEM) of *Mesenchytraeus lusitanicus*: lateral view of the anterior region.

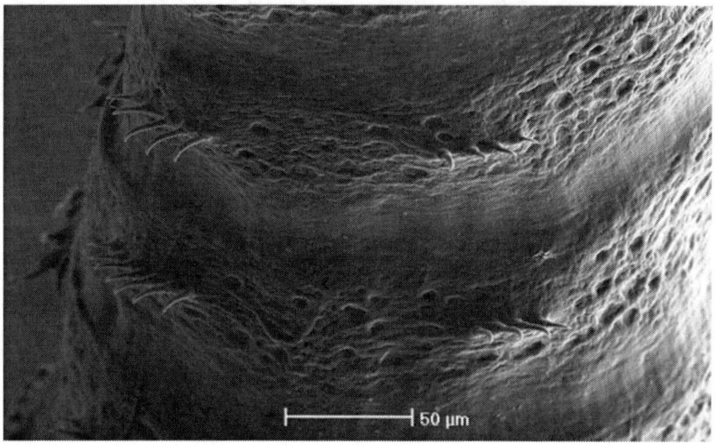

Figure 8b. Microphotograph (SEM) of *Mesenchytraeus lusitanicus*: closed view of the setae bundles.

Two pairs of primary pharyngeal glands are present in segments 4 and 5 plus two or three pairs of secondary glands in segments 5, 6 and/or 7. There are four pairs of preclitellar nephridia from segments 6/7 to 9/10 (Schmelz pers. comm.). The dorsal blood vessel originates from segment 13. The spermathecae are free from the oesophagus, with an ectal duct shorter in length than the club-shaped ampulla [30, 47]. These worms are commonly found in acid and wet soils [47] and together with its reported geographical distribution explain why this *Mesenchytraeus* species was the most abundantly recorded in Serra do Xistral, with no significant differences being detected between the two localities (Table 3).

Genus Cognettia Nielsen and Christensen 1959

This genus includes both small (< 6 mm) and large specimens (> 20 mm) with two or three sigmoid setae per bundle which lack nodulus. Its oesophagus, without appendages, is not joined to the simple spermathecae. Most *Cognettia* species usually have a brain with a posterior incision, but it could also be straight in other specimens [30].

***Cognettia sphagnetorum* (Vedjoský 1877)**

These enchytraeids (Figure 9a) measure about 10-25 mm in length (40-65 segments) and the setae are arranged as two or three per bundle (Figure 9b). Most individuals have one pair of primary pharyngeal glands dorsally separated located in 3 or more segments, from 4 to 8; however, there is certain variation between individuals and thus, some have only three pairs in 4-6, in other cases the anterior lobes can be dorsally connected or there could be some kind of ventral glands (which are really slight swellings on the afferent fascicles) in segments 6-9.

The dorsal blood vessel usually originates from segments 15-16. Because this species usually reproduces by fragmentation, the first pair of nephridia is located in segments 8/9 or have a posterior location, whereas the sexual organs (and the clitellum if present) are displaced forwards three or four segments, with the male pore usually located on segment 9 (instead of 12). The seminal vesicle is well developed in most cases or even absent in others [30, 31, 42]. This enchytraeid species has a wide distribution but it prefers organic, acid and moist soils [30, 43, 44] where it could become the dominant species [37, 40, 48-50].

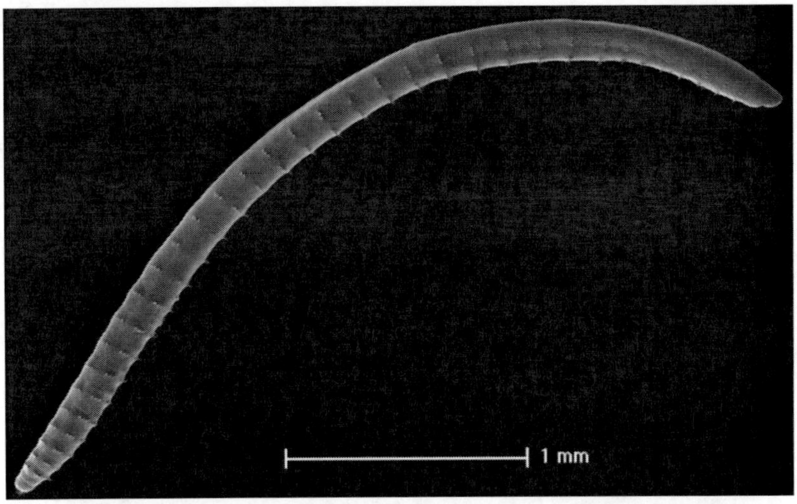

Figure 9a. Microphotograph (SEM) of *Cognettia sphagnetorum*: full view of one individual.

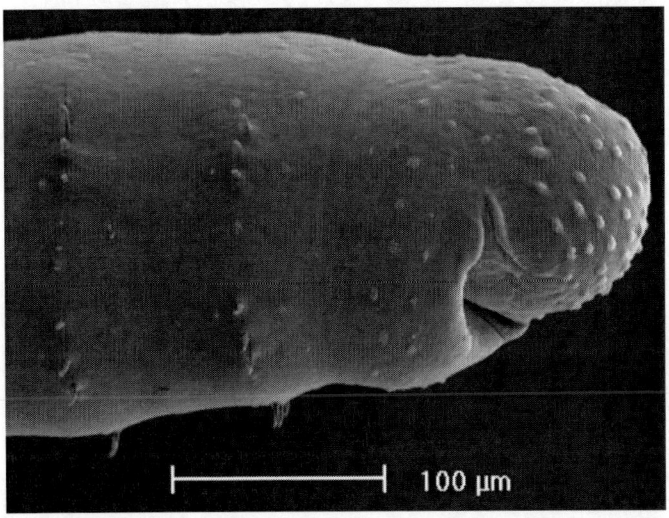

Figure 9b. Microphotograph (SEM) of *Cognettia sphagnetoru:* lateral view of the anterior region.

Accordingly, *C. sphagnetorum* was also the most abundant enchytraeid species in Serra do Xistral, with an average population density of about 39,000 individuals m^{-2} at the most oceanic locality, whereas half this number ($p < 0.05$) was recorded at Chan do Eume (Table 3).

Cognettia hibernica Healy 1975

This small sized species (< 5 mm in length and 34 segments) has only two setae per bundle across all its entire body (Figures 10a, b). The coelomocytes are scarce, small, with fine granules inside. The lobes of the first and second pairs of pharyngeal glands (located in segments 4 and 5) are broadly connected, whereas the two lobes of the third pair (in segment 6) are separated, elongated and bigger in size. In addition, there are small secondary glands in segments 5 and 6. The preclitellar nephridia are placed in segments 6/7 and 7/8 (sometimes in

10/11), the dorsal blood vessel becomes evident from segment 13 and there is no seminal vesicle [30, 47]. The geographical distribution of *C. hibernica* seems to be limited to Ireland and northern Spain, where it is usually found in soils with high water content (70-80%) and low pH (below 5.5), typical conditions of peatlands, coniferous forests and acid grasslands, but it has also been collected in streams, neutral grasslands, deciduous woodlands or marshes [30, 33, 43, 47]. In Serra do Xistral, *C. hibernica* was more abundant at the most continental locality than at the oceanic one ($p < 0.05$; Table 3), with more than 15,000 individuals per m^2 and thus, reaching a similar density to *C. sphagnetorum* at this particular drier site.

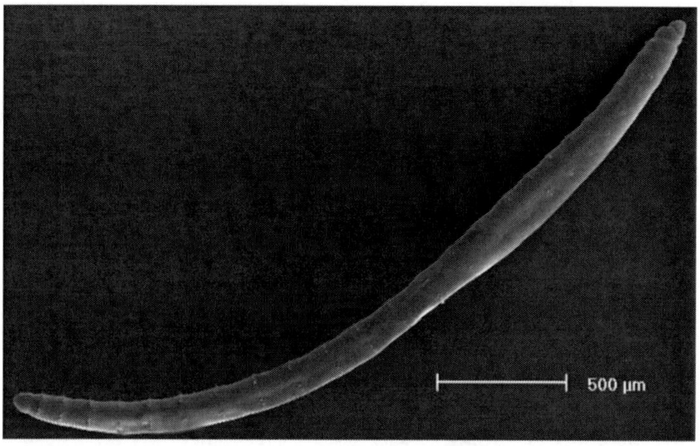

Figure 10a. Microphotograph (SEM) of *Cognettia hibernica*: full view of one individual.

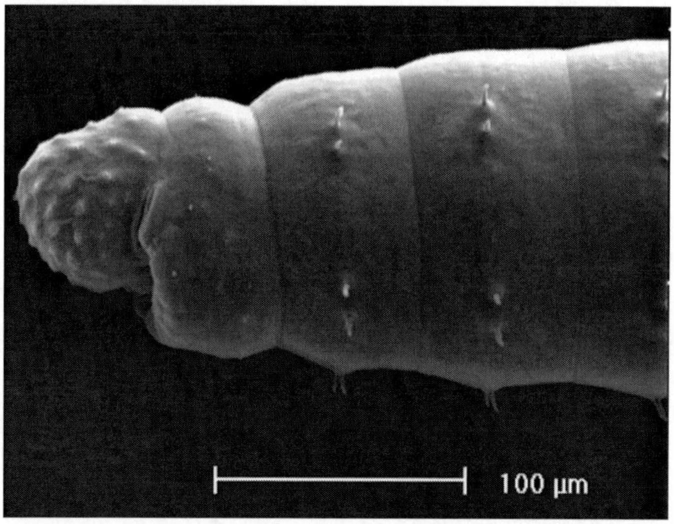

Figure 10b. Microphotograph (SEM) of *Cognettia hibernica*: lateral view of the anterior region.

Cognettia cognettii (Issel 1905)

This species is smaller in size than *C. sphagnetorum* but larger than *C. hibernica*, with length 6-11 mm and segment number 30-35 (Figure 11a).

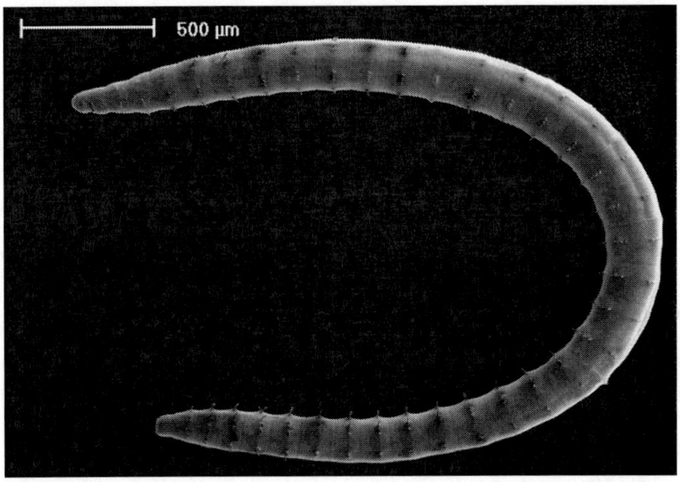

Figure 11a. Microphotograph (SEM) of *Cognettia cognettii*: full view of one individual.

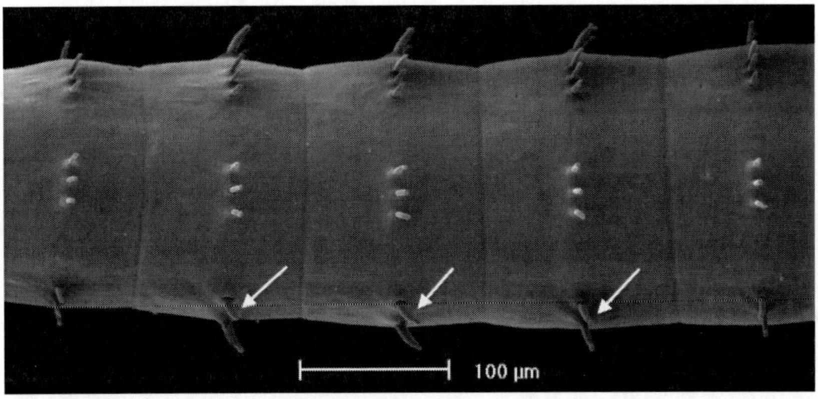

Figure 11b. Microphotograph (SEM) of *Cognettia cognettii*: ventral view of segments III-VII with the arrows indicating the lateral enlarged chaetae in segments IV, V and VI.

Although according to the original description [51] *C. cognettii* has three sigmoid yellowish setae per bundle, the revised description of this species provided by Cernosvitov [52] indicates that most of the lateral bundles only contain two setae, being characteristic the presence of enlarged setae (about one third longer than average) in two or three lateral bundles on segments 3-7 (Figure 11b). Three pairs of primary pharyngeal glands are present in segments 4, 5 and 6 (each one with only one dorsal lobe, but two lobes could also be present in segment 4) and three pairs of secondary ones in segments 5, 6 and 7 (the latter bigger in size than the two previous ones). There are four pairs of preclitellar nephridia (from 6/7 to 9/10). The coelomocytes contain fine, pale, occasionally slightly yellowish or opaque granules and the dorsal blood vessel originates from segment 14. The spermathecae have a long and thin ectal duct in segments 5 and 6 and become wider when it reaches the sac-shaped ampulla at its distal end [30, 52]. *C. cognettii* is widely distributed, although its greatest densities have been recorded in slightly acid soils with 50-70% of soil water content [30, 43, 44]. Because of this preference for drier soils, this species was more abundant ($p < 0.05$) at the most continental site than in Chan do Lamoso (Table 3).

Cognettia clarae Bauer 1993

The external morphology of this species (Figure 12a) is similar to that of *C. cognettii* but larger in size, measuring from 10 to 16 mm and having 55-67 segments. The description provided by Bauer [53] based on Austrian populations indicates that it has only two setae per bundle on every segment; however, the specimens collected in Serra do Xistral belong to the "occidental group", being characterised by the presence of three setae in the ventral bundles from segments 2/3 to 5/6 (Figure 12b) [30]. The coelomocytes are oval and filled up with small translucent granules. Each of the three primary pharyngeal glands, located in segments 4, 5 and 6, has two lobes which are dorsally separated. Three secondary lobes are also present in segments 5, 6 and 7. Their nephridia are unique because the first pair is located in a very anterior position, in segments 2/3 and the nephridial vesicle is large. In addition, the seminal vesicle is absent [30, 53].

C. clarae has a Holarctic distribution, but it is more frequent in cold mountain regions of Central and North Europe [30, 53]. Its presence in the more temperate region of Serra do Xistral was rather scarce and only very few individuals were collected at Chan do Eume (Table 3).

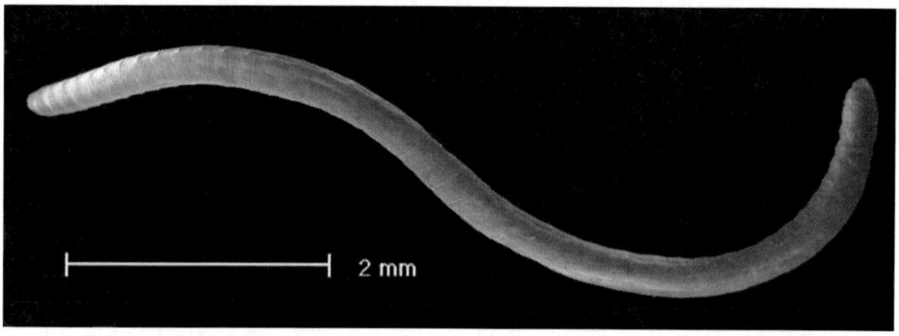

Figure 12a. Microphotograph (SEM) of *Cognettia clarae*: full view of one individual.

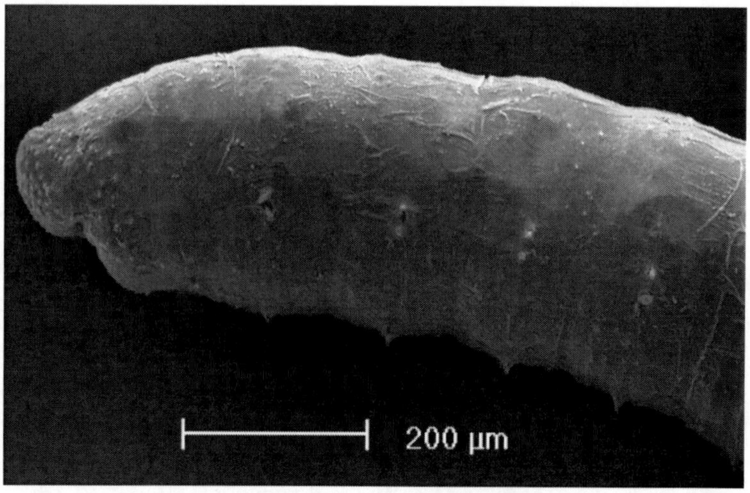

Figure 12b. Microphotograph (SEM) of *Cognettia clarae*: lateral view of the anterior region.

Conclusion

The Oligochaeta community from the upland peatlands within Serra do Xistral was mainly represented by two families, Lumbricidae and Enchytraeidae. However, only two species of lumbricids were collected in very low densities (about 50 individuals per m^2) at both localities, whereas enchytraeids showed a far greater diversity (up to 11 species were identified) and their population numbers reached, on average, more than 60,000 individuals per m^2. In addition, although at both sites the enchytraeid community was dominated by one species (*Cognettia sphagnetorum*), its structure showed marked differences between localities. Overall, enchytraeid numbers were significantly higher at the oceanic site (Chan do Lamoso) due to the greater abundance of *C. sphagnetorum, Cernosvitoviella atrata, Marionina filiformis* and *Mesenchytraeus sanguineus*, whereas the drier conditions at Chan do Eume significantly favoured higher population numbers of five species and in particular those of *C. hibernica* and *Achaeta* spp. which reached similar densities to those of the dominant *C. sphagnetorum* at this site. Finally, *M. flavus* and *M. lusitanicus* showed similar abundances at both sites. Our results highlight the close link between enchytraeids and soil water content in these Galician peatland habitats. Due to the key role these oligochaetes play on soil organic matter decomposition and nutrient cycling, any alterations in the abiotic factors are expected to have important implications in the functioning of these systems.

Acknowledgments

We thank Dr. Esther Barreal, Dr. Pedro Pablo Gallego and Dr. Pablo Ramil-Rego for their assistance in the field. We are also indebted to Dr. Rüdiger Schmelz for his valuable help in identifying enchytraeid worms.

References

[1] Z. Yu, D. W. Beilman, S. Frolking, G. M. MacDonald, N. T. Roulet, P. Camill and D. J. Charman, *Eos* 92, 97 (2011).

[2] H. Čížková, J. Kvét, F. A. Comín, R. Laiho, J. Pokorný and D. Pithart, *Aquatic Sciences* DOI: 10.1007/s00027-011-0233-4 (2011).

[3] P. Ramil-Rego, M. Rodríguez Guitián and J. Rodríguez Oubiña in *Avances en la Reconstrucción Paleoambiental de las Áreas de Montaña Lucenses*, pp. 165-187, Excma. Diputación Provincial de Lugo, Lugo (1996).

[4] J. B. Cragg, *The Journal of Ecology* 49, 477 (1961).

[5] L. Brussaard, D. K. Aanen, M. J. I. Briones, T. Decaëns, G. B. De Deyn, T. M. Fayle, S. W. James and T. Nobre, in *Soil Ecology and Ecosystem Services*, pp. 563-570, Oxford University Press, Oxford (2012).

[6] P. Lavelle and A. Martin, *Soil Biology and Biochemistry* 12, 1491 (1992).

[7] L. Brussaard, V. M. Beham-Pelletier, D. E. Bignell, V. K. Brown, W. A. M. Didden, P. Folgarait, C. Fragoso, D. W. Freckman, V. V. S. R. Gupta, T. Hattori, D. L.

Hawksworth, P. Lavelle, D. W. Malloch, J. Rusek, B. Söderström, J. M. Tiedje and R. A. Virginia, *Ambio* 26, 563 (1997).
[8] G. G. Brown, I. Barois and P. Lavelle, *European Journal of Soil Biology* 36, 177 (2000).
[9] P. M. C. Bruneau, D. A. Davidson, I. C. Grieve, I. M. Young and N. Nunan, *FEMS Microbiology Ecology* 52, 139 (2005).
[10] P. Lavelle, T. Decaëns, M. Aubert, S. Barot, M. Blouin, F. Bureau, P. Margerie, P. Mora, and J.-P. Rossi, *European Journal of Soil Biology* 42, S3-S15 (2006).
[11] K. Hedlund and A. Augustsson, *Soil Biology and Biochemistry* 7, 905 (1995).
[12] M. L. Rantalainen, H. Fritze, J. Haimi, O. Kiikkilä, T. Pennanen and H. Setälä, *Biology and Fertility of Soils* 39, 200 (2004).
[13] S. Moody, M. J. I. Briones, T. G. Piearce and J. Dighton, *Soil Biology and Biochemistry* 27, 1209 (1995).
[14] D. C. Coleman, D. A. Jr. Crossley and P. F. Hendrix, *Fundamentals of Soil Ecology* 2nd. ed., Academic Press, New York (2004).
[15] M. J. I. Briones, P. Ineson and A. Heinemeyer, *Global Change Biology* 13, 2252 (2007).
[16] S. Rivas-Martínez and J. Loidi Arregui, *Itinera Geobotanica* 13, 41 (1999).
[17] M. A. Rodríguez Guitián and P. Ramil-Rego, *Recursos Rurais* 4, 19 (2008).
[18] M. A. Rodríguez Guitián, P. Ramil-Rego, C. Real, R. Díaz Varela, J. Ferreiro da Costa and C. Cillero, in *Botánica Pirenaico-Cantábrica en el siglo XXI*, pp. 633-654, Universidad de León, León (2009).
[19] F. B. O´Connor, *Nature* 175, 815 (1955).
[20] J. C. Coulson and J. B. Whittaker in *Production Ecology of British Moors and Montane Grasslands*, pp. 52-93, Springer Verlag, Berlin (1978).
[21] J.E. Peachey, *Pedobiologia* 2, 81 (1963).
[22] M. J. I. Briones, P. Ineson and T. G. Piearce, *Applied Soil Ecology* 6, 117 (1997).
[23] R. J. Blakemore in *A series of searchable texts on earthworm biodiversity, ecology and systematics from various regions of the world. 2nd Edition and Supplement*, COE Soil Ecology Research Group, Yokohama National University, Japan (2007).
[24] M. J.Swift, O. W. Heal and J. M. Anderson, *Decomposition in terrestrial ecosystems*, Blackwell Scientific Publications, Oxford (1979).
[25] M. B. Bouché, *Lombriciens de France, Ecologie et Systematique*, Publ. Institut National de la Recherche Agronomique, Paris (1972).
[26] R. W. Sims and B. M. Gerard, *Earthworms: Notes for the identification of British species, synopses of the British fauna (New Series)*, Linnean Society, London (1999).
[27] J. A. Svendsen, *Journal of Animal Ecology* 26, 411 (1957).
[28] J. Álvarez, *Los oligoquetos terrícolas de la Peninsula Ibérica,* Publicaciones de la Facultad de Ciencias de la Universidad Complutense, Madrid (1971).
[29] E. Rota and C. Erséus, *Annales Zoologi Fennici* 34, 89 (1997).
[30] R. M. Schmelz and R. Collado, *Soil Organisms* 82, 1 (2010).
[31] C. O. Nielsen and B. Christensen, *Natura Jutlandica* 8-9, 1 (1959).
[32] R. M. Schmelz, *Taxonomy of Fridericia (Oligochaeta, Enchytraeidae). Revision of species with morphological and biochemical methods.* Abhandlung des Naturwissenschaftlichen Vereins, Hamburg (2003).
[33] B. Healy, *Journal of Life Sciences Royal Dublin Society* 1, 39 (1979).

[34] J. A. Springett, *Journal of Animal Ecology* 39, 725 (1970).
[35] V. Standen, *Journal of Animal Ecology* 42, 219 (1973).
[36] J. A. Springett and P. M. Latter, *Journal of Animal Ecology* 46, 959 (1977).
[37] H. Lundkvist, *Journal of Applied Ecology* 20, 873 (1983).
[38] G. A. van Duinen, T. Timm, A. J. P. Smolders, A. M. T. Brock, W. C. E. P. Verberk and H.Esselink, *Hydrobiologia* 564, 143 (2006).
[39] K. Maraldo, I. K. Schmidt, C. Beier and M. Holmstrup, *Soil Biology and Biochemistry* 40, 1765 (2008).
[40] M. Prendergast-Miller, V. Standen, R. Rees, J. Parker, Leith I. and L. Sheppard, *European Journal of Soil Biology* 44, 101 (2008).
[41] J. Chalupský, *Zoologica Scripta* 21, 133 (1992).
[42] E. Rota, *Italian Journal of Zoology* 62, 183 (1995).
[43] B. Healy, *Pedobiologia* 20, 159 (1980).
[44] Beylich and U. Graefe, *Annelid coenoses of wetlands representing different decomposer communities*, G.Broll, W. Merbach and E. Pfeiffer, Berlin (2002).
[45] R. M. Schmelz, R. Collado and M. J. I. Briones, *Verhandlungen des Naturwissenschaftlichen Vereins in Hamburg (Neue Folge)* 44, 23 (2008).
[46] B. Christensen and K. Dózsa-Farkas, *Polar Biology* 29, 454 (2006).
[47] R. Collado, E. Martínez-Ansemil and N. Giani, *Annales de Limnologie* 29, 129 (1993).
[48] T. Klungland, *Pedobiologia* 22, 112 (1981).
[49] H. Lundkvist, *Pedobiologia* 23, 21 (1982).
[50] E. Uhía and M. J. I. Briones, *Acta Oecologica* 23, 349 (2002).
[51] R. Issel, *Zoologische Jahrbücher, Abteilung für Systematik* 22, 451 (1905).
[52] L. Cernosvitov, *Proceedings of the Zoological Society of London* 114, 524 (1945).
[53] R. Bauer, *Linzer Biologische Beitraege* 25, 685 (1993).

In: Invertebrates: Classification, Evolution and Biodiversity ISBN: 978-1-62618-789-4
Editor: Rafael Riosmena-Rodriguez

Chapter 3

Marine Zooplankton Grazing Studies in Mexico: An Overview

Jorge García-Pámanes[1] and José Rubén Lara-Lara[2]
[1]Departamento de Biología Marina, UABCS, La Paz, México
[2]Centro de Investigación Científica y de Educación Superior de Ensenada, México

Abstract

The first work in Mexico on macrozooplankton grazing was published in 1989 and that on microzooplankton appeared in 1991. Only six reports exist on microzooplankton grazing processes; in all of them the dilution method was used. Seven works have been published on macrozooplankton grazing; three of them were carried out in Todos Santos Bay, B.C., at sites with different dynamic characteristics.In two of them, ^{14}C was used as a tracer, one in an upwelling area and another in an estuarine environment. The method of change in chlorophyll-a concentration was used to develop a third survey in a stable hydrographic area with oceanic influence. The fourth report, carried out in the central region of the Gulf of California, where ^{14}C was also used as a tracer. Results on daily variations indicate that grazing rates are up to 75% higher during the night-time that during the day. The perspective of the future scenario is not encouraging for the development of this research line; only a reduced group of researchers are working currently on this topic in Mexico. It is necessary to realise the importance of having information on *in situ* consumption rates of the zooplankton; it is only by having this information that we will be able to estimate the ecological role of the zooplankton and its function in the ecosystem. This situation is changing due to the efforts several authors or co-authors of the above papers who are actively working on this topic or are in the stage of preparation of articles on specific grazing rates and filtration for micro- and macrozooplankton, using information derived from previous work. Moreover, the number of working groups that address this issue are increasing gradually.

Introduction

Studies of the carbon cycle are closely related to the understanding of the flux of organic matter in the ocean [1]. The knowledge of the food habits of species of zooplankton is a step toward assessing the role of these species in pelagic food webs. Consumption of phytoplankton by zooplankton is called grazing or herbivory, is one of the causes of mortality of phytoplankton, along with the collapse of the euphotic zone cells [2] and viral diseases [3], but is the factor which has been considered of greatest significance in regulating phytoplankton populations. Commonly in studies of zooplankton, grazing is divided into two classes: macrozooplankton (greater than 200 micron organisms) and microzooplankton (organisms smaller than 200 microns). The macrozooplankton (or mesozooplankton) group is composed mainly of cladocerans, copepods, theropods, euphausiids, chaetognaths and larvae of various groups [4]. Of these, the group of copepods is primarily responsible for the grazing of phytoplankton. In contrast, the microzooplankton group is composed of protozoa and larvae, the nauplius stages of metazoans. Two suborders of protozoan ciliates, the tintinnids and oligotrichios, are ubiquitous and abundant in pelagic ecosystems and usually dominate the fauna microzooplanctonica [5]. The range of food particles potentially available to an organism is a function of the morphology of their food collector device [6]; therefore, it is necessary to separately perform measurements of consumption and the macrozooplankton/ microzooplankton, since the two groups affect different fractions of phytoplankton.

Background Studies on Grazing by Zooplankton in Marine Environments of Mexico

The first Mexican investigator who addressed the topic was Dr José Rubén Lara-Lara, when he undertook his doctoral studies at Oregon State University, USA. Returning to Mexico prompted the development of this line of research, directing the master's thesis of Jorge Garcia-Pámanes, which was the first work investigating grazing in the country, and that of Mary Eugenia Matus-Hernandez, years later, who directed the doctoral thesis of the same investigator, which resulted in two publications related to the topic.

Since then, other researchers have been interested in the issue of grazing, including Antonio Trujillo-Ortiz, who led the master's thesis of Mary Moramay Badillo-Padilla [7, 8, 9, 10, 11, 12]. Our series of publications has contributed to the general understanding of this process in the Ocean using examples from the Mexican coast.

Grazing by Marine Microzooplankton in Mexico

We found six references on this subject: the master's thesis of Matus-Hernández [13] was the first paper on the subject in the country. Garcia Pámanes PhD proyect [14], in which discuss grazing activity by microzooplankton, published in March 2001 [15]; Palomares-García [7]; Noriega-Cañar and Palomares-García [9] Herrera-Moro-Chao et al. [11] and Hernandez-Trujillo [12].

Ensenada B.C.

Matus-Hernández [13] performed their work in a coastal town in the Bay of Todos Santos at Ensenada B.C (Figure 1). Sampling was carried out daily for 15 days in July 1990. The objective was to evaluate grazing rates of the dominant herbivores (micro and macrozooplankton) on phytoplankton biomass. The microzooplankton grazing rates were determined by the dilution method [16], quantifying chlorophyll-a content by fluorimetry. Phytoplankton populations were dominated (>75%) by cells <2.6 microns, and the integrated concentration of chlorophyll-a values ranged from 50 to 500 mg m^{-2} and the primary productivity from 300 to 5600 mg C m^{-2} d^{-1}. Microzooplankton (200 microns) consisted mainly of ciliates, with an average surface abundance of 427 org. l^{-1}, whose grazing rates ranged from undetectable to 2.88 d^{-1}. Even when environmental conditions were stable during the study period, the grazing rates of microzooplankton showed high daily variability (CV => 170).

Gulf of California

García-Pámanes [14, 15] performed his work in the central Gulf of California in 1995(Figure 1). Grazing rates of microzooplankton (organisms <200 microns) were determined for two stations located on the sill of the Guaymas Basin (U) and the other in the central Gulf (RC) in three seasons (spring, summer and fall), using the dilution technique. García-Pámanes found that phytoplankton abundance was different in the sill and in the RC during the study period. In the U, diatoms were always dominant, while in the RC they were present during the spring and autumn, but were the dominant flagellates only during the summer. The abundance of flagellates and diatoms presented an inverted pattern in the two sampling regions. The microzooplankton was dominated by protozoa, with dinoflagellates being the largest group; the most frequent genera were *Amphidinium, Oxytoxum, Gyrodinium* and *Peridinium.* The ciliates were the second largest group, with the most common genera being *Mesodinium, Strombidium* and *Strombilidium.*

This author reported that the average growth rates of phytoplankton were 1.58, 1.64 and 1.82 d^{-1} for spring, summer and autumn, respectively, and the average grazing rates were 1.25, 1.93 and 1.84 d^{-1} for spring, summer and autumn, respectively (Table 1). No statistically significant differences were found between the rates of grazing seasons ($p > 0.05$), suggesting that both grazing and phytoplankton growth remained very constant throughout the year. It was estimated that the microzooplankton can consume between ~90, 106 and 100% of the daily primary production potential during the spring, summer and autumn, respectively (Table 1), which agrees well with estimates made in other temperate regions in both coastal and oceanic waters. Their results suggest that in the Guaymas Basin there are two seasons in a year: spring, where a relatively small fraction of nanophytoplankton fails to be removed and is possibly exported to the ocean, and summer-autumn, where the entire nanophytoplankton biomass is consumed in the pelagic system.

Palomares-García et al. [7] carried out their work in Bahia Concepcion B.C.S. (Figure 1), using the dilution method to estimate the rates of microzooplankton grazing by determining the change in pigment concentration by high-performance liquid chromatography (HPLC).

They found that concentrations of chlorophyll-a (Chl-a) varied widely (from 0.34 to 3.32 mg L^{-1}) showing two peaks in late spring and autumn, associated with the transition between mixed and stratified conditions. Growth rates of phytoplankton varied seasonally, with the lowest rates in summer (range 0.01-2.55 d^{-1} of Chl-a; 0.00-3.84 d^{-1} of Chl-b, 0.26 to 3.29 d^{-1} for fucoxanthin; 0.00 to 6.27 d^{-1} for peridinin; and 0.00 to 4.35 d^{-1} for zeaxanthin).

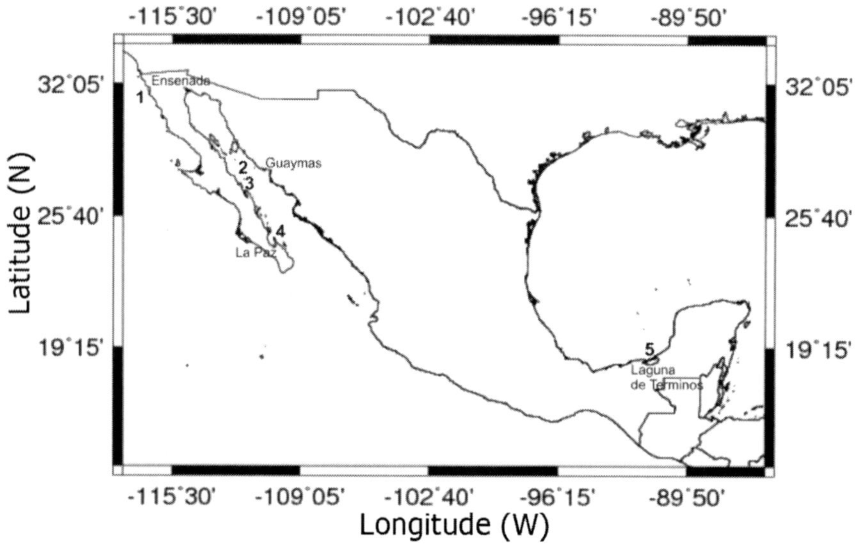

Figure 1. Location of the study areas.

Table 1. Microzooplankton grazing rates for the central zone of the Gulf of California [14, 15]

Date	Station	r^2	k	Phyto doubling	Phyto Potential Production	Real Prod. Phyto	g	% Take off. Potential
(1995)			(d^{-1})	(d^{-1})	($mgCm^{-3}d^{-1}$)	($mgCm^{-3}d^{-1}$)	(d^{-1})	**(d^{-1})**
Mar. 28	Drift	0.0	- 0.2 ns				0.1 ns	
Apr.02	Sill	0.6	1.6	2.3	55.7	5.6	1.25	90
Apr.02	Sill	0.2	1.1 ns				0.7 ns	
Aug.30	Sill	0.8	1.5	2.1	113.2	-11.3	1.9	110
Aug.30	Sill	0.6	1.5	2.2	122.2	- 9.0	1.9	107
Sep. 01	Drift	0.9	1.9	2.7	99.2	- 2.1	2.0	102
X			1.6±0.1	2.3±0.2	111.5±6.7	-7.5±2.8	1.9±0.05	106±2
Dic.01	Drift	0.2	1.1ns				1.3 ns	
Dic.06	Sill	0.8	1.8	2.6	78	- 0.3	1.84	100
Dic.06	Sill		- 0.4 ns				0.9 ns	

Phytoplankton growth rate *(k)*; phytoplankton doublings *(k* / In 2); potential production $[P_p = (P_0 e^{kt}) - P_0]$; actual production = $[P_a = (P_0 e^{[k-g]t}) - P_0]$; grazing rate *(g)*; % potential production grazed $[(P_p - P_a) / P_p]$.

Mean ± std error ($^{S}/_{\sqrt{n}}$); x = mean; ns = not significant.

Grazing by microzooplankton was a major loss process (range 0.00-1.89 d^{-1} of Chl-a; 0.00-3.12 d^{-1} of Chl-b; 0.26-3.29 d^{-1} for fucoxanthin; 0.00-2.03 d^{-1} for peridinin; and 0.00-3.51 d^{-1} of zeaxanthin). They suggested that some species of heterotrophic dinoflagellates, included in the genera *Gyrodinium* and *Protoperidinium*, which were abundant during the study period, may be responsible for grazing. Grazing rates represent the average of ~68-89% of the estimated average rates of pigment-specific phytoplankton growth. They conclude that their results indicate a tight coupling between community grazing microzooplankton and phytoplankton growth rates and do not reflect a selective feeding behaviour. These results suggest that microzooplankton play a fundamental role and can significantly modify the availability and efficiency of energy transfer at higher trophic levels.

Noriega-Cañar and Palomares-García [9] also worked in Bahia Concepcion, BCS, used chlorophyll-a levels, estimated by HPLC and microscopy, to determine rates of phyto-plankton growth; and microzooplankton grazing obtained by the dilution technique, in the middle region of Bay. They found that the growth rates (μ) of nano and microphytoplankton, estimated by microscopy, in different seasons were: 1.04 and 1.06 d^{-1} in the spring, 1.2 and 1.04 d^{-1} in the summer, 1.1 and 1.03 d^{-1} in the autumn, and 1.05 and 1.00 d^{-1} in the winter. Also, the consumption of nano- and microphytoplankton by microzooplankton (g) were: 1.06 and 1.12 d^{-1} in the spring, 1.2 and 1.07 d^{-1} in the summer, 1.03 and 1.10 d^{-1} in the autumn, and 1.14 and 1.02 d^{-1} for the winter.

The growth rates (μ) and grazing (g) of Chl-a evaluated by HPLC were: 2.47 and 1.33 d^{-1} in the spring, 1.31 and 1.00 d^{-1} in the summer; 2.55 and 1.89 d^{-1} in the autumn, and 0.29 and 0.05 d^{-1} in the winter. In the microscopic analysis, there was no clear preference for size class for the consumption of primary production potential removed (PPr). Estimates of phytoplankton growth rates (μ) and grazing by microzooplankton (g) determined with microscopy and HPLC, PPR were similar (98.0% for nanophytoplankton, 97.2% microphytoplankton, and 99.7% for HPLC).This indicates that tintinnidos and heterotrophic dinoflagellates, which were abundant during the study period, and in particular species belonging to the genera *Gyrodinium* and *Protoperidinium*, may be responsible for grazing.

Bay of La Paz, B.C.S.

Hernandez-Trujillo [12] estimated phytoplankton growth and the impact of grazing by microzooplankton in the fall and winter of 2007 and spring and summer of 2008 at a fixed station in the Bay of La Paz, BCS. The grazing experiments were performed using the method of dilution with surface water samples. The time series obtained showed a seasonal pattern in which grazing was highest in the winter, coinciding with the maximum growth of phytoplankton; this tend to reduce starting in winter and reached its minimum in the summer. The impact of grazing was higher in the winter when consumption was estimated up to 55% d^{-1} of phytoplankton production and markedly decreased along the year until a minimum of 3% in the summer. No correlation was seen between growth rates and grazing with temperature, but ANOVA detected differences in seasonal consumption due to the concentration of chlorophyll-a. Although microzooplankton herbivores had a definite seasonal pattern, this requires a greater number of observations including factors such as the type of phytoplankton and its influence on the variation in the rate of grazing.

Gulf of Mexico

The only reference found on grazing on the east coast of Mexico was the work of Herrera-Moro-Chao et al. [11], which determined the grazing pressure of microzooplankton on major phytoplankton groups in Laguna de Terminos, Campeche (Figure 1); the dilution method was used during the dry seasons of 2009 and 2010. Diagnostic pigments quantified different phytoplankton groups and the different size fractions were analysed by HPLC. Parallel-collected phytoplankton samples were fixed in formalin for later identification. The phytoplankton was dominated by diatoms with a total of 50 species.

This author found that higher grazing pressure was exerted on the picofitoplanctonica fraction with a value of 65% and 55% d^{-1} for the dry seasons of 2009 and 2010, respectively. Grazing analysis using diagnostic pigments showed that higher grazing pressure was exerted on the group of Cyanobacteria during the two seasons (35% and 51% d^{-1}, respectively) followed by the group of Bacillariophytas. They conclude that their results confirm that microzooplankton is an important mechanism of control on phytoplankton in coastal lagoons and can consume a significant portion of the total biomass, and that this predation is exerted on small fractions.

Grazing by Marine Macrozooplankton in Mexico

We found seven references to works regarding macrozooplankton grazing in Mexico, three of which were in the region of Ensenada, Baja California and four in the Gulf of California; one was in the central area and three in the Bay of La Paz.

Ensenada B.C.

García-Pámanes [17, 18] was the first work on grazing that took place in Mexico. The work consisted of a series over a period of 15 days, from 26th June to 10th July 1986. Samples were collected at a station located 5 km southwest of Punta Banda, Baja California (Fufure 1), designated as an area of intense upwelling. ^{14}C was used as a tracer and experiments were run in the laboratory.

Table 2. Daily average filter rates *(ml cop.$^{-1}$* day^{-1}) and specific grazing rates *(h^{-1})* of *Calanus pacificus* and *Acartia tonsa* in Punta Banda, Baja California [17, 18]

		Calanus pacificus				*Acartia tonsa*				
		Filtration rate		Grazing rate		Filtration rate			Grazing rate	
Date	*N*	X	CV	X (x 10^{-3})	CV	*N*	X	CV	X (x 10^{-3})	CV
27/VI/86	18	51.4±74.3	146.4	8.6±12.4	146.5	12	23.2±48.0	212.8	3.9±7.9	208.0
28/VI/86	30	28.6±52.9	186.6	4.7±8.8	187.0	20	29.3±26.0	90.0	4.1±3.6	90.1
29/VI/86	19	13.3±19.5	147.8	2.2±3.2	147.8					
30/VI/86	30	31.1±51.4	166.7	4.2±5.7	137.0					
01/VII/86	30	35.5±48.1	136.7	5.9±8.0	136.7					

		Calanus pacificus				*Acartia tonsa*				
		Filtration rate		Grazing rate		Filtration rate			Grazing rate	
Date	*N*	X	CV	X (x 10^{-3})	CV	*N*	X	CV	X (x 10^{-3})	CV
02/VII/86	30	28.5±45.7	161.9	5.0±7.6	152.0					
03/VII/86	30	26.7±33.9	128.0	4.4±7.6	128.1					
04/VII/86	30	11.6±31.7	275.3	1.9±5.3	279.1	18	263.0±792.1	305.3	43.8±132.0	305.3
05/VII/86	30	19.8±32.6	166.2	3.3±5.4	166.1	18	40.4±52.3	131.2	6.7±8.7	131.1
06/VII/86	30	43.6±78.2	180.7	7.3±13.0	179.7	18	34.1±31.7	94.2	5.7±5.3	94.2
07/VII/86	30	13.8±16.2	118.9	2.3±2.7	119.1	18	15.8±24.6	157.7	2.6±4.1	157.7
08/VII/86	30	12.7±18.3	144.5	2.0±2.9	150.5	18	76.0±88.3	117.8	12.7±14.7	117.6
09/VII/86	30	10.2±8.2	80.7	1.7±1.4	80.5	18	13.5±20.2	152.1	2.5±3.4	138.4
10/VII/86	30	2.5±6.7	276.0	0.4±1.1	275.6	18	1.4±2.9	202.8	0.2±0.5	196.9

N = number of data; X = mean; S = standard deviation; CV = variation coefficient corrected by number of data.

Table 3. Day to day variability of the macrozooplankton grazing rates and potential production of chlorophyll-a consumed by the macrozooplankton in the Todos Santos Bay [19]

Day	G	Productivity
July	(d^{-1})	consumed (%)
14	0.08	15
15	0.35	57
16	0.14	25
17	0.06	11
18	0.02	4
19	0.01	2
20	0.08	15
21	0.14	25
22	----	0
23	0.10	18
24	2.46	178
25	0.06	11
26	----	0
27	----	0
28	0.03	6

Table 4. Grazing and filtration rates of *Acartia californiensis* during the time series August 11th to 23rd 1994 in the Punta Banda estuary [20]

	Grazing Rates (h^{-1})			Filtration Rates (ml org.$^{-1}$ día^{-1})		
Day						
	Md	*X*	EE	Md	*X*	EE
August	(1 x 10^{-4})	(1 x 10^{-4})	(1 x 10^{-4})			
11	6.47	6.77	2.36	3.11	3.26	1.14
12	2.33	2.35	9.30	1.13	1.13	0.44
13	3.18	4.03	3.93	1.19	1.19	0.70

Table 4. (Continued)

	Grazing Rates (h^{-1})			Filtration Rates (ml org.$^{-1}$ día^{-1})		
Day						
	Md	*X*	EE	Md	*X*	EE
August	(1 x 10^{-4})	(1 x 10^{-4})	(1 x 10^{-4})			
14	7.00	1.31 x 10^{-3}	1.67 X 10^{-3}	3.36	3.90	2.07
15	8.56	9.37	4.11	4.11	4.50	1.95
16	7.50	7.62	3.45	3.60	3.65	1.66
17	6.22	6.82	3.88	2.98	3.28	1.85
18	4.48	4.79	2.65	2.14	2.29	1.28
19	2.79	3.38	2.39	1.34	1.62	1.14
20	3.53	5.37	5.78	1.69	2.58	2.77
21	2.03	2.75	2.06	0.97	1.32	0.98
22	3.25	3.34	4.42	1.56	1.60	0.89
23	3.75	4.88	4.29	1.79	2.34	2.07
Md total	4.17			1.99		
X		5.74	1.28		2.51	0.39

Md = moda; *X*= mean; EE = standard error.

Table 5. Macrozooplankton grazing rates (G d^{-1}) and primary production consumed (% PpR) in some localities of the central region of the Gulf of California [14, 21]

station	autumn 1994		spring 1995		summer 1995		autumn 1995	
	G	$\%P_pR$	G	$\%P_pR$	G	$\%P_pR$	G	$\%P_pR$
1	1.87	101	1.16	87	0.67	60	nd	
2	0.68	59	2.40	115	0.95	76	nd	
3	1.49	92	nd		0.69	61	nd	
4	nd		nd		1.57	98	nd	
5	0.96	74	2.30	113	1.74	102	1.73	98
6	1.14	81	1.32	92	4.75	122	1.01	76
7	nd		1.47	97	2.69	115	4.84	118
8	nd		1.04	81	0.31	33	1.89	101
9	nd		0.49	49	1.83	103	1.72	98
21	0.34	34	0.73	65				
22	0.13	15	0.35	37				
X	0.94±0.23	65±12	1.25±0.24	82±9	1.69±0.45	86±10	2.24±0.67	98±7
CV	66%	48%	58%	33%	81%	34%	67%	15%
Drift 1	0.20	22	1.88	107	1.02	79	0.92	72
Drift 2	4.18	nd	0.43	44	4.19	nd	1.11	80
Drift 3	0.71	60	0.40	42	1.08	nd		
Drift 4	0.55	50	0.71	64	1.75	102		
Drift 5	1.53	94	0.40	42	3.42	nd		
Drift 6			0.48	48	2.17	nd		
Drift 7			1.36	94				
X	1.44±0.72	57±15	0.81±0.22	63±10	2.27±0.53	91±11	1.01±0.10	76±4
CV	112%	53%	72%	43%	57%	18%	14%	18%

station	autumn 1994		spring 1995		summer 1995		autumn 1995	
	G	$\%P_pR$	G	$\%P_pR$	G	$\%P_pR$	G	$\%P_pR$
red	*n* = 12	*n* = 11	*n* = 16	*n* = 16	*n* = 15	*n* = 11	*n*=7	*n*=7
X	1.15±0.32	62±9	1.06±0.17	74±7	1.92±0.34	86±8	1.89±51	92±6
CV	96%	48%	64%	38%	69%	32%	72%	18%
Sill 11			1.30	92	1.63	99	0.85	68
Sill 12			1.08	83	0.24	26	0.85	68
Sill 13			0.74	66	0.56	52	1.11	80
Sill 14			0.61	58	5.12	122	0.21	23
Sill 15			0.58	55	1.32	90		
Sill 16			0.38	40				
X			0.78±0.14	66±8	1.77±0.87	78±17	0.75±0.19	60±13
CV			44%	29%	110%	49%	51%	42%
Sill 21							2.25	107
Sill 22							0.46	44
Sill 23							0.83	67
Sill 24							0.27	28
X							0.95±0.45	62±17
CV							95%	55%

X = mean; CV = coefficient of variation (standard deviation/ mean)x 100; ± = Standard error (S/√n); red = average drift stations including; Sill10 = spring tides; Sill20 = neap tides, nd = not determined.

For the experiments, phytoplankton and zooplankton were collected from the wild. The phytoplankton was collecting the day before the experiment and was left for 24 h for marking with ^{14}C, while the zooplankton was collecting on the day of the experiment, during early morning. During the study period, the dominant group was comprised of the zooplankton copepods, alternating between *Calanus pacificus* dominance in the first half of the study period and *Acartia tonsa* in the second. Filtration rates and specific grazing varied from day to day; those of *C. pacificus* were found in the range of 2.5 to 51.4 ml copepod^{-1} d^{-1} and 0.4 x 10-3 to 8.6 x 10-3 d^{-1}, respectively, while those of *A. tonsa* ranged from 1.4 to 263.0 ml copepod^{-1} d^{-1} and 0.2 x 10-3 to 43.8 x 10-3 d^{-1}, respectively (Table 2).

Upwelling events that occurred during the study period heavily influenced the filtration rates and specific grazing, with low rates coinciding with high variability during the outcrops and variable high rates during the relaxation. The results showed diurnal variations, as *C. pacificus* feeds preferably overnight, showing the highest rates of filtration and grazing between 20:00 h and midnight and the lowest rates at midday, with nightly rates up to 68% higher than daytime rates. Carbon consumption by the zooplankton community was low, being between 0.2 and 7.2% of the primary production of the first 5 m of the water column (50% of the euphotic zone), or from 0.4 to 24.7% of the primary production of the fraction greater than 8 microns, but could be up to 68% higher if the results of the diurnal variations were applied to the entire study period. The studies draw attention to the large variations that occur in filtration rates and specific grazing overnight, and highlight the need for more precise determinations, to generate time series data on the process of grazing in order to understand the influence of changes in the environment on zooplankton feeding processes.

Matus-Hernandez [13, 19] performed a series over 15 days, from 14th to 28th July 1990. The sampling location was the Todos Santos Bay, BC, which has a strong oceanic influence.

The variation of the temperature of the sea in this area has a close relationship with the California Current, and shows the effect of some local events (upwelling, surface heating and mixing of shallow areas). The method used was analysis of changes in the concentrations of chlorophyll a. Two types of experiments were performed, one feeding the organisms with the total phytoplankton and the other feeding them phytoplankton less than 20 microns. During our study, omnivorous organisms, copepods and euphausiids predominated over carnivores. Within this, omnivorous copepods comprised the dominant group, constituting 68 to 98% of total zooplankton. Grazing rates of the community presented great variability from day to day, with values from non-detectable to 2.5 d^{-1}, with an average value of 0.25 d^{-1} (Table 3); there was no statistically significant difference between the rates of grazing animals fed total phytoplankton and those fed less than 20 microns. Filtration rates also showed the same pattern of daily fluctuations, with values from zero to 8.8 ml $org.^{-1}$ d^{-1}, with an average of 1.07 ml $org.^{-1}$ d^{-1} (Table 3); these filtration rates are comparable to those reported for other regions. The percentage of the potential production of chlorophyll consumed by macrozooplankton values ranged from zero to 178%, averaging 25% (Table 3).

An outstanding fact is the great variability presented for daily filtration rates and grazing. The author mentions a similar phenomenon as recorded by Matus-Hernández [13] in microzooplankton populations in the same sampling station and time of year and Garcia-Pámanes [17, 18] in a nearby station but characterised by intense mixing processes; however, the variability they reported was much higher, possibly due to the intense mixing in the study area. Therefore, this group concludes that regardless of the environmental conditions in the region (water column stable or mixed), other physiological factors influence food activities of the populations of both microzooplankton and macrozooplankton. The rate of grazing by macrozooplankton and the percentage of potential chlorophyll a output consumed on average (25%) were about half of those registered by Matus-Hernández [13] for microzooplankton (43%) in the same season; this showed that the greatest impact on phytoplankton populations was due to microconsumers at this time.

Badillo-Padilla [20] carried out their work in the Punta Banda estuary, located at the southern end of the Bay of Todos Santos, Baja California. Two types of experiments were performed to determine the filtration rate and the specific rate of *Acartia californiensis* grazing. The first was a time series of 13 days, from 11 to 23 August 1994. The second was a diurnal variation *in situ* on the 27th and 28th of the same month. The organisms for experiments were separated by sex and the effects of different densities of organisms were observed in experimental flasks; ^{14}C was used as a tracer. The results indicated that specific grazing rates of filtration and showed the same behaviour. The grazing rate had an average value of $5.74 \times 10^{-4} \pm 1.28 \times 10^{-4}$ h^{-1}. The highest rate of grazing was filed on August 15th with an average value of $9.37 \times 10^{-4} \pm 4.11 \times 10^{-4}$ h^{-1} and the minimum was on 21st August with an average value of $2.75 \times 10^{-4} \pm 2.06 \times 10^{-4}$ h^{-1} (Table 4). The statistical results show significant differences between sampling days, specifically between the days where there were the highest and the lowest rates. The specific rate of grazing females showed an average value of $6.58 \times 10^{-4} \pm 2.35 \times 10^{-4}$ h^{-1} and $4.62 \times 10^{-4} \pm 9.23 \times 10^{-5}$ h^{-1} for males, while the filtration rate showed an average value of 2.92 ± 0.62 ml org^{-1} d^{-1} for females and 2.11 ± 0.42 ml org^{-1} d^{-1} for males.

The results of different densities of organisms put into experimental flasks indicated that specific grazing and filtration rates decreased with an increase in the number of animals.

Diurnal variation took place during neap, representing a low dynamic water column. The highest filtration rate was presented at 20:00 h with an average value of 5.68 ± 2.28 ml org^{-1} d^{-1} and the lowest at 04:00 h with an average of 0.97 ± 0.24 ml org^{-1} day^{-1}. Only 24.39% of the total carbon produced by phytoplankton was consumed by *A. californiensis*, the rest was available for use by other trophic levels or to be exported from the lagoon.

Gulf of California

García-Pámanes [14, 21] made four cruises to the central Gulf of California: in the fall of 1994, and in the spring, summer and autumn of 1995, during which the experiments were conducted. The community grazing rate was determined using ^{14}C as a tracer. Sampling was performed in two areas: the sill of the Guaymas Basin (U), characterised by the occurrence of highly dynamic processes, high power dissipation, hydraulic jumps, internal waves and tides, and the region of the basin (RC), which was less dynamic sill. In the RC a network of nine stations was established. On the sill and just in one network station, on every cruise, there was a diurnal variation of ~36-48 h with sampling every six hours. Grazing rates of the community (macrozooplankton > 200 microns) varied between 0.13 and 1.87 d^{-1} in the fall of 1994; between 0.35 and 2.40 d^{-1} in the spring, 0.31 and 4.75 d^{-1} in the summer and 1.01 and 4.84 d^{-1} in the autumn of 1995 (Table 5). Spatial heterogeneity was found in grazing rates; this variation could be due to the patchy distribution of phytoplankton and zooplankton or the time of sampling sampling. The results of the estimation of the grazing rate of the community during the diurnal variations indicate that rates during the day are about 75% lower than during the night. Statistically significant differences ($p < 0.05$) were only found between autumn 1994 and autumn 1995. The macrozooplankton presented an average consumption of primary production potential daily of 65% in autumn 1994; 82% in the spring, 86% in the summer and 98% in the autumn of 1995, representing a significant biological sink for primary production.

Bahía de La Paz, B. C. S.

The work by Hernandez-Trujillo et al. [8] consisted of a series of time from 16th to 24th October 2002, and involved measuring the daily feeding rate of mesozooplankton, and estimating changes in the concentration of chlorophyll-a in a station in the Bay of La Paz. They determined the growth rate of phytoplankton (k), the ratio of consumption or grazing rate (g), the rate of clearance of water (F) and the rate of intake (I) by *Acartia lilljeborgii* as a baseline for the area of study. The grazing rate ranged between 0.72 and 2.74 ind d^{-1}, while the growth rate of phytoplankton was 0.03 to 0.76 d^{-1}. The clearance rate of water was determined between 11.9 and 45.6 ml ind^{-1} d^{-1} and the rate of intake between 5.2 and 18.4 mg Chl-a ind^{-1} d^{-1}. The mesozooplankton could remove from 1.0 to 43% of the potential daily production (Table 6), and the clearance rates of intake had a similar daily variation throughout the study. However, the growth rate showed that grazing was lower during the time of our study. The results showed that mesozooplankton can consume about 17% of the daily production of phytoplankton in the study area.

Band-Schmidt et al. [10] tested different diets of phytoplankton in *Acartia clausi* and *A. lilljeborgii* of Bahia de La Paz, to determine their effect on survival, egg production and intake rate. Copepod females were fed with strains of diatoms (*Chaetoceros* sp., *Cylindrotheca closterium*, and *Odontella longicruris Dytilum brightwelli*), dinoflagellates (*Scrippsiella* sp., *Gyrodinium* sp., *Prorocentrum micans* and *P. rhathymum*) and a rafidofita (*Chattonella* sp.). They found that after 24 h of incubation in darkness at 24°C, with the different diets, survival was greater than 91%. When just feeding *A. clausi* with *Gyrodinium* sp., survival was low (44.5%), whereas dinoflagellate diets showed the highest egg production. *A. clausi* showed higher egg production when feeding with *P. rhathymum*, *P. micans*, *Gyrodinium* sp. *Scrippsiella* sp. and *Chattonella* sp, while *A. lilljeborgii* had higher egg production with *Scrippsiella* sp., *P. micans, Gyrodinium* sp. and *Chaetoceros* sp. Ingestion rates were significantly higher in both copepods when fed *Chaetoceros* sp., *Odontella longicruris*, *Prorocentrum rhathymum* and *Chattonella* sp. In both copepods, ingestion rates were lower (below 300 ng C copepod^{-1} h^{-1}) when fed with *Cylindrotheca closterium, Dytilum brightwelli, Scrippsiella* sp. and *Gyrodinium* sp. Ingestion rates were different with a diet of *Prorocentrum micans*; *A. lilljeborgii* showed an ingestion rate that was higher than *A. clausi*. Their results suggest that both *Acartia* species respond to the nutritional quality of phytoplankton over a short period of time. The highest rates of intake did not necessarily result in greater egg production, suggesting that the relations hip between the rate of intake and egg production could be related to food quality rather than the quantity ingested. The response in egg production for both copepods appears to be specific for each species, but, in general, dinoflagellate diets seem to have a higher nutritional value, compared with the diatoms with the exception of *Chaetoceros* sp.

The work of Hernandez-Trujillo [22] consisted of performing two time series, from 15th to 19th October and from 3rd to 7th December 2007, in a station at Ensenada of La Paz, which is located in the south-western region of the Bay of La Paz. Bioassays were conducted with specimens of *Acartia clausi*, *A. lilljeborgii* and *Centropages furcatus*, the grazing on natural food species where mesured calculating the change in chlorophyll-a concentration at the beginning and end of each experimental period, and the nominal production of eggs per female was determined. In October, zooplankton biomass showed a downward trend from the first to the fifth day of sampling and the values were 13 to 12 ml biomass m^{-3}. Grazing experiments showed that the nominal production of eggs per female was 3 to 11 eggs h^{-1}, while the egg hatching rate of the species ranged between 60 and 90%. In December, zooplankton biomass showed mixed variation by reducing 3 to 2.8 ml biomass m^{-3} in the first three days, then increasing to levels close to 3.2 ml biomass m^{-3}. Statistically, the variation was not significant. Experiments showed that the nominal production of eggs per female fluctuated between 1 and 11 and eggs h^{-1}, hatching rate varied between 41 and 77%. Between the two time series there was a noticeable change of the sea surface temperature, decreasing by almost 7°C in December. Similarly, zooplankton volume in December showed a reduction of almost 77% compared to October, but the number of eggs produced per female showed no significant change between the months in which the bioassays were conducted.

Table 6. Results of experiments performed on the instantaneous growth rate of phytoplankton (k), grazing rate (g), clarification rate (F), intake rate (I) and percentage of potential production consumed [22]

					% Cl *a* potential Production consumed by zooplankton
Date	k (d^{-1})	g ($ind \cdot d^{-1}$)	F ($ind \cdot d^{-1}$)	I (Cl $a \cdot$ ind d^{-1})	(d^{-1})
17/10/2002	0.25±0.06	1.67±0.20	26.06±3.42	9.07±1.25	16.80±4.13
	85.43%	38.02%	41.53%	43.89%	77.83%
19/10/2002	0.32±0.06	1.16±0.29	19.43±4.86	10.69±0.85	27.88±1.68
	26.51%	35.81%	35.37%	10.77%	8.52%
21/10/2002	0.48±0.27	1.81±0.03	21.16±0.37	7.02±0.22	26.94±15.63
	80.18%	2.72%	2.47%	4.43%	82.06%
23/10/2002	0.26±0.13	2.53±0.21	42.16±3.46	5.31±0.01	10.74±6.03
	70.71%	11.73%	11.62%	2.66%	79.43%
25/10/2002	0.04±0.01	1.78±0.42	29.69±7.0	14.69±3.95	2.88±1.69
	47.14%	33.36%	33.36%	38.02%	82.98%
Whole	0.25±0.06	1.67±0.20	26.1±3.42	9.1±1.26	16.8±4.14
period	85.43%	38.02%	41.52%	43.89%	77.83%

Mean ± standard error, Coefficient of variation (standard deviation / mean x 100).

Conclusion

Importantly, the approaches used for studies on hervibory are varied, as has been evidenced in the work outlined. García-Pámanes and Matus-Hernández [13, 17, 18, 19] oriented their studies towards the effect of the water dynamics on feeding mechanisms, García-Pámanes et al. [14, 15, 21] focused on the carbon fluxes, Badillo-Padilla [20] on physiology and ecological impact, and Band-Schmidt et al. [10] determined the survival and egg production. In contrast, the working group of Jose Bustillos-Guzman, David López-Cortés and Francisco Hernández (personal communication), at the Centre for Biological Research of the Northwest, investigated the impact on populations of zooplankton on phytoplankton and the physiology of shrimp.

Moreover, although the works reported here were conducted in areas with different environmental conditions (upwelling areas, high and low dynamics, estuarine and oceanic), they agree on the influence that the dynamics of water have on the grazing rate. This includes the daily variability and day-night period, which should be factors taken into account in future works. They also suggest that because of the large variability in daily rates of zooplankton grazing in coastal ecosystems, it is necessary to generate time series at least in the frequency range of days and that, because of the large variability in the contribution by fractions zooplankton, studies should include, wherever possible, both fractions (macro and micro) of consumers, to get a better understanding of the effects of the process of grazing on phytoplankton populations.

Since the first formal study of grazing was in 1986, 26 years have passed, during which there have been thirteen researches on grazing by zooplankton, which equates to an investigation every two years; this is a very low figure considering the extent of the coasts of Mexico. The interest in this type of research has increased over time, as shown by the results of the studies that have been carried out. Studies in this period can be divided into two groups: from 1986 to 1995 and more recently from 2006 to the present. Moreover, it should be noted that all but one of these took place in the north-western region of the country.

The future looks promising; today, several of the authors or co-authors of the above papers are actively working on this topic or are in the stage of preparation of articles on specific grazing rates and filtration for micro and macrozooplankton, with information derived from previous work. On the other hand, the numbers of working groups that are addressing this issue are increasing gradually. Is urgent to realise the importance of studies that provide not only more but also accurate information on consumption rates of zooplankton *in situ*, for the Pacific coast as well as the Atlantic and Cribean areas, only by generating this information we will be able to understand the ecological role played by zooplankton in the ecosystem.

References

[1] Lampitt R. S., T. Noji and B. von Bodungen, *Marine Biology,* 104, (1990).

[2] Raymont J. E. G., Pergamon Press, Oxford, (1980).

[3] Suttle, C. A., Chan A. M. and Cottrell M. T., *Nature* 347, (1990).

[4] Omori M. and Ikeda T., John Wiley & Sons, (1984).

[5] Gifford D. J., *Marine Ecology Progress Series,* 47, (1988).

[6] Longhurst A. R. and Harrison W. G., *Progress in Oceanography,* 22, (1989).

[7] Palomares-García J. R., Bustillos-Guzmán J. J. and López-Cortés D., *Journal of Plankton Research* 28, 12, (2006).

[8] Hernández-Trujillo S., Zarate-Villafranco A., Pacheco-Chávez R., Esqueda-Escárrega G., Hernández-Alfonso J. R. and Acevez-Medina G., *Hidrobiológica,* 17, 3, (2007).

[9] Noriega-Cañar A. M. and Palomares-García J. R., *Hidrobiológica* 18, (Supplement 1), (2008).

[10] Band-Schmidt C. J., Pacheco-Chávez R. and Hernández-Trujillo S., *Hidrobiológica,* 18, (Supplement 1), (2008).

[11] Herrera-Moro-Chao I. R., Ponce-Manjarrez E. J., Segura-Soto Y. I., Varona-Cordero F., Bustillos-Guzmán J. and Gutiérrez-Mendieta F. J., *XVII Reunión de la SOMPAC,* 9-13 April 2012 (Chetumal, Quintana Roo, México).

[12] Hernández-Trujillo S., *Estimación de la producción secundaria del zooplancton en la Bahía de La Paz.* http://ecologiadelplanctonmarino.blogspot.mx/ (in preparation).

[13] Matus-Hernández M. E., *Tasas de pastoreo diarias del zooplancton en una localidad costera de Baja California, México,* Centro de Investigación Científica y Educación Superior de Ensenada, Tesis de Maestría (1991).

[14] García-Pámanes J., *Flujo del carbono dentro de la zona epipelágica de la región central del Golfo de California,* Centro de Investigación Científica y Educación Superior de Ensenada, Tesis Doctoral, (1999).

[15] García-Pámanes J. y Lara-Lara J. R., *Ciencias Marinas* 27, 1, (2001).

[16] Landry J. R. and Hassett R. P. *Marine Biology,* 67, (1982).

[17] García-Pámanes J., *Variación día-día de la tasa de pastoreo zooplanctónico frente a Baja California, México,* Centro de Investigación Científica y Educación Superior de Ensenada, Tesis de Maestría, (1989).

[18] García-Pámanes J., Lara-Lara J. R. and Gaxiola-Castro G., Estuarine, *Coastal and Shelf Science,* 32, (1991).

[19] Lara-Lara J. R. and Matus-Hernández M. E., *Ciencias Marinas,* 23, 1, (1997).

[20] Badillo-Padilla M. M., *Variación de la tasa de pastoreo de Acartia californiensis Trinast (Copepoda: Calanoide) en el estero de Punta Banda, Baja California, México.* Universidad Autónoma de Baja California, Facultad de Ciencias Marinas, Instituto de Investigaciones Oceanológicas, Tesis de Maestría, (1996).

[21] Hernández de la Torre B. y Gaxiola Castro G., *Carbono en Ecosistemas Acuáticos de México* (2007). García-Pámanes J., Lara-Lara J. R. and Bazán-Guzmán C., Pastoreo por el mesozooplancton en la región central del Golfo de California: un estudio estacional. Instituto Nacional de Ecología, SEMARNAT, Centro de Investigación Científica y de Educación Superior de Ensenada, México, pp141–155.

[22] Hernández-Trujillo S., *Pastoreo del microzooplancton durante un ciclo estacional en la Bahía de La Paz,* B. C. S. http://ecologiadelplanctonmarino.blogspot.mx/ (in prepara- tion).

In: Invertebrates: Classification, Evolution and Biodiversity ISBN: 978-1-62618-789-4
Editor: Rafael Riosmena-Rodriguez

Chapter 4

Invertebrates: Structure of the Community and Biodiversity Associated to Rhodolith- Sponge Complex at Magdalena Bay. South Baja California

Pamela Herrera- Martínez[*1], *Sergio Álvarez-Hernández*[1a], *María del Carmen Méndez-Trejo*[2,3] *and Rafael Riosmena-Rodríguez*[2]

[1]Departamento de Hidrobiología,
Universidad Autónoma Metropolitana-Iztapalapa, México
[1a]Laboratorio de Ficología Aplicada, Departamento de Hidrobiología, Universidad Autónoma Metropolitana- Iztapalapa, México
[2]Programa de Investigación en Botánica Marina,
Departamento de Biología Marina,
Universidad de Baja California Sur. La Paz, Baja California Sur, México
[3]Centro de Investigación Científica del Mar (CICIMAR),
Instituto Politécnico Nacional, La Paz, Baja California Sur, México

Abstract

One of the most exceptional and multifunctional habitats for cryptic and epibenthic fauna has been established in the costal zones of Magdalena Bay conforming an interesting interaction between rhodoliths and marine sponges. These function primarily as refuges, where the rhodolith beds (red coralline algae) grow in association with a high biodiversity of marine sponges and form functional complexes to counteract the hydrodynamic effect generated by the current flow. The main objective in this chapter was to highlight the community structure of cryptic, epibenthic, and sessile fauna with a multifactorial analysis including: the ecological analysis via alpha diversity, the

* Email address: herrera.martinez.pamel@gmail.com.

oceanographic dynamics of the Bay, and the intrinsic complexity of the habitat, including the chemical-biological interaction between sponges and rhodoliths comparing those variables between two different localities (CFE and PED) subjected to different dynamics at the Lagoon System of Magdalena Bay.

The sample area covered five hundred square metres at each site, covering a total area of $250m^2$, reviewing a total of 70 rhodoliths with 7,223 associated invertebrates; these belonged to 52 orders and 105 families, with a total of 110 species (42 determined to species and 68 to genera level) analysed. The morphological arrangement of the rhodolith varies as a response to the physical regimens presented by our results revealed that higher number of individuals is supported in PED (52.48± 29.86) organisms cm^2 when rhodoliths present 7 to 4 branches in a volume of 250 to 200 ml. The cryptic and epibenthic mobile groups include: Crustaceans that contributed 67% of the total abundance in both localities, Molluscs at levels of 5.3% in CFE and 8.3% in PED, representing the higher richness taxa, Polychaetes at levels of 19.6% in CFE and 14% in PED and Ophiurida at levels of 8.1% in CFE and 9.9% in PED. In terms of the richness, molluscs, crustaceans and polychaetes are the principal components of the 42 species in CFE and 54 species in PED (this last with a high Simpson ds= 6.24). The epibenthic sessile fauna was integrated by: Cnidaria that was monospecific in both localities and represent more than 87% of the sessile abundance, Ascidacea provided 3% of the total abundance and the Porifera taxa representing only 8% of abundance but with a higher coverage and richness. This dovetailing provides a complex habitat represented by a high biodiversity. It is important to highlight that this chapter presents the results of an extensive study of the invertebrate community associated with rhodoliths in the Mexican region of the eastern Pacific.

Introduction

The Rhodoliths beds are red coralline algae (Corallinaceae, Rhodophyta) that grow free with a crusty shape and are at the mercy of current movements and the physical forces of waves and currents that maintain them [11]. Those coralline beds "build communities on benthic substrate made up branching or rounded thalli that collectively create a fragile, structured biogenic matrix over coarse or fine carbonate sediment" [26]. They also set up different types of substrate, including: clayey sands and calcareous rocks with sands, typical of costal zones at Magdalena Bay. In this region is where the rhodoliths create an exceptional and multifunctional habitat for cryptic and epibenthic fauna, including the amazing association with a high diversity of marine sponges. There are others research that has documented the ecological interaction between rhodoliths and invertebrates species like echinoderms [16], free living corals [20], and new cryptochiton species [6], and other studies that have documented the association of sponges with other rhodophytes including: the research of [24] that describe 39 species founded in rhodoliths at Ria Arousa(Galicia, Spain), *Haliclona cymaeformis-Ceratodyction spongosium* in the biggest Australian Reef [8, 31, 32], a similar association of *H. caerulea-Jania adherents* described at Mexican Pacific Coast located at Mazatlan, Manzanillo, Colima, Huatulco and Oaxaca [4, 5], and the multiple researches of the associated fauna with rhodoliths at the Gulf of California [15, 26]. However, there have been few investigations of this topic in the Mexican Pacific region, taking into account the preliminary study of the sponge community associated with rhodoliths at Magdalena Bay [2].

This chapter evaluates the community structure on rhodolith beds, but in order to understand the up-growth and success of the algae reefs like rhodoliths beds, it is necessary to analyse the biology and ecology of both rhodoliths and all of the symbiotic groups associated. This includes their principal sessile epibenthic component, the sponges, because it is known that a variety of factors can influence the diversity, distribution and abundance of organisms associated with rhodoliths beds [26], like seasonal patterns of light and water temperature and the complexity of the rhodolith thalli branching [12, 15, 26, 33]. This methodological research was focused on multifactorial variables that impact over the stability on this complex, under natural conditions; this chapter considers the oceanographic patron incidents on each areas, comparing two localities of Magdalena Bay, and also the intrinsic complexity given by the up-growth of the rhodoliths units, evaluated with the number of branches and the dimension of this units measured in terms of volume. Those factors explain the variation in size, shape and branching patterns influenced by variations in water motion [3]. The other factor provided was the chemistry-biology interaction with sponges evaluated by the structural function and their role as biological filters, as sponges are highly efficient filter-feeders [8]. For this reason, the processes that occur inside and outside the rhodolith-sponge units are particularly interesting for understanding the symbiosis between them, because, as Davy *et al.* [8] mentioned, "the nutritional interaction in sponge-alga symbiosis are still poorly defined". Previously studies have indicated that the nitrogen required for algae growth can potentially be provided by sponge catabolism supplying excretory DIN to its algae partner [8]. Based on the above, this preliminary research contemplates the results of these variable undernatural conditions. All of the factors had been integrated at the interpretation of the changes in the species abundance, showing a patron distribution, evaluated by ecological analysis. Therefore, the relevance of this chapter generates an important contribution to the Mexican Pacific region, with an extensive study of the invertebrates' biodiversity and their function and interaction, in order to describe the rhodolith-sponge complex as a habitat characteristic at the costal zones of Magdalena Bay.

Material and Methods

Study Area

The study area is located in Magdalena Bay, which is a Lagoon system located between 24° 15′N 111° 30′W and 25° 20' N 112° 15' W. This region is known to be one of the most productive zones of the Mexican Pacific Coast and is an anti-estuarine Lagoon system, possessing 50 km of longitude and several intertidal low zones boarded with narrow channels of mangrove with an average depth of 3 m [29]. The tidal regimen is mixed semidiurnal with periods of high tide and low tide, producing the biggest flow rate during low tide [27, 28]. Circulation sources are modulated by the interactions that maintain with the open sea through the tidal dynamics [27], showing that this bay is strongly influenced by the tidal regimen.

Sampling was conducted in July, which represents the warmest month and the transition month between the warm and cold season, according to the dynamic oceanographic study in Magdalena Bay, [7]. The rhodolith beds were selected taking account the previous research in the area [1], selecting two regions of the bay that might be subject to different dynamics based on their location.

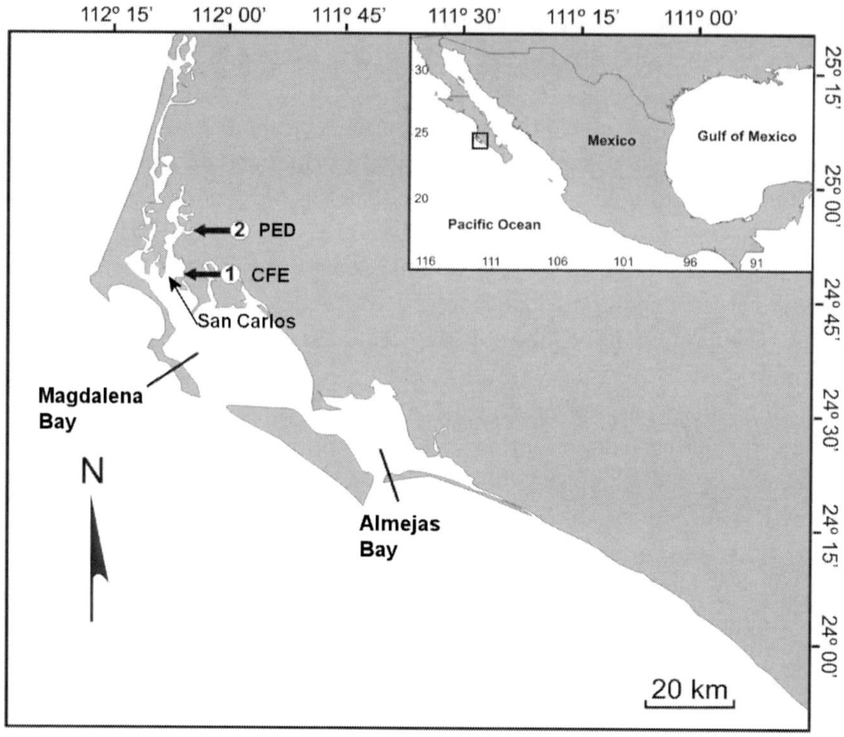

Figure 1. Study sites in Bahia Magdalena.

The two rhodolith beds were located within the Estero Banderitas, designed as CFE (locality 1); 24°48'45"N; 112°05'59"W and PED (locality 2); 24°54'12''N; 112°05'24''W (Figure 1). CFE is part of Lagoon Tlaque, which is bordered with mangroves and composed of clay soils with a high accumulation of detritus; the flow is low but subject to the tidal rhythms. In contrast, PED is located in the middle part of Estero Banderitas within the dendritic channel, named "Estero San Fernando" (Figure 1), surrounded by mangrove. The bottom is comprised of calcareous rocks and sands that correspond to a biggest current flow, generating a continuous flow and constant circulation. The environmental parameters water temperate (°C), salinity psu and water column depth (m) were recorded at each side, using a mercury thermometer, a refractometer, and a barometer, respectively. Also, the concentration of Ca-Mg (mg/l) in water was evaluated with volumetric methods (APHA,1980), and analysed by the Laboratory of Water Quality at CIBNOR (Centro de Investigaciones Biológicas del Noroeste),in order to describe the habitat associated to the Rhodolith-Sponge complex.

Evaluation of Nutrients in Situ and the Concentration in Biological Filters

The concentration of ammonium in the sediment (NH_4) was evaluated in samples extracted manually with plastic syringes in each locality; the concentration of ammonium was determined using the method of Solórzano [23]. In the same manner, the ammonium concentration was evaluated in the samples of interstitial water extracted with syringes (5 ml) as a principal sessile components associated with rhodoliths (ascidians and marine sponges), taking two samples of two sponges and one of ascidians. The selection was made based on the following criteria: the most abundant species and other species that were exclusive to that locality. The samples were analysed using the method of Solórzano [23] by the Water Quality Laboratory at CIBNOR.

Coverage of Sponges in the Beds: CFE and PED

The density and coverage of sponges was estimated in two 25 m transects, with each one perpendicular to the shore line of coast. The density was estimated by counting the number of sponges associated with rhodoliths (classified as complex) and all of the individual rhodoliths that did not have visible sponges associated, inside a PVC frame of 1 m^2 that was laid randomly on top of or below the transect, every 5 m. There was a total of 5 quadrants with transect lines (1m x 5 quadrants x 25 m transect = 125m^2), covering a total sampling area of 250 m^2 (2 transects x 125 m^2) in each locality (CFE and PED). The coverage was expressed as a percentage, by counting the total number of rhodoliths and the number of complexes by quadrant in each locality in order to estimate the coverage of the habitat (rhodolith-sponge).

Intrinsic Complexity of the Rhodolith

The structural complexity of the rhodoliths was analysed in order to understand how the variation in the number of thalli and the size could influence the fauna ensembles in abundance terms. In the area, we conducted an intensive sampling counting the number of branches in 1 cm^2 and measuring the displacement volume (ml) in each rhodolith.

Richness and Abundance of the Sponges Associated to the Rhodolith Beds

The richness was determined by scuba diving photographs and the collection of 1 or 2 samples species of the different sponge morphologies, along with analysis of the species that were present along the two transects line perpendicular to the shore (25 m each). The specific abundance was determined by counting the number of sponges over the number of rhodoliths. These units were collected from five quadrants of 0.0625 m^2 (25cm x 25 cm) placed in the superior corner of the quadrant of 1 m^2, along the 2 transect lines of 25 m in each locality. The samples were placed into plastic bags and frozen at -10°C for laboratory identification using specific literature, considering the previous identification by Ávila [2].

Interactions: Richness and Biodiversity

The cryptofauna, epifauna, and other sessile components associated were manually separated, classified by taxa and identified to the lowest possible taxonomic level. For each rhodolith collected from 0.0625 m^2 (25cm x 25 cm), samples were put in plastic bags and sealed under water and then frozen (-10°C) for transportation to the laboratory. A total of 70 rhodoliths were examined removing epifauna from the surface of branched thalli and also removing the cryptofauna that had burrowed into the coralline thalli. In order to make a complete revision of internal organisms, the rhodoliths were fragmented into small pieces, using a mesh of 250 μm, by a method involving pressure and rinsing, removing, separating, counting and identifying all of the invertebrates larger than 250 mm; all of the microscopic species were retained in the mesh. Data of species abundance and richness were examined using PAST software applying Shapiro Wilk [14] for normality ($\alpha = 0.05$).

The richness, abundance and diversity were evaluated by taxa divided into two groups: a) criptofauna and epibenthic mobile fauna and b) epibenthic sessile fauna; this was according to ecological classification and the relationship with the habitat. All of the invertebrates species were reported considering their mean abundance and standard deviation; the organisms not identified until genera level were not considered for the analysis of richness, abundance and diversity by taxa.

Alpha diversity and the index applied by taxa included: richness (S) dominance (λ Simpson1949), evenness Simpson (ed) and diversity of Simpson (ds); finally, two way ANOVA [14] was applied in order to compare the abundance of the two localities.

Results and Discussion

The primary causes of natural disturbance in rhodolith reefs are water motion and sedimentation, as reported by [25]. This is an intrinsic consequence of dynamics in the system, as presented in Magdalena Bay, but this chapter revealed that the extreme changes in temperature, salinity, dissolved metal ions ($CaCo_3$) and nutrient (NH_4) regimens could directly affect rhodoliths and indirectly affect the associated community. For the rhodoliths, the presence of Ca^+ and Mg^+ as structural components is fundamental and triggers the photosynthesis activity; "recent studies have shown that red coralline algae record both elemental and isotopic signals, which contain environmental information within the calcite skeleton" [17]. We considered the previous cited investigation for a preliminary evaluation, measuring the concentration of $CaCo_3$ in water. Our results showed that compounds changed dramatically between localities and differed with the depth, recording the highest concentration in superficial water in CFE (7,704 mg/l) compared with PED (6,272 mg/l), which presented the highest values at 3 m (6,953 mg/ l) (Table 1).

Those changes might be related to the temperature stratification, anthropogenic entrance to the water surface and biogenic efforts resulting from the biodegradation of carbonate structures. The most important factor found to influence the concentration of $CaCO_3$ is temperature, which causes the precipitation of this compound. This variable is determinant, because it has been shown that temperature in Magdalena Bay presents a pronounced seasonal variation [18, 27, 28]; however, according to Gómez-Gutierrez *et al.* [13], the variability of temperature in spring tides and neap tides is of equal or greater magnitude than the seasonal variability with temperature mean >23.9°C reported by Cervantes-Duarte *et al.* [7], which corresponds to the local variation of temperature at CFE and PED (Table 1).

Table 1. Comparative table of the physic-chemical factors at Magdalena Bay

Oceanographic variables	Seasonal cycle research [7]			Measure of local condition (July 2011)		Biological filters NH_4 (μM)		
	Season			Locality		Locality		
	Cool	Warm		CFE	PED		CFE	PED
	Nutrients					Nutrients at interstitial water		
WATER			Depth (m)	WATER				
	—	—	0	7,704	6,272			
$CaCo_3$ (mg/l)	—	—	3	6,032	6,953	Sponge		
NH4 (μM)	3.6 ± 3.0	2.4±5.6	0	—	—			
SEDIMENT				SEDIMENT		*Scopalina* sp.	5.185	0.5
NH_4 (mg/kg)	—	—	3	22.2	6.3	*Haposclerida* sp.1	9.205	—
Physico-Chemical				Physico-Chemical		*Haliclona* sp.	SV—	80.9
	18.8±2.2	23.9±2.6	0	28	25	Ascidians		
Temperature (°C)			3	25.3	21.5			
	34.9±0.7	34.9±0.8	0	36	35.5	*Trididemnum* sp.	12.423	—
Salinity (‰)			3	37.5	33.5	*Didemnidae* sp. 1	—	104.33

Robinson *et al.* [21] showed that "the most current dynamics and possible transport of nutrients occurs during periods of spring tide"; these values were found in the same month of sampling as the work by Cervantes-Duarte *et al.* [7]. The principal amount of nutrients enters the Bay via the current tides. The superficial salinity between localities presented greater stability in contrast with temperature, but showed a considerable change in the water column, reflected by spatial patterns of circulation during the period of sampling. CFE presents positive circulation (psu 0m<psu 3m), whereas PED presents negative circulation (psu 0m >3m). Both localities are euryhaline systems (Table 1), and those changes are associated with the local characteristics derived from the position into the estuary and the local water dynamics at the zone.

Chemical and Biological Interaction between Rhodoliths and Sponge

Nitrogen compounds are the main source of nutrients in estuarine-lagoon systems; therefore, the last stage in the nitrogen cycle was evaluated with regard to ammonium (NH_4) in the three phases of the ecosystem: column water (taking as reference the study of Cervante-Duarte [7]) and two benthic phases: a) interstitial water in the biological filters and b) the sediment content (Table 1). Lower concentrations of NH_4 are present in the water during the warm season (2.4± 5.6) compared with the cool season (3.6 ± 3.0), presenting a higher standard deviation as a result of temporal variations and local conditions (Table 1).It is clear that primary producers are reduced in biomass during this season, as a result of the nutrient reduction.

The benthic phase reflects biological processes as a result of local conditions, because comparing the sediment values between localities, CFE presented higher ammonium concentrations (22.2 mg/kg), unlike PED (6.3 mg/kg). Higher concentrations in the soil at CFE are the result of large amounts of litter generating a greater accumulation of detritus. However, in the benthic phase, the function of biological filters is the key; according to our results, during the warm season the sponged sampled presents high concentration of NH_4 in interstitial water compared with the water column values excepting *Scopalina* sp. with lower concentration at PED Our results suggest that not every sponge filters at an equal rate; some are better ammonium filters, such as *Haliclona* sp. with 80.4 μm NH_4 (Table1), compared with other species that, although they have a higher abundance, such as *Scopalina* sp., their capacity as a filter is reduced. For that reason, they increase their coverage areas in the rhodolith beds and are probably more related to a structural function or have another symbiotic relationship with the rhodoliths in Magdalena Bay. This might happen with other sponge species found in our study. Therefore, future studies to increase the knowledge in this line of investigation are recommended because another interesting discovery is the function of ascidians that filter ammonium, such as Didemnidae sp. (104.33 μm), which is another frequent component of sessile fauna (Figure 6,E; Figure 7, D) (Table 1). Both type of filters sponges and ascidians register higher concentrations at PED. These biological components could be indicators of health in this ecosystem, because, according to [8], "the nitrogen requirement for algae growth can potentially be provided by sponge catabolism alone". Our results suggested that the sponge supply by excreted DIN (dissolved inorganic nitrogen) to the rhodolith and "the sponge beneficed with structural support" [5, 19, 31, 32].

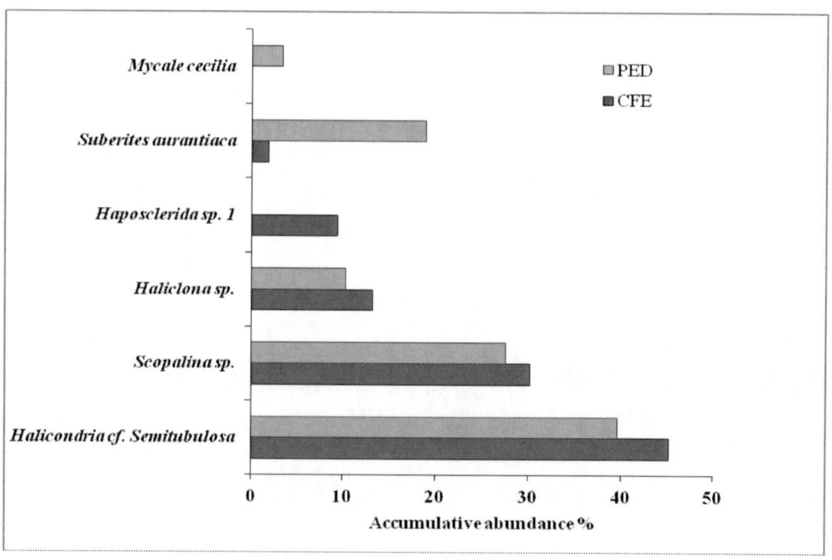

Figure 2. Percentage contribution of sponges associated with rhodoliths comparing localities of sampling.

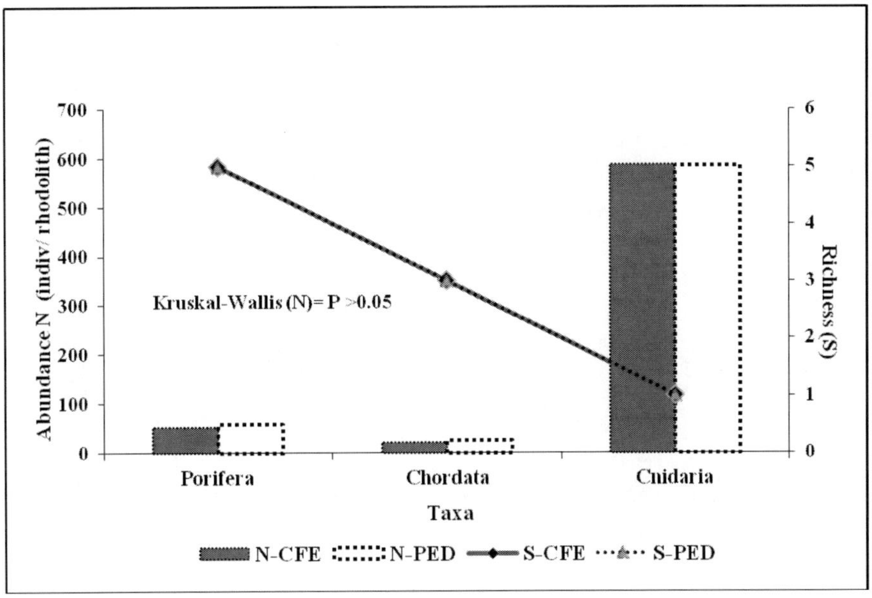

Figure 3. Richness and total abundance of the epibenthic sessile fauna by taxa: comparing localities of sampling.

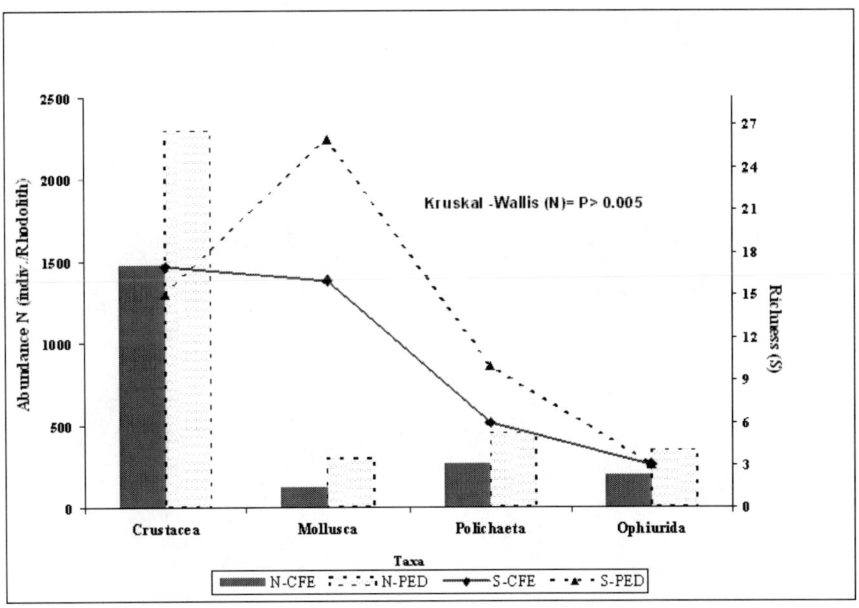

Figure 4. Richness and total abundance of the cryptic and mobile epibenthic fauna by taxa: comparing localities of sampling.

It would be essential for future studies to calculate the nitrogen requirements of algae and the amount of nitrogen source excreted as ammonium by sponge and ascidians in natural conditions, because these could be excellent indicators of water quality.

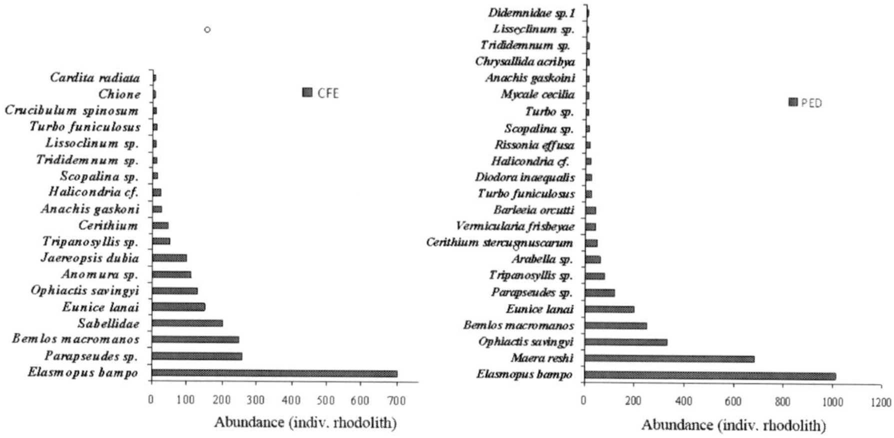

Figure 5. Abundance of dominant species by location: Assuming (dominance species > average abundance by taxa).

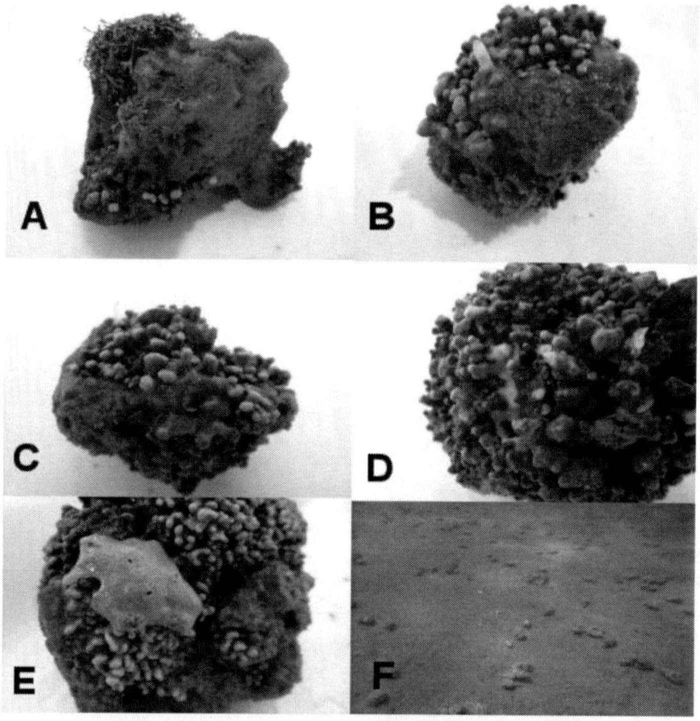

Figure 6. Interactions: A) PED: *Suberites aurantiacus* agglomerating to types of Rhodophyta algae: *Lithophillum margarite* and *Amphiroa* sp. B) CFE: Different types of sponge grow: *Halichondria* cf. *semitubulosa* (branching grow) and *Scopalina* sp. (incrusting) C) CFE: *Ophiactis savingy* (ophiuroidea) embracing to *Lissoclinum* sp. (Ascidia), to the left *Scopalina* sp. and back an atrium of *Halichondria* cf. *semitubulosa* D) CFE: *Trididemnum* sp. (incrusted ascidia) and tube worm *Eunice* sp. associated to rhodolith E) PED Sessile components from right to left: *Antopleura* sp. (cnidaria), Didemnidae (Ascidia), *Scopalina* sp. (sponge) F) CFE: foresia between hermit crabs and rhodoliths.

Figure 7. Localities. A) CFE: shows the epibentic components molluscs and hermit crabs, with a higher abundance of cnidarians that covers large area, this space is dominated by sponges B) Community associated to rhodoliths at CFE show the predominate sessile specie *Scopalina* sp. accompanied with the colonial ascidia *Lissoclinum* sp. C) PED: Conglomerate unit integrate by incrusting sponge, ascidians, molluscs and hermit crabs in association with rhodoliths and other Rhodhophyta species. D) Another type of multi-species algae balls conformed by *Acanthopora* balls with incrusted Didemnidae (ascidia) and *Halichondria* cf. *semitubulosa* (sponge) intercalated in rhodolith beds, this structure only was founded in PED.

Coverage, Richness and Abundance of Sponges Associated with Rhodolith Beds

Sponges are the principal component of sessile fauna associated with rhodoliths. Coverage was estimated from a visual census covering 10 m^2 (2 transects x 5m^2) of the total area of sampling in the coralline beds by locality. In the rhodoliths, 59% of CFE was superficially covered by sponges, with a total of five species that were encrusting sponges (Figure7, A). *Scopalina* sp. (30%) was the most striking for its bright orange colour and the coverage area over rhodolith units (Figure 7, B). Although 40.9% appeared to not possess sponges, upon laboratory review it was verified that sponges were growing into the rhodolith units and agglutinating over and between the rhodolith branches without reaching the surface; for example, *Halicondria* cf. *Semitubulosa* was the most abundant species on rhodoliths (45.20%), followed by *Haliclona* sp., *Haposclerida* sp. and in some case *Suberites aurantiaca* (Figure 2).

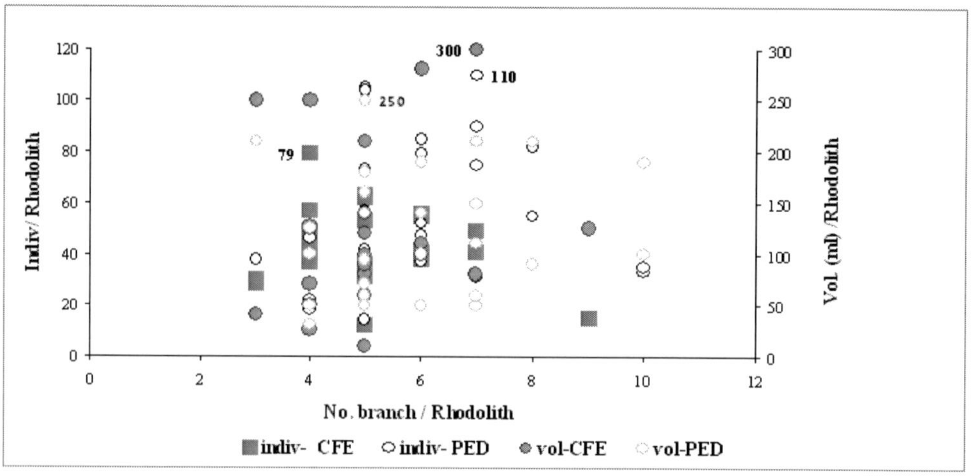

Figure 8. Change in total number of individuals and the increasing rhodolith volume (cm^2) on the basis of the number of branches.

This explains why we did not observe the presence of sponges in the visual census. In PED, 69% of rhodoliths beds were covered by sponge and only 31% appeared to be rhodoliths alone; as for CFE, the sponge was found to be growing into the rhodolith structure, and the rhodoliths without sponge were poorly developed rhodoliths, most probably up-growth. The dominant species was the same as CFE, *Halicondria* cf. *Semitubulosa* (39.6%), followed by *Scopalina* sp., *Suberites aurantiaca*, *Haliclona* sp., and *Mycale cecilia* (Figure 2). From those results, and when comparisons with other studies [31] are made, it is evident that the algae may acquire structural reinforcement with the presence of sponges contra-resting the abrasive hydrodynamics changes, and expanding their photosynthetic surface (10).

Description of the Structural Arrangement of Other Sessile Groups Associated with Rhodolith Beds

Epifauna restricted to the rhodolith beds included species requiring hard substrates, such as anemones, tunicates and sponge, and all epibenthic filters taxa. The Cnidaria taxon, represented by the Anthozoa class, was monospecific in both localities and represented more than 87% of the total sessile abundance and more than 580 individuals by taxa in both localities.

Cnidarian abundance was the result of small size and a great ability to adapt over the hard surface in rhodoliths. Porifera taxa represented up to 8% of the abundant sessile group and more than 50 individuals by taxa with a slight increase in PED (Figure 3), although of the three taxa that integrated the sessile fauna, the sponge presented the greater coverage and richness in coralline beds. This highlights their relevance as a structural component of this microhabitat. The Ascidacea represented only 3% of the sessile abundance, and was increased in PED composed predominantly of colonial ascidians, belonging to the families Didemnidae and Polyclinidae, which are frequently confused with sponges (Figure 7 A, B and Figure 6 D).

Table 2. Spatial analysis and ecological community by taxa

	SESSIL FAUNA							
	Richness (S)		Simpson Diversity (ds)		Simpson (ed)		Simpson (λ)	
	CFE	PED	CFE	PED	CFE	PED	CFE	PED
Porifera	5	5	3.23	3.72	0.60	0.69	0.31	0.27
Ascidiacea	3	3	2.23	3.13	0.47	0.68	0.45	0.32
Total	8	9	1.26	1.30	0.14	0.14	0.79	0.77
	CRYPTIC AND EPIBIONTIC FAUNA							
	Richness (S)		Simpson Diversity (ds)		Simpson (ed)		Simpson (λ)	
	CFE	PED	CFE	PED	CFE	PED	CFE	PED
Polychaeta	6	10	2.63	3.84	0.43	0.38	0.38	0.26
Crustacea	17	15	3.45	3.33	0.20	0.22	0.29	0.30
Mollusca	16	26	5.15	10.26	0.28	0.36	0.19	0.10
Ophiurida	3	3	1.87	1.09	0.62	0.36	0.54	0.92
Total	42	54	6.24	6.53	0.84	0.85	0.16	0.15

According to the Shapiro Wilk index, the abundance of sessile groups did not have a normal distribution ($p<0.05$), which is commonly seen in communities that present a marked dominant species. In spite of the marked differences between localities (as a result of the different local conditions), the sessile community has the same opportunities of colonisation, according to the Kruskal Wallis analysis, which showed no significance differences ($p>0.05$) in abundance terms between localities. A true indicator of the biodiversity index is represented by ecological taxa analysis, because this correlated the conditions of established species with the environmental conditions presented by locality. According to this, the better environmental conditions at PED generated an increased biodiversity of the Porifera taxon followed by the Chordata taxon, which presented a lower biodiversity compared with the sponges (Table2). This was as a result of a bigger equitability in the average abundance species (Table 3), and a smaller number of dominant species (Figure 5).

These are indicators of the response to environmental gradients, because they function as biological filters and they are better represented under homogenous ambient conditions inferred by the values of evenness of lambda index (Sponges ed= 0.69 at PED and Ascidians ed= 0.68 at PED) (Table 2), which explained the better conditions for the establishment of the benthic community at PED induced by the continuous current flow. Compared with other similar studies of the Gulf of California, specifically at Concepcion Bay [15, 26], where the dominant epibenthic components were composed of different species of foliose macroalgae and mats of tunicates/sponge, the coverage was increased during the winter [26]. Another community study of rhodoliths performed in Brazilian algae reefs reported a different composition of the epibenthic community associated with rhodoliths, where the dominant fauna taxa are represented by Bryozoan and Crustacean, which are primarily barnacles, and sponge was the least important group (Villas-Boas pers. com.). According to this, the community of sessile invertebrates at the studied zone of Magdalena Bay is integrated by filters, where the sponge has the principal role. The change in faunal composition could be related mainly to the environmental impacts of the zone caused by the current flow and the circulation of water at Coastal zones that change dramatically between regions such as the Pacific, Gulf of California and the Southwest Atlantic zones.

Structure of the Community and Interaction: Criptofauna and Epifauna

The cryptic and epibenthic community was four to six times more diverse than the sessile community; the taxa included in this group were: Crustacea, Mollusca, Polichaeta and Ophiurida. According to the Shapiro Wilk test, these do not have normal distribution ($p<0.05$) in terms of abundance, as a result of the dominance of Crustacean species that contribute more than 67% of the total abundance for the cryptic and epibenthic group in both localities and present a higher abundance of taxa in PED (N= 2,293 indiv. by taxa). This was followed by the Polychaetes, which supply 19.6% in CFE and 14% in PED to the most abundant group and present a higher abundance in PED (N= 447 indiv. by taxa), then the Ophiurids, with 8.1% in CFE and 9.9% in PED (N= 348 indiv. by taxa) and finally the Molluscs, with the fewest in abundance terms with 5.3% in CFE (N= 126 indiv. by taxa) and 8.3% in PED (N= 292 indiv. by taxa). However, Molluscs was the highest richness taxa, with 26 species in PED and 16 in CFE (Figure 4).

In spite of the differences in terms of richness and abundance, the capacity of colonisation between CFE and PED are equal according to Kruskal Wallis test, the results of which show no significance differences ($p>0.05$) in abundance and richness terms between localities (Figure 4) as a result of the habitat plasticity. The multifactorial variables previously described in the chapter are related to: a) local condition, b) chemical and biological interaction between principal components, and c) the complexity of the habitat. Those impact on the structural arrangement to taxa level and particularly on the composition and distribution of the associated species, described in this section through the ecological analysis. According to this, a high diversity in PED was showed by the Mollusca taxon (ds= 5.15), followed by the Polychaetes group (ds= 3.84). Crustaceans and Ophiuroida had a higher biodiversity at locality one (Table 2). In terms of richness, Molluscs, Crustaceans and Polychaetes are the principal components of a great diversity of cryptic and epibenthic species, although the dominance and composition of species are very different between localities (Figure 5). Molluscs are highly represented by microcrustacean species, principally the families Cerithiidae and Turbinidae. The Crustaceans are dominated by a great richness of micro crustaceans, represented by Amphipoda, Isopoda, Tanaidacea and Decapoda orders, the Polychaetes are dominated principally by the Sabellidae, Eunicidae and Syllidae families, and the Ophiurids are formed of two families: Ophionereididae and Ophiactidae. Molluscs and polychaetes presented a homogeneous distribution with a higher richness of species at PED and a greater dominance at CFE, represented by *Cerithium stercusmuscarum* CFE (1.35 ±2.01); PED (1.49 ±2.67), *Anakis gaskoini* CFE (0.79 ±1.57); PED (0.28 ±0.75) and *Turbo funiculosu s*were more abundant in locality two CFE (0.35 ±0.98); PED (0.71 ±1.38) (Figure 5 and Table 3).

Table 3. Taxonomic list of the invertebrates associated with rhodoliths

		Abundance M(SE)	
Phylum / Locality		CFE (n=35)	PED (n=35)
SESSIL FAUNA		14.08	15.92
PORIFERA		8.83	9.67
Class	Desmospongia		
Order	Halichondrida		
family	Dyctionellidae		
	Scopalina sp.	0.47 (0.51)	0.46 (0.51)
Family	Halicondriidae		
	Halichondria cf. *semitubulosa*	0.71 (0.46)	0.64 (0.48)
Order	Poecilosclerida		
Family	Mycalidae		
	Mycale Laubenfels, 1936	NR	0.31 (0.47)
Order	Hadromerida		
Family	Suberitidae		
	Suberites aurantiacus (Duchassaingand & Michelotti, 1864)	0.03 (0.17)	0.06 (0.24)
Order	Haposclerida sp. 1	0.15 (0.35)	NR
Family	Chalinidae		
	Haliclona sp. Grant, 1836	0.21 (0.41)	0.17 (0.38)
CHORDATA		5.25	6.25
Class	Ascidiacea		
Order	Enterogona		
Family	Didemnidae sp. 1	NR	0.17 (0.38)
	Lissoclinum sp.	0.24 (0.43)	0.26 (0.44)
	Trididemnum sp.	0.35 (0.49)	0.29 (0.45)
Family	Polyclinidae		
	Pseudodistoma sp.	0.03 (0.17)	NR
CNIDARIA			
Class	Anthozoa		
Family	Actiniidae		
	Anthopleura sp.	17.71 (13.21)	16.71 (13.47)
CRYPTIC AND EPIBIONTIC FAUNA		187.90	284.16
ANNELIDA		42.55	44.64
Class	Polychaeta		
Order	Canalipalpata		

Table 3. (Continued)

		Abundance M(SE)	
Phylum / Locality		CFE (n=35)	PED (n=35)
Family	Sabellidae sp.	5.91 (6.73)	1.25 (1.8)
Order	Phyllodocida		
Family	Polynoidae		
	Halosydna glabra Hartman, 1939	0.71 (0.97)	0.11 (0.53)
Order	Palpata		
Family	Eunicidae		
	Marphysa sp.	NR	1.09 (1.38)
	Eunice cedroensis Fauchald, 1970	0.06 (0.24)	0.46 (1.67)
	Eunice lanai Carrera-Parra y Salazar, 1998	4.44 (3.62)	5.71 (7.71)
Order	Aciculata		
Family	Oeninidae		
	Arabella sp.	1.06 (1.67)	1.71 (2.11)
Family	Syllidae		
	Tripanosyllis sp.	1.5 (2.57)	2.26 (2.23)
Family	Nereididae		
	Ceratonereis sp.	NR	0.69 (1.3)
	Laeonereis culveri (Webster, 1879)	NR	0.17 (0.45)
	Neanthes sp.	0.09 (0.38)	0.31 (0.72)
Order	Capitellida		
Family	Maldanidae		
	Isocirrus papilatus Berkeley and Berkeley, 1939	NR	0.26 (0.61)
Arthropoda			
Subphylum CRUSTACEA		76.19	113.10
Class	Malacostraca		
Order	Amphipoda		
Family	Amphitoidae		
	Ampithoe plumulosa Shoemarker, 1938	0.6 (1.35)	0.3 (0.95)
Family	Aoridae		
	Bemlos macromanus Shoemaker, 1925	25 (25.24)	24.8 (20.06)
Family	Colomastigidae		
	Colomastix sp.	0.2 (0.63)	3.4 (10.06)
Family	Ischyroceridae		
	Ericthonius brasiliensis (Dana, 1853)	NR	1 (3.16)
Family	Leucothoidea		

		Abundance M(SE)	
Phylum / Locality		CFE (n=35)	PED (n=35)
	Leucothoe sp.	4.9 (4.18)	3.7 (3.33)
Family	Lissianasidae		
	Lisianopsis sp.	1.9 (3.6)	NR
	Lysianassa sp.	5.8 (5.55)	1.7 (2.58)
Family	Mellitidae		
	Elasmopus bampo J. L. Barnard, 1979	70 (39.79)	101.2 (37.8)
	Melita sp. Leach, 1814	0.6 (1.35)	NR
	Maera reshi J.L. Barnard, 1979	0.4 (1.26)	68.2 (28.93)
Family	Phoxocepalidae		
	Foxiphalus sp. J. L. Barnard, 1979	0.8 (2.2)	NR
Order	Isopoda		
Family	Anthuridae		
	Mesanthura occidentalis Menzies and J. L. Barnard, 1959	0.2 (0.42)	0.3 (0.67)
Family	Corallanidae		
	Excorallana tricornis(Hansen, 1890)	1 (1.63)	0.2 (0.42)
Family	Jaeropsididae		
	Jaeropsis dubia Menzis, 1951	9.8 (13.7)	10.5 (15.79)
family	Paranthuridae		
	Colantura bruscai Poore 1984	NR	0.6 (1.35)
Order	Tanaidacea		
Family	Leptocheliidae		
	Leptochelia dubia (Krøyer, 1842)	0.4 (1.26)	0.9 (1.6)
Family	Parapseudidae		
	Parapseudes sp.	25.7 (30.63)	12 (16.4)
Order	Decapoda		
Class	Malacostraca		
Family	Palaemonidae		
	Periclimeneus sp.	0.09 (0.38)	NR
	Alpheus sp.	0.06 (0.24)	0.14 (0.43)
Infraorden	Anomura sp.	3.29 (10.6)	2.03 (5.06)
	Brachyura sp.	0.29 (0.62)	0.31 (0.53)
MOLLUSCA		4.50	10.43
Class	Gastropoda		
Order	Archaeogastropoda		
Family	Fissurelladae		
	Diodora inaequalis Sowerby, 1835	0.11 (0.33)	0.71 (0.89)

Table 3. (Continued)

		Abundance M(SE)	
Phylum / Locality		CFE (n=35)	PED (n=35)
Family	Haliotidae		
	Arene sp.	NR	0.71 (0.45)
Family	Turbinidae		
	Turbo funiculosus Kienner,1847	0.35 (0.98)	0.71 (1.38)
	Turbo sp.	0.03 (0.17)	0.34 (0.76)
Order	Neogastropoda		
Family	Collumbellidae		
	Anachis gaskoini Carpenter, 1857	0.79 (1.57)	0.28 (0.75)
	Mitrella sp.	NR	0.11 (0.32)
Family	Muricidae		
	Muricopsis sp.	NR	0.03 (0.17)
Family	Pseudomelatomidae		
	Pseudomelatoma penicillata (Carpenter, 1864)	NR	0.03 (0.17)
	Zetekia gemmulosa (C.B. Adams, 1852)	NR	0.14 (0.55)
Order	Neotaenioglossa		
Family	Barleiidae		
	Barleeia orcutti Bartsch, 1920	NR	1.2 (3.52)
Family	Caecidae		
	Caecum carpenteri (Bartsch, 1920)	NR	0.17 (0.51)
Family	Cerithiidae		
	Cerithium stercusmuscarum Valeciennes, 1833	1.35 (2.01)	1.49 (2.67)
Family	Epitoniidae		
	Epitonium mitraeforme (Sowerby, 1844)	NR	0.03 (0.17)
Family	Litiopidae		
	Alaba supralirata Carpenter, 1857	NR	0.23 (0.65)
Family	Caliptreidae		
	Crepidula arenata (Broderip, 1834)	NR	0.06 (0.24)
Family	Turritelidae		
	Vermicularia frisbeyae McLean, J.H., 1970	0.09 (0.29)	1.26 (1.96)
Order	Neritopsina		
Family	Neritidea		
	Theodoxus luteofaciatus Miller, 1879	0.12 (0.33)	0.09 (0.28)
Order	Ctenobranchia		

		Abundance M(SE)	
Phylum / Locality		CFE (n=35)	PED (n=35)
Family	Rissoinidae		
	Rissonia effusa Mürch, 1860	0.09 (0.29)	0.51 (1.09)
Order	Heterostropha		
Family	Piramidellidae		
	Chrysallida acrybya Dalland Bartsch,1909	NR	0.23 (0.57)
Class:	Bivalvia		
Order	Carditoidea		
Family	Carditidae		
	Cardita radiata Sowerby, 1833	0.15 (0.70)	0.06 (0.24)
Order	Littorinimorpha		
Family	Calyptraeidae		
	Crucibulum spinosum (G. B. Sowerby I, 1824)	0.24 (0.55)	0.06 (0.24)
Order	Mytiloidea		
Family	Mytilidae		
	Lithophaga aristata (Dillwyn, 1817)	0.03 (0.17)	NR
Order	Ostreoida		
Family	Ostreidae		
	Ostrea palmula Carpenter,1857	0.06 (0.34)	0.14 (0.43)
	Ostrea sp.	NR	0.06 (0.24)
Order	Tellinoidea		
Family	Tellinidae		
	Tellina straminea Deshayes,1855	0.03 (0.17)	0.08 (0.37)
Order	Veneroida		
Family	Lucinidae		
	Lucina prolongata Carpenter, 1857	0.03 (0.17)	0.03 (0.17)
Family	Veneridae		
	Chione guatulconensis Hertleinand Strong, 1948	0.21 (0.41)	0.03 (0.17)
Class	Poliplacophora		
Order	Chitonida		
Family	Ischnochitonidae		
	Radsiella sp.	0.03 (0.17)	NR
Echinodermata		64.67	116.00
Class	Ophiuroidea		
Order	Ophiurida		
Family	Ophionereididae		

Table 3. (Continued)

		Abundance M(SE)	
Phylum / Locality		CFE (n=35)	PED (n=35)
	Ophionereis annulata Le Conte, 1851	0.15 (0.56)	0.11 (0.32)
Family	Ophiactidae		
	Ophiactis sp.	1.76 (3.84)	0.31 (0.83)
	Ophiactis savingy (Müller and Troschel, 1842)	3.79 (4.07)	9.51 (10.16)

Those epibenthic species are very frequent in rhodolith beds, and like the rest of the Molluscs, are composed of microcrustacean species, which can be hypothetically explained by larval settlement preferences for coralline algae refuges because the morphology of structures can be protected. The more abundant Polychaetes species were: the Sabellidae family CFE (5.91 ±6.73); PED (1.25 ±1.8), *Eunice lanai* CFE (4.44 ±3.62); PED (5.71 ±.71) and *Tripanosyllis* sp.CFE (1.5 ±2.57); PED (2.26 ±2.23) (Figure 5, Table 3). Some motile organisms such as Polychaetes worms and Crustaceans can be associated with multiple sub-habitats (Foster, 2003). In this research, most of the Polychaetes species recorded were tubeworms; nine of the eleven species (Table 3) are adapted with rhodoliths creating their own microhabitat in the tube to counteract environmental effects. In contrast, the Crustacean and Ophiurids present a higher diversity in CFE and great dominance in PED represented by *Elasmophus bampo* CFE (70± 39.79); PED(101.2± 37.8), which is a relevant microcrustacean, and other abundant species like: *Bemlos macromanus* CFE(25 ±25.24); PED (24.8 ± 20.06), and *Parapseudes* sp. CFE(25.7 ±30.63); PED(12 ±16.4) (Figure 5, Table 3). Those cryptic species roll a primordial paper and probably regulate the community structure in locality two, which isthe same as Anomuran: very frequent in rhodolith beds highly adapted to the conditions of this habitat and in accordance with that observed in field maintain an amazing intraspecific relationship between anomuran-rhodolith and between the mollusc *Turbo funiculosus*-rhodolith, (Figure 6, F). Those relationships are known as "foresia"[1] and "tanatocresis"[2] observed between anomura and the gastropod *Cerithium stercusmuscarum* (obs. pers). Ophiurids present a higher dominance at locality two represented by *Ophiactis savingyi* PED (9.51 ±10.16); CFE (3.79 ±4.07).In accordance with this, the community responds to the environmental gradient in terms of the distribution and dominance of species, but the assemblage is maintained because the associated populations integrate in a persistent habitat that could be adapted to natural conditions in space and time. Comparing with other studies of communities associated with rhodoliths in the Gulf of California [15, 26], some species were common, but there is a clear difference in the abundance of taxa. The Echinodermata (primarily *Ophiactis simplex* and *Ophiactis savingy),* Arthropoda (mainly amphipods) and Annelida contributed the greatest percentage of abundance by taxa at the Gulf of California [26]. In the case of Crustaceans: *Bemlos macromanos* and the common genera *Elasmopus* are dominant species for the taxa. The Ophiuran *Ophiactissavingy* and *Ophiactys simplex* are common species in this study. This

[1] It is an interspecific relation between two species one of this transports passively to the other.

[2] Interspecific relation where one specie takes the remains of another species, like: skeletons or corpses in order to protect or use it as a tool

observation supports the hypothesis that rhodoliths could support a wide geographical distribution of species, highlighting the evident ecological and biological importance of this habitat, as a refuge of great diversity for cryptic and unique species, as well as for other species that have not been reported for the Pacific zone.

Intrinsic Complexity of the Habitat Rhodolith-Sponge

The dominant rhodolith-forming species in the sample areas was *Lithophyllu mmargaritae,* which presents fruticose and foliose growth forms, reported previously [1]. In this survey, the number of branches of rhodoliths in one square centimetre and the volume displaced (ml) were evaluated in order to evaluate the complexity of the habitat related to the number of individual rhodoliths. According to the results in CFE, the branch density ranges from 9 to 3 cm^2, volume from 300 to 10 ml, and the individual number of invertebrates ranges from 79 to 12 individuals; compared with PED, where the branch density ranges from 10 to 3 cm^2, volume from 250 to 30 ml, and the individual number of invertebrates ranges from 110 to 11 (Figure 8). A multiple correlation analysis revealed that the criptofauna abundance does not show a linear correlation with the number of branches or the rhodolith volume, because those present an asymptotic arrangement (Figure 8). However, under certain conditions, in terms of complexity (branching density) and space available (thallus volume), the abundance increases, coinciding with [26]. As the results revealed, higher number of individuals are supported in PED (52.48 ± 29.86 organisms cm^2), where the rhodoliths have 7 to 4 branches and their volume is 250 to 200 ml, compared with CFE ($43.33\pm$ 14.40 organisms cm^2) presented under the following structural arrangement: 4 to 7 branches and a volume of 200 to 210 ml (Figure 8). At both localities the sponges present a high coverage by thalli which is implicit in the displaced volume (ml) measured in this survey and groupings of more than two rhodoliths over the rhodolith beds were recorded in some cases. The agglutination was frequently shownin the presence of *Suberitesauranticus* (Figure 6, A). The rhodoliths were interspersed by another spherical configuration: the *Acantophora* sp. algae balls presented in PED (Figure 6, A); this kind of morphological arrangement appears to be a response to the physical regimens presented in this locality and the different degrees in sphericity observed in rhodoliths display different stages of growth. Both algae balls (*Lithophylum margarite* and *Acantophora* sp.) are capable of harbouring very large numbers of faunal associates with the maze of their branching fronds [8] and with the increased area provided by this sponge partner, there are increased resources and possibly habitats and niches [2].

Conclusion

The rhodolith beds at Magdalena Bay form individual ecosystem units where the sponge is the structural component of these microhabitats, harbouring large amounts of species. The plasticity to withstand oceanographic conditions in coastal areas is provided by their interaction and structural conformation that is maintains by their massive structural component: the sponges. For this reason, it is important to conserve this unique habitat with the purpose of maintaining the structure and trophic flow in the community and highlighting the importance of this symbiotic association resulting in functional units for nutrient recycling, which probably have a significant contribution on the net productivity of Coastal Lagoons. This chapter provides new knowledge of this habitat, but the real goal is to strengthen the research between physical processes and the response of the biota. This work contributes to stimulate the conservation of these biological habitats as a source of diversity in all dynamic and productive ecosystems, such as Magdalena Bay.

Acknowledgments

We wish to thank M. Biol. B. Moreno (Programa de BotánicaMárina, UABCS) and Biol. T. Pineda (Laboratorio de Sistemática y Ecología de Equinodermos ICMyL) for their valuable contribution on taxonomy identification assistance. We also thank Dr J. López-Vivas for his help in field work and Dr G. Hinojosa for his collaboration and providing us with facilities at Bahía Magdalena. Pamela Herrera wishes to thank Dr Rodolfo Rioja sector project (108464), SEMARNAT-CONACyT for the scholarship during this project.

References

[1] Ávila E., N.I. Blancas-Gallangos, R. Riosmena-Rodriguez and L. Paul-Chavez, (2009). *Journal of the Marine Biological Association of the United Kingdom*, vol. 90, pp.193-202.

[2] Ávila E., and R. Riosmena-Rodriguez, (2011), *Brazilian Journal of Oceanography*, Vol. 59, no.4, (online) consulted 8 February of 2012, http://dx.doi.org/10.1590/S1679-87592011000400007.

[3] Bosence D. W.J., (1976), *Paleontology*, Vol. 19, pp. 365-395.

[4] Carballo J.L., and E.Ávila, (2004), *Marine Ecology Progress series*, Vol. 279, pp. 93-104.

[5] Carballo, J.L., E.Ávila, S.Enríquez, and L.Camacho, (2006), *Marine Biology*, Vol. 148, pp. 467-478.

[6] ClarkR.N., (2000), *Memorial-Occasional papers of the Delaware Museum of Natural History,* Vol. 43, no. 43, pp. 1-20.

[7] Cervantes-Duarte R., S. López-López, E.Gonzales-Rodríguez and S.Futema-Jimenez, (2010), *CICIMAR Oceánides*, Vol. 25, no. 2, pp. 111-120.

[8] Davy S.K., D.A. Trautman, M.A.Borowitzka, and R.Hinde, (2002). *Journal of Experimental Biology,* Vol. 205, pp. 3505-3511.

[9] Freiwald A., (1995). *Beitr. Paläont.,* Vol. 20, no.7, pp. 7-19.

[10] Enríquez S., J.L. Carballo, and E. Ávila, (2009). *Journal of Phycology* vol. 45, pp.81-90.

[11] Foster M.S., L.M. McConnico, L. Lundsten, and T. Wadsworth, (2007), *Ciencias Marinas,* Vol. 33, Issue4: 367-384.

[12] Grall J., and M. Glemarec, (1977), *Vie et Miliea,* Vol. 47, pp. 339-349.

[13] Gómez-Gutierrez J., and C.J. Robinson, (2006), *Marine ecology Progress series,* vol. 320, pp. 215-231.

[14] Hammer Ø.H.D.,and P.D. Rast, (2001). En: *Palaeontologia Electronica.* version 2.15 ed. Paleontological Software for Education and Data Analysis, p. 9.

[15] Hinojosa-Arango G., and R. Riosmena-Rodriguez, (2004), *Marine ecology,* Vol.25, issue. 2, pp. 109-127.

[16] James, D.W., (2000), *Marine Biology*, Vol. 137, no. 5-6, pp. 913-923.

[17] Kamenos N.A., M. Cusack, T.Huthwelker, P.Lagarde, and R.E. Scheibling, (2009), *ELSEVIER*, Vol. 73 Issue. 7, pp. 1901-1907.

[18] Lluch-Belda D., M.E. Hernández-Rivas, R.Saldierna-Martínez and R. Guerrero-Caballero, (2000). Oceánides, vol.15 pp. 1-23.

[19] Reyes-Bonilla H., R. Riosmena-Rodriguez, and M.S. Foster, (1997), "Hermatipic corals associated with rhodolith beds in the Gulf of California, México". *Pacific Science*, Vol. 51, Issue3, pp. 328-337.

[20] Robinson C.J., S. Gómez-Aguirre, and J. Gómez-Gutiérrez, (2007), *Journal Fish Biology,*Vol. 71, pp. 1-19.

[21] Rützler K., (1990), In RüetzlerK.,(ed.). *New perspectives in sponge biology.* Washington DC: Smithsonian Institution Press pp. 455-466.

[22] Strickland J.D.H., and T.R. Parsons, (1972), *A practical handbook of seawater analysis.* Fish.Res. Bd. Canada bulletin 167 (second edition).

[23] Solórzano L., (1969) *Limnology Oceanographic*, Vol. 14, pp.799-801.

[24] Solórzano M., F.Cristobo, and V.Urgorri (1991), Boletín de la Real Sociedad Española de Historia Natural, Vol. 87, Issue.1-4, pp. 175-185.

[25] Steller D.L., and Foster M.S., (1995), *Journal of Experimental Marine Biology and Ecology*, Vol. 194, pp. 201-212.

[26] Steller D., (2003), *Aquatic Conservation Marine Freshwater Ecosystem,* Vol. 13, pp. S5-S20.

[27] Sánchez-Montante O., (2004), Unpublished Ph D. thesis CICIMAR-IPN, 288 pp.

[28] Sánchez-Montante, O., O.Zaitev, and M. Saldivar-Reyes, (2007), "In: Funes-Rodriguez, R, J. Gómez-Gutiérrez and R. Palomares-García (Eds.). *Estudios Ecológicos en Bahía Magdalena.* Instituto Politécnico Nacional.

[29] Schweers T.M., Wolff, V. Koch, and F.Sinsel-Duarte, (2006), *Revista de Biología Tropica,* Vol. 54, pp. 1003-1017.

[30] Van Soest R.W.M, N. Boury-Esnault, J.N.A. Hooper, K. Rützler, N.J. de Voogd, B. Alvarez de Glasby, E. Hajdu, A.B. Pisera, R. Manconi, C. Schoenberg, D. Janussen, K.R. Tabachnick, M. Klautau, B. Picton, M. Kelly, and J. Vacelet, (2011),WorldPoriferadatabase. Available online at http://www.marinespecies.org /porifera. Consulted on 2011-07-30.

[31] Trautman D.A., R.Hinde, andand M.A. Borowittzka, (2000), Dordrecht, the Netherlands: Kluwer Academic Publishers pp. 521-537.

[32] Trautman D.A., R. Hinden, and M.A.Borowitzka, (2002), *J. Exp. Mar. Biol. Ecol.* Vol. 244, pp. 87-105.

[33] Weber-Van BoseA., andM.Foslie, (1904), *Leiden Bulletin Reports,*Vol. 61, pp. 1-110.

In: Invertebrates: Classification, Evolution and Biodiversity ISBN: 978-1-62618-789-4
Editor: Rafael Riosmena-Rodriguez

Chapter 5

When Rhodoliths Do Not Roll: Invertebrate Community Responses to Environmental Stability and Natural Disturbance

Gustavo Hinojosa-Arango[1], Christine A. Maggs[2], Michael P. Johnson[3] and Rafael Riosmena-Rodríguez[4]

[1]Centro para la Biodiversidad Marina y la Conservación A.C. (CBMC). Benito Juárez. La Paz, BCS. México

[2]School of Biological and Food Sciences.
Queen's University Belfast (QUB). Belfast, Northern Ireland

[3]Ryan Institute. National University Ireland (NUI). Galway, Ireland.

[4]Programa de Botánica Marina.
Universidad Autónoma de Baja California Sur (UABCS).
La Paz, BCS, México

Abstract

Rhodolith beds are aggregations of free-living, non-geniculated coralline algae with a worldwide occurrence. Their distribution and survivorship is regulated by many factors, from which light, temperature, and sedimentation are considered as main restrainers of growth and structure. Rhodoliths are bioengineers and provide a three-dimensional habitat for associated species. Distinctive faunal and floral assemblages of high diversity are harbored by rhodolith thalli by virtue of their branching and interlocking nature. Rhodolith communities are of ecological and economical importance around the world and subject to periodical events disturbance. Periodic rotation is necessary, but does not need to be frequent, to prevent sedimentation and to allow light reaching all surfaces of the thalli. This rotation is a natural disturbance, stochastic phenomena recognized as an important determinant of presence and abundance of associated species. After a disturbance event the associated community with any substratum starts a recovery process that might eventually return it to what it used to be. This process can take from days up to years. In this chapter, the role of disturbance in structuring the invertebrate assemblages associated with two rhodolith beds was studied. We artificially stabilized rhodolith thalli for periods of 6 months to analyze the response of the associated invertebrate community to stability and disturbance. Samples from untied thalli (natural) were compared to artificially stabilized rhodoliths. Results demonstrated that the influence of disturbance on invertebrate assemblages is not simple and will strongly depend on the characteristics of the disturbance as well as the physiological, morphological and behavioral capabilities of the species present at any given time. Factors such as the time of the year when disturbance events happen can modify the recovery patterns of the assemblage diversity. The patterns of the species associated with natural and artificially stabilized rhodoliths emerge at three scales: a) between stability periods (two, four and six months), b) between levels of stability (natural and stabilized thalli), and c) among assemblages in experiments started on different dates. The value of rhodoliths as a unique biotope is under threat by different kinds of human activities around the world. The lack of regulation on small and large scale trawling, reduction of water quality, modification of water circulation, ocean acidification, and any other activities that affect rhodolith stability and disturbance are of special concern. Invertebrate assemblages on rhodolith beds provide recruits for many commercial fisheries around the world. It is necessary to take into consideration the response of these assemblages to both natural and artificial disturbance in the development and implementation of coastal management plans around the world.

Introduction

The diversity of the marine assemblages reflects a balance between environmental conditions and the degree of disturbance they experience [47, 13]. Disturbance is considered one of the main influences affecting diversity and structure of marine communities [41, 30]. However, the way that diversity responds to environmental disturbance is not consistent or predictable [27]. Both empirical observation and current theory suggest that increasing levels of disturbance may either decrease or increase diversity, or even have no effect [50]. Disturbance events are followed by succession processes during which biodiversity and biomass generally increase, reach a peak in the transitional stage, and then decline in late stages [32, 12]. The Intermediate Disturbance Hypothesis (IDH) proposed by Connell in 1978 [7] has been used to explain community response to environmental states that disturbance can affect assemblages on different scales. Environments with low levels of disturbance have lower species richness due to competitive exclusion while high levels of disturbance eliminate those species incapable of rapid colonization. Therefore assemblage richness peaks in environments with moderate disturbance levels [36, 55, 38, 23]. Different variables, such as frequency, intensity, and duration of disturbance, together with the time since the last event as well as interactions between two or more disturbance attributes result in variations of the community assemblages in a given habitat [35]. The essential elements of the IDH are: 1) there must be repeated local disturbance; 2) disturbance must be frequent enough that competitive exclusion does not occur over the whole area, yet not so frequent that most species are eliminated; 3) the frequency of disturbance has to be seen in the context of generation time; and 4) there must be a trade-off between a species' ability to colonize and to compete [56].

The most likely scenarios for the assemblages associated with artificially stabilized rhodoliths in comparison with natural beds as a result of disturbance are represented as richness curves (Figure 1). The associated assemblages with artificially stabilized rhodolith may initially show a positive response to the increased stability, with higher richness than that of non-stabilized rhodolith samples. In accordance with the IDH, it is expected that the biological interactions become more intense as the period of stability increases resulting in lower richness relative to natural rhodolith samples. Richness values (inserted graphs representing dissimilarity calculated from ANOSIM analysis) are believed to increase with time until it reaches the highest values at intermediate levels of disturbance when samples present the maximum diversity (Figure 1a). The second scenario is that the induced rhodolith stability will result on either a positive or negative effect on the richness of the assemblage composition in relation to the natural rhodolith (Figure 1b). Dissimilarity (R-values) may be high from the beginning and those differences will be maintained throughout the experiments as a consequence of the observed effect of stabilization on the richness of the assemblages. In the third state, stability is not an important factor in recruitment and survival. Increased periods of stability will have no significant effect on the diversity values of artificially stabilized samples (Figure 1c). Dissimilarity will be low between stabilized and natural samples.

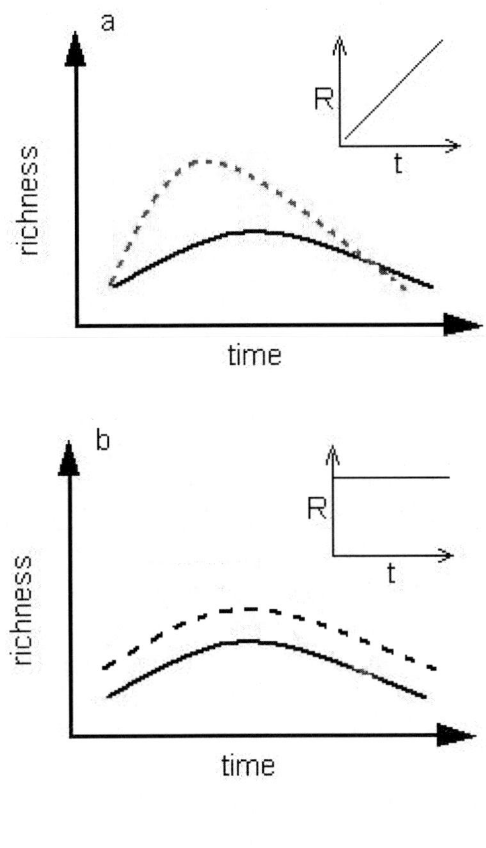

Figure 1. (Continued).

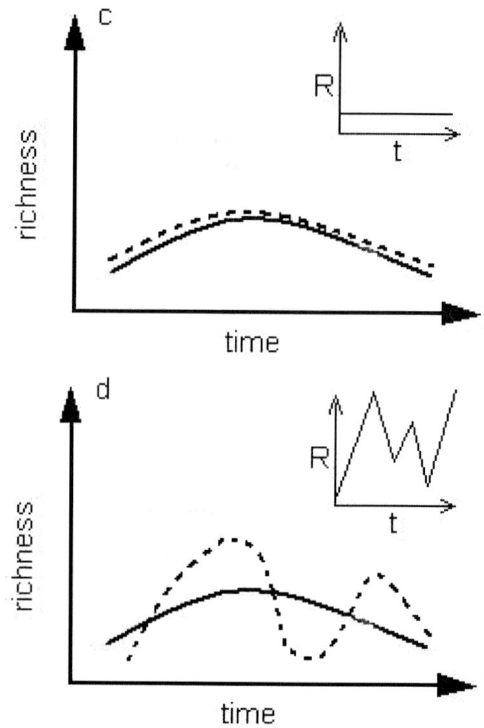

Figure 1. Most likely responses to disturbance of the assemblages associated with artificially stabilized rhodolith (doted line) and natural rhodolith (continuous line). a) Development in accordance with the IDH. b) Artificial stabilization has a positive/negative effect allowing the development of more/less diverse assemblages than on natural rhodolith regardless of time since last disturbance. c) No effect due to reduction on disturbance frequency (neutral effect). d) Idiosyncratic model with no defined pattern of development. Models compiled and adapted from Warwick and Clarke [51], Wilkinson [55], Shea *et al.* [38], and Johst and Huth [23]. Inserted graphs represent dissimilarity between assemblages calculated as R-values from ANOSIM analyses.

The fourth possible scenario is an idiosyncratic model with no defined pattern of richness variations as a result of stabilization, with the assemblages changing unpredictably (Figure 1d). Dissimilarity will increase or decrease randomly over time, making it hard to predict the recovery of the associated community after a disturbance event. The precise shape of the curves in all cases will depend on the number of species that can be added before disturbance and competition become intense enough to reduce diversity.

Studies of disturbance effects on community structure have concentrated on single "pulse disturbances", focussing on intermediate and short-term effects [1, 59, 54]. Nevertheless, long-term studies of succession cycles and associated biotic changes are necessary to evaluate the ecological effects of various disturbance regimes, especially those due to annual or seasonal factors. Rhodolith communities with their high diversity are of ecological importance and the particular environmental conditions in which they develop should provide a suitable system to study diversity-disturbance relationships. Experiments on mobile habitats are scarce and only a few artificial stabilization studies have been conducted on cobbles [41] and boulders [28], nevertheless this kind of analysis has never been made for rhodoliths. The role of disturbance in structuring the assemblages associated with two rhodolith beds was studied to test if the observed richness of associated invertebrate species followed the patterns predicted by the IDH or other scenarios explained in figure 1.

It is generally agreed that the result of colonization and stabilization experiments depends on the time of the year when sampling surfaces are deployed [4, 40, 33]. This is the first study in which the relationship between disturbance and species richness has been experimentally examined for rhodolith assemblages. The present work addressed the following questions: 1) Is there any effect on the assemblages associated with rhodolith due to increased periods of stability in relation to natural disturbance conditions? and 2) Do the assemblages develop in the same way if experiments are started on different dates?

Study Area

Strangford Lough on the east coast of Northern Ireland is a narrow arm of the sea (Figure 2). It is approximately 24 km in length with an average width of 5 km and a maximum depth of 60 m. This body of water is connected to the Irish Sea by a narrow strait less than 1 km in width. Tidal currents can reach up to 16 k/hr, but it is more common to observe speeds of 9.26 to 12.96 km/hr.

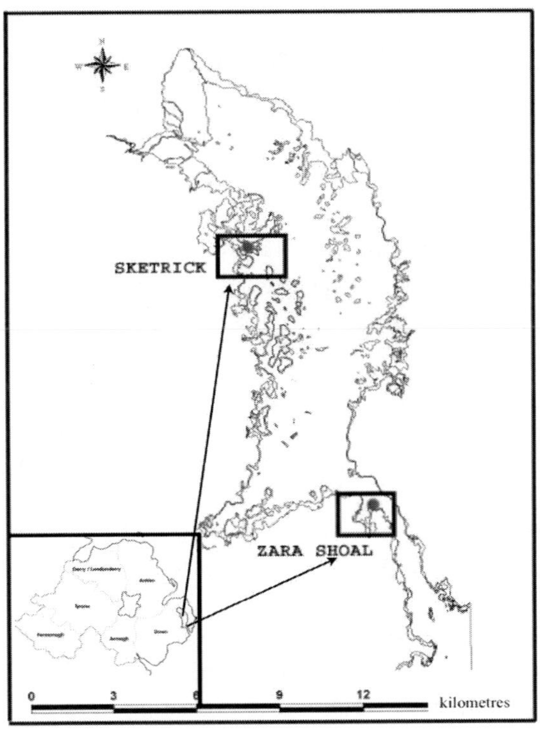

Figure 2. Rhodolith beds location in Strangford Lough, Northern Ireland. The Zara Shoal bed in the south en of the Lough and is mainly composed by free-living thalli of the coralline algae *Phymatolithon calcareum*, it experiences a fast flow because is situated in the narrowest portion of the Lough. Sketrick is a shallow rhodolith bed in the north of Strangford, composed mainly of sub fossil *Lithothamnion glaciale* that dates to 4,200 YBP (Present-calibrated radiocarbon date, [3].

The Zara Shoal maerl bed is situated in Castleward Bay (Figure 2). The bed is mainly composed of living *Phymatolithon calcareum*, a very common widespread rhodolith species, at a mean depth of 8 m. Most of the sea bottom is covered by thalli, covering up to 90% of the total area; in some places rhodoliths are mixed with boulders and cobbles where large macroalgae, such as *Laminaria saccharina* and *Saccorhiza polyschides*, can develop giving added shelter for associated species.

This rhodolith bed is exposed to wind from the south and southeast, and the sea state becomes moderate to strong in more than Force 4 winds. The sampling site is closer to the narrows channel and consequently the tidal currents can reach velocities of 5.55 to 11.11 km/hr.

The Sketrick Island rhodolith bed is located in the middle section of the Lough, with depths ranging from the intertidal to 5 m deep (Figure 2). It is located in the channel between Sketrick Island and mainland that protects it from bad weather that occurs in the main body of the Lough.

This bed consists primarily of sub fossil thalli of *Lithothamnion glaciale* mixed with sparse live *Phymatolithon calcareum* thalli. In shallower areas, boulders can be found with large macroalgae, such as *Laminaria saccharina* and *Sargassum muticum*, growing to lengths over 2 m.

Design to Test Invertebrate Assemblages Response to Stability and Natural Disturbance

A. Experimental Sampling

The effects of stability on the assemblages associated with artificially stabilized and free-living rhodoliths were tested at two beds, Sketrick Island and Zara Shoal, in Strangford Lough (Figure 2). In the case of the Zara Shoal bed, the rhodolith-forming species *Phymatolithon calcareum* was used for the experiments. 18 plastic meshes of 15 x 15 cm (225 cm^2) completely covered with rhodolith thalli attached by regular fishing line were prepared for deployment (Figure 3). At the Sketrick Island rhodolith bed, sub fossil thalli of *Lithothamnion glaciale* were artificially stabilized and compared to loose thalli of the same species.

Meshes were attached onto three concrete paving slabs placed at both locations at the beginning of each of the three experiments (Figure 4). Every two months, three plastic meshes were retrieved from both localities to evaluate abundance and richness of the assemblages over different periods of time. Natural rhodolith (referred to as loose rhodolith) was collected from quadrats with similar dimensions (volume and shape) as the plastic meshes at the same time as meshes were retrieved, in order to provide comparative data for natural rhodolith assemblages for each sampling date. Samples were enclosed gently underwater in separate plastic bags to be transported to the laboratory.

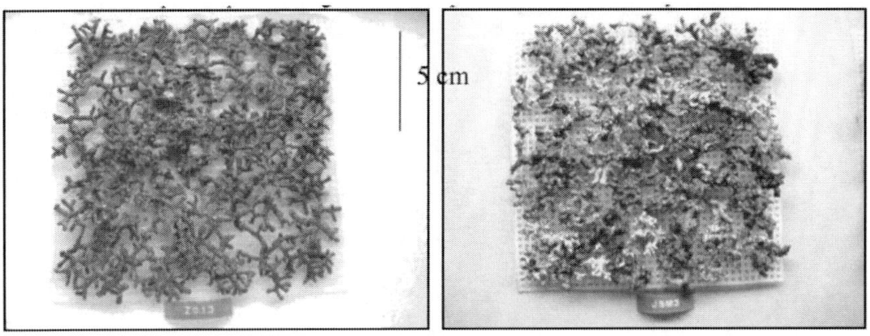

a) b)

Figure 3. Artificially stabilized thalli of rhodolith-forming species *Phymatolithon calcareum* (a) and sub fossil *Lithothamnion glaciale* (b). Thalli were tied onto plastic meshes and deployed onto concrete paving slabs for different periods of time.

(a)

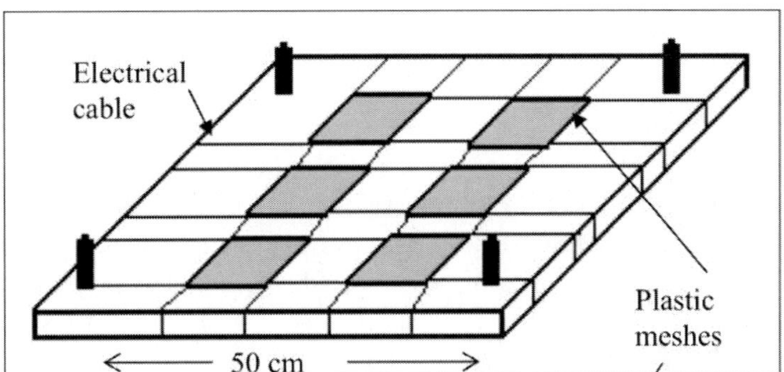

(b)

Figure 4. (Continued).

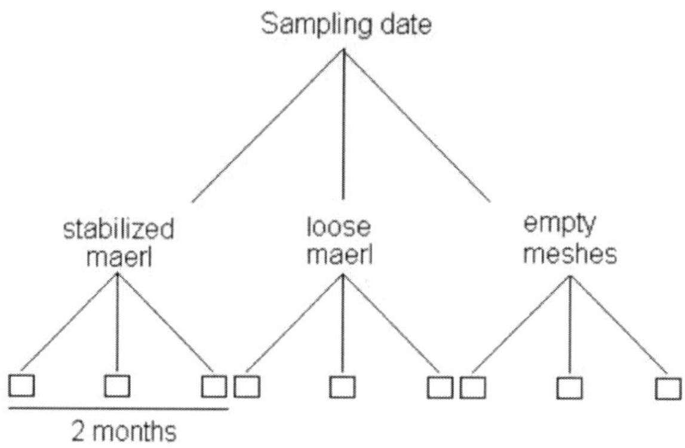

c)

Figure 4. Example of a concrete paving slab used for experimental set up. (a) Diagram of the paving slabs deployed at the two rhodolith beds, showing positions of plastic meshes with tied thalli and empty meshes onto these slabs for different periods of time depending on the experiment being carried out. (b) Underwater of rhodolith-covered meshes at experimental sites. (c) Diagram of the experimental design to evaluate the effects of disturbance on the assemblages associated with rhodolith beds.

B. Laboratory Procedures

Plastic meshes and quadrats were photographed before and after sampling to compare differences in algal coverage. Meshes and rhodolith from quadrats were shaken vigorously to extract all the vagile animals. Sessile organisms such as ascidians and sponges, and some molluscs, were removed using pincers. The water from each sample was sieved through a 0.5 mm mesh to collect loosened epifauna. The retained sediment was sorted under a dissecting microscope to extract any remaining epifaunal organisms. Organisms were identified to species, genus or to the highest level possible and preserved in a 70% alcohol solution. Encrusting colonial invertebrates were counted as one individual for each rhodolith piece they were attached to, independently of the number of colonies present on each thallus.

C. Statistical Analyses

Data were tested for normality and heterogeneity of variances using Kolmogorov-Smirnov and Cochran's *C*-tests. Data were analysed using two-way Analyses of Variance with MINITAB©. The time since stabilization (Date: two, four or six months) and stability of rhodolith (two levels: stabilized or loose) were used as fixed factors. Each one of the three experiments was analysed for the richness of the assemblages, including epifloral and epifaunal species. Tukey tests were used to determine the significance of all possible differences between groups.

Two-way crossed ANOSIM analyses with PRIMER©, based on Bray-Curtis similarity for the untransformed abundance and presence/absence data, were used to compare the assemblages associated with artificially stabilized and natural rhodolith samples [6]. The dissimilarity of richness between rhodolith assemblages on the different sampling dates was plotted as Global R-values from the ANOSIM analyses. R-values of 0 indicate similar assemblages, while R=1 indicates completely different assemblages. Multidimensional scaling plots (MDS) were calculated to visualize the similarity between species richness and abundance. The most important species were identified by SIMPER analyses and their contribution to the similarity between assemblages was calculated (cut-off = 90%).

A final comparison using a two-way ANOVA analysis was carried out to evaluate the effect of the starting date of the three experiments (first source of variation: May 2002, March 2003, and September 2003) and the time elapsed since artificial stabilization (second source of variation: two, four, and six months) on the assemblages associated with the artificially stabilized rhodolith.

Response of the Associated Invertebrate Community with Rhodoliths to Stability and Disturbance

1. Assemblage Response to Increased Rhodolith Stability at the Sketrick Island Bed

At Sketrick Island, the richness of the assemblages associated with artificially stabilized *Lithothamnion glaciale* differed from those of the natural rhodolith. The richness followed different patterns in response to the two sources of variation analysed, level of stability and time since stabilization. The 2-way ANOVA analyses were not always significant for the effects of artificial stabilization.

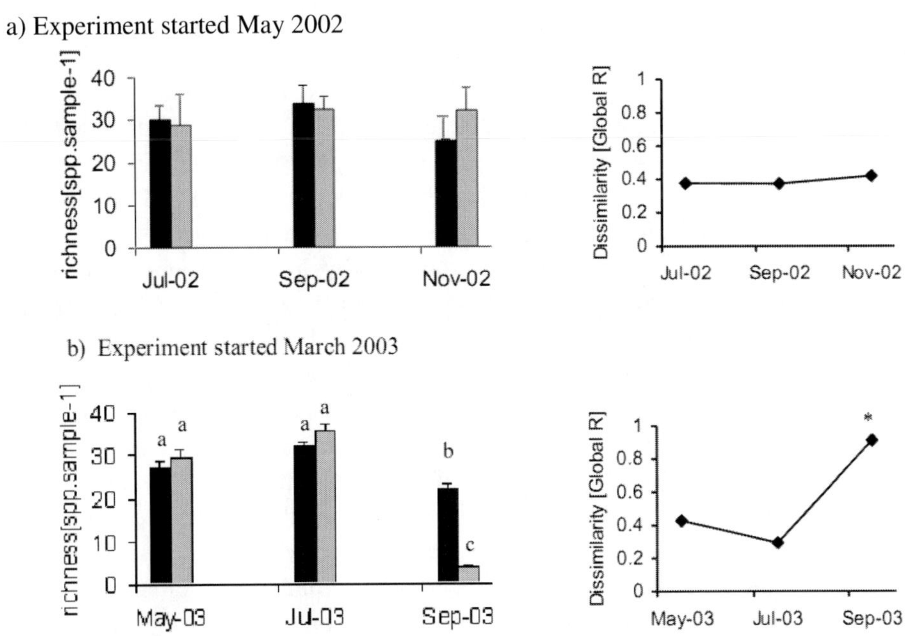

c) Experiment started September 2003

Figure 5. (Continued).

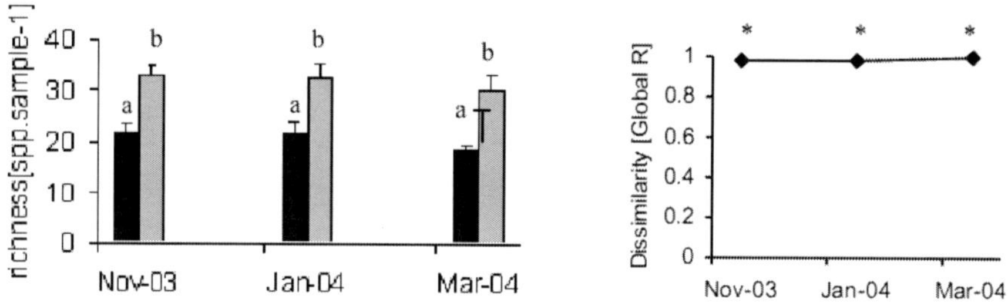

Figure 5. Graphs on left correspond to the total richness (means ± standard deviations) associated with loose (black) and stabilized (grey) *L. glaciale* at Sketrick Island (n= 3). Dates show time since stabilization with different starting dates for each experiment. Stabilized rhodolith was retrieved after 2, 4 and 6 months, corresponding to the first, second and third sampling dates shown for each of the experiments. Graphs on right represent dissimilarity between stability regimes, calculated as Global R values from the ANOSIM analysis. * Denotes significant difference ($p < 0.05$) between assemblages on stabilized and natural rhodolith at that sampling date using a Tukey test. Letters denote statistically different groups according to Tukey test.

In the first experiment, started in May 2002, species richness associated with artificially stabilized rhodolith presented no significant differences relative to the natural samples or between the different periods of stability (Figure 5a and Table Ia). These assemblages responded to disturbance as in the third predicted scenario with a no-effect model (Figure 1c) rather than the Intermediate Disturbance Hypothesis. The ANOSIM analysis showed that the dissimilarity in the number of species between samples was low, not significant and relatively stable in relation to the period of stability (Figure 5a, Table I appendix).

In the experiment started in March 2003, the assemblages of stabilized and natural rhodolith followed the Intermediate Disturbance Hypothesis pattern (Figure 5b). Species richness differed significantly between periods of stability. Whether assemblages developed on artificially stabilized or loose rhodolith was significant by the sixth month after stabilization. The interaction of both factors had a significant effect on the assemblages (Table Ib Total richness: Date, Stability and Interaction). The richness increased by the second sampling date, becoming higher in stabilized samples, but decreasing significantly after six months, the decline in richness being bigger for stabilized samples. The dissimilarity of the assemblages remained relatively constant after four months, but increased significantly due to reduction of species richness on stabilized rhodolith by the sixth month (Table Ib).

The experiment started in September 2003 represented the second possible scenario for the response of diversity to disturbance and the artificial stabilization in relation to the patterns described previously (Figure 1b). The richness differed significantly between stable and loose rhodolith assemblages, always being higher for artificially stabilized samples (Table Ic Stability). According to the ANOSIM analysis, the assemblages were significantly different by the second month of induced stability. Such differences were recorded as dissimilarity values close to 1 and were maintained throughout all the experiment (Figure 5c).

In accordance with the 2-way ANOVA analyses, PRIMER detected significant differences between the stabilized and natural rhodolith assemblages. Additionally, SIMPER analyses reported low levels of similarity between stabilized and natural samples. The species that contributed the greatest abundance to assemblage dissimilarity in the three different experiments were four amphipod species, *Lysianassa ceratina*, *Corophium sextonae*, *Aora gracilis* and *Jassa falcata*, which were the most important species after two months of rhodolith stability. The importance of these species shifted at different stages for each experiment when their abundance decreased and some other species increased in importance.

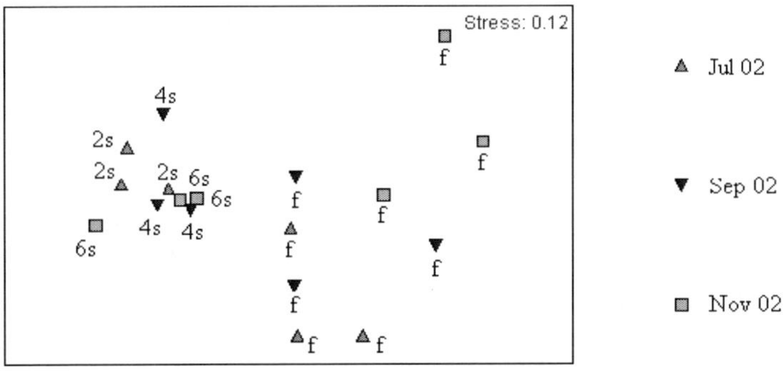

Figure 6. MDS ordination plot for the effect of disturbance on stabilized and natural rhodolith samples (PRIMER Bray-Curtis similarity, untransformed data). Assemblages were compared after two (2s), four (4s), and six months (6s) of artificial stabilization to the assemblages associated with loose rhodoliths (f) collected on the same dates. Plot represents the similarities of the assemblages for the experiment started in May 2002 at the Sketrick Island bed. The other two experiments presented similar distributions and therefore were not included.

In the second and third experiments the amphipod *Microdeutopus versiculatus* increased in abundance to values similar to those of the other amphipod species after four months of stabilization. The multi-dimensional scaling plots showed significant differences between artificially stabilized and natural assemblages (Figure 6). This pattern was seen in all three experiments carried out. The similarity within stabilized samples was higher than between the natural rhodolith assemblages. The stabilized samples are grouped more closely than the natural rhodolith samples. The ANOSIM presented similar results whether the abundance of the species or only the presence/absence data were considered. Such similarity in results indicated that assemblages were very similar over time and that changes in the distribution of the abundance between species were responsible for the differences observed.

2. Assemblage Responses to Increased Rhodolith Stability at the Zara Shoal Bed

The variations in species richness for the assemblages associated with artificially stabilized and natural *P. calcareum* at the Zara Shoal rhodolith bed followed different patterns for the three experiments carried out between May 2002 and April 2004 (Figure 7). None of the experiments followed an Intermediate Disturbance Hypothesis distribution, but rather an idiosyncratic or no-effects model. The assemblage richness in the first and third experiments, May-02 and September-03, presented a pattern for both the artificially stabilized and the loose rhodolith (Figure 7a and b), which fits the predictions of an idiosyncratic model (Figure 1d). The assemblages differed between periods of stability and the interaction effect of the two sources of variation (level of stability and time elapsed since artificial stabilization). Richness in March 2003 was no different between stabilized and natural samples (Table III). The ANOSIM analysis showed that the similarity between assemblages decreased in later phases of the experiment (Figure 7b). Molluscs and crustaceans were the most important groups in natural samples. Both groups also dominated the assemblages on artificially stabilized rhodolith, with shifts in the importance of the dominant species over time. The mussel *Mytilus edulis* was one of the most important components of the assemblages. In addition another mollusc, *Musculus costulatus* and the amphipod species *Corophium sextonae*, *Lysianassa ceratina* and *Jassa falcata* made a high contribution to the assemblage composition. The amphipod *C. sextonae* increased in abundance with time in the three stabilization experiments, although those increases were not correlated with the total richness of the assemblages.

Assemblages associated with artificially stabilized rhodolith differed from those of natural samples, which are represented as two distinct groups in the multi-dimensional scaling plot (Figure 8). This pattern was seen in all three experiments carried out. Natural assemblages presented larger differences between sampling dates. The similarity within artificially stabilized samples was higher and they were grouped more closely in the MDS plots. According to the ANOSIM analyses, the differences observed are due to changes in the species abundance. The analyses of presence/absence indicated that assemblages are very similar in composition but there were changes in the distribution of the abundance between the different species.

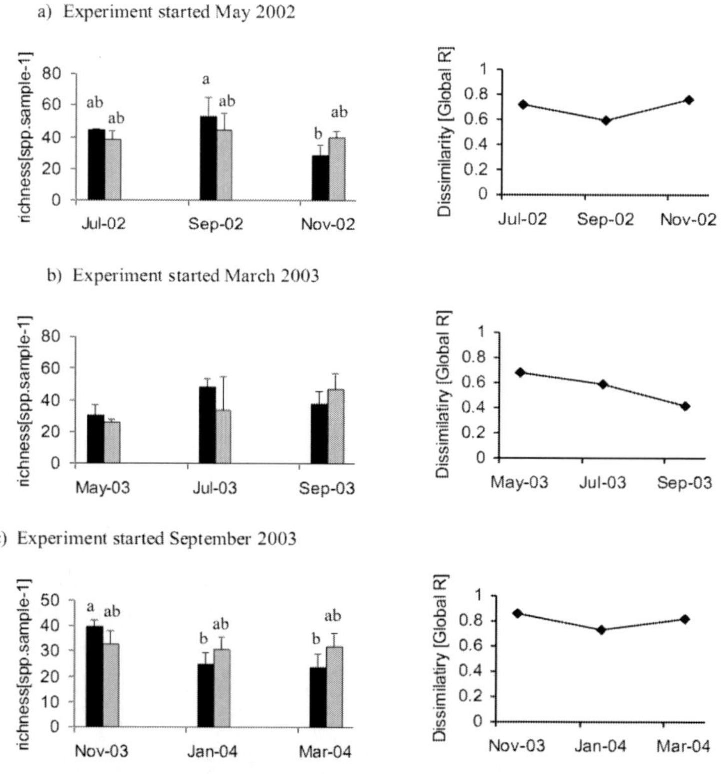

Figure 7. Left graphs correspond to the total richness (means ± standard deviations) associated with loose (black) and stabilized (grey) *P. calcareum* at Zara Shoal (n= 3). Dates show time since stabilization with different starting dates for each experiment. Stabilized rhodolith was retrieved after 2, 4 and 6 months, corresponding to the first, second and third sampling dates shown for each of the experiments. Graphs on right represent dissimilarity between stability regimes, calculated as Global R values from the ANOSIM analysis. Letters denote statistically different groups according to Tukey test ($p < 0.05$).

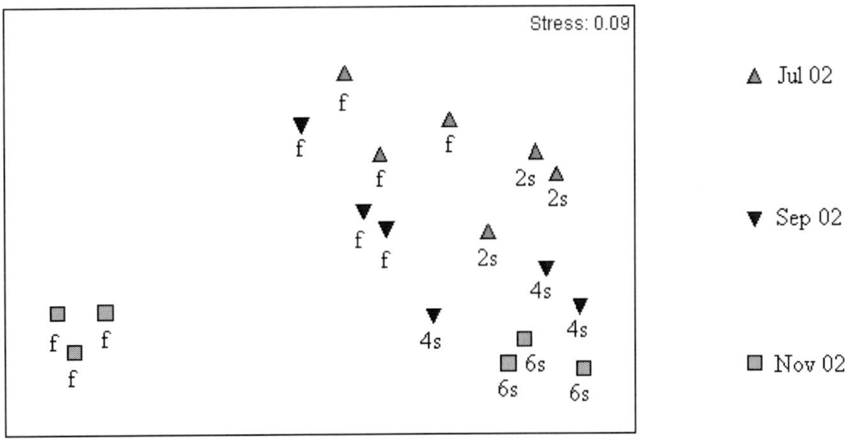

Figure 8. MDS plot for the effect of disturbance on stabilized and natural rhodolith samples (PRIMER, Bray-Curtis similarity, untransformed data). Assemblages were compared after two (2s), four (4s), and six months (6s) of artificial stabilization to loose rhodolith samples (f) collected on the same dates. Plot represents the similarities of the assemblages for the experiment started in May 2002 at the Zara Shoal bed. The other two experiments presented similar distributions and therefore were not included.

3. Effects of Different Starting Date on the Assemblages Associated with Artificially Stabilized Rhodolith

A comparison between the assemblages associated with stabilized samples was carried out to determine if there were any effects due to the date when samples were deployed. The response of the assemblages to artificial stabilization was of different magnitudes depending on the time of year when the experiments started. The richness of the assemblages associated with stabilized rhodolith at Sketrick Island differed significantly. The experiments started in May 2002 and September 2003 were more similar than with the assemblages from the experiment started in March 2003. The 2-way ANOVA test found significant differences between the periods of stability, starting date and the interaction of the two sources of variation (Table III). Therefore, the effect of the time elapsed since artificial stabilization on the resulting assemblages is highly dependent on the starting date of the experiments (Figure 9).

The richness of the assemblages associated with the artificially stabilized rhodolith differed significantly depending on the date when samples were deployed (Table III Starting date). The period that samples were kept stable did not have a significant effect on the diversity of the assemblages. The interaction of the two sources of variation, starting date and period of stability, affected significantly the number of species associated with the stabilized rhodolith (Table II Starting date and interaction); therefore, the effect of the period of stabilization on the resulting assemblages depends on the starting date of the experiments (Figure 10).

The experiment started in March 2003 was the most different in comparison to May 2002 and September 2003 experiments that presented higher similarity.

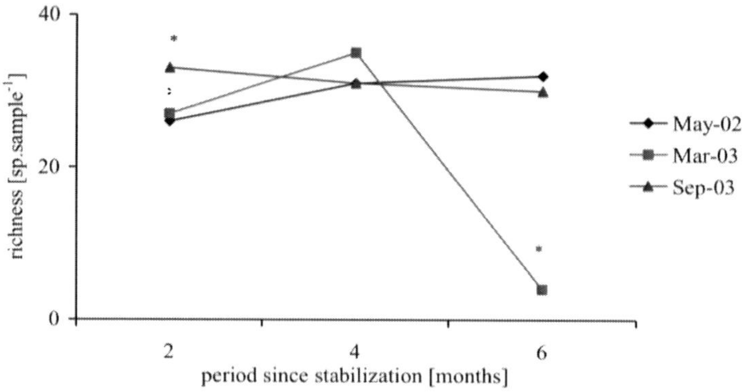

Figure 9. Interactions plot for the richness of the assemblages associated with artificially stabilized rhodolith at Sketrick Island. The factors considered were: the starting dates (3 levels: May-02, March-03, and September-03) and the time elapsed since rhodolith thalli were stabilized (3 levels: two, four, and six months). * Denotes significant differences ($p<0.05$) according to Tukey test.

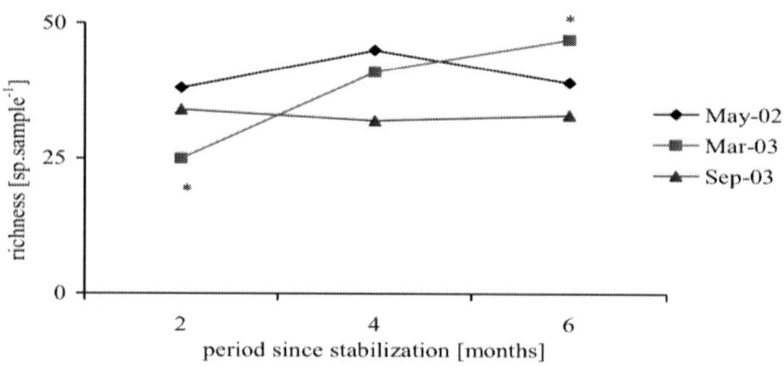

Figure 10. Interactions plot for the richness of the assemblages associated with artificially stabilized rhodolith at Zara Shoal. The factors considered were: the starting dates (3 levels: May-02, March-03, and September-03) and the time since rhodolith was stabilized (3 levels: two, four, and six months). * Denotes significant differences ($p<0.05$) according to Tukey test.

The patterns observed within rhodolith beds are highly dependent on the starting date of the experiments. At Sketrick Island, the assemblages started in March presented different temporal dynamics of species richness. If experiments started in September, assemblages differed highly due to disturbance. The experiment started in May was not affected by increased stability; the same result was observed for the experiment started in March at Zara Shoal. The other two starting dates at Zara Shoal did not have a clear effect on the assemblages, which presented an idiosyncratic pattern (Table IV appendix).

Discussion

Effects of Disturbance on Rhodolith Assemblages

The stabilization experiments conducted for the present study demonstrated that the influence of disturbance on assemblage dynamics of associated organisms to rhodolith beds is not simple and will depend on the characteristics of the disturbance as well as the physiological, morphological and behavioural capabilities of the species present at any given time. Furthermore, factors such as the time of the year when disturbance events happen can modify the recovery patterns of the assemblage diversity. The differences in richness between natural and artificially stabilized rhodolith assemblages were mainly an effect of decreased movement disturbance. Nevertheless changes in environmental and biological variables, together with the starting date of the experiments, accounted to different extents for the patterns observed. Previous studies have demonstrated that deployment of artificial substrata at different times of the year leads to the development of communities dominated by different taxa [1, 5, 40]. Artificial substrata may initially be occupied by taxa that pre-empt space and affect subsequent recruitment of other organisms [1] and some species recruit only after initial settlers have increased the complexity of the surface, making it suitable for secondary settlement [33]. For rhodolith assemblages the two main factors controlling diversity are the frequency and intensity of disturbance. Both factors directly affect the number of species that can successfully settle and survive under certain environmental conditions. It is also important to consider the timing of disturbance and their synchronization with availability of species dispersal propagules. Apparently, rhodolith environments in deeper water like the Zara Shoal bed are more stable and the differences between natural and artificially stabilized rhodolith are reduced. The growth patterns of *Phymatolithon calcareum* thalli at the studied beds (flat like forms as observed in Figure 3a) are likely to be held in place by overlying rhodolith reducing movement to a level closer to that of the total stabilization, therefore reducing the differences observed between stabilized and loose rhodoliths. It is important to note that two of the experiments at Zara Shoal followed an idiosyncratic model, which reflects the importance of biological interactions in species composition on rhodolith assemblages. Richness differed between dates because these experiments incorporated samples with diverse disturbance histories and regimes. If disturbance alone were the most important controlling influence on assemblage development, samples should exhibit a similar sequence of richness variations irrespective of the actual date of mesh immersion. Thus the patterns in species richness and diversity associated with natural and artificially stabilized rhodolith emerge at three scales: between stability periods (two, four and six months), between levels of stability (natural and stabilized rhodolith), and among starting dates (May 2002, March 2003 and September 2003). The data obtained do not allow the identification of which of the disturbance attributes: frequency, intensity, extent or duration, are responsible for the differences observed in the assemblages associated with rhodolith. Like in many other ecosystems [8, 9, 11, 17] there are three sets of environmental factors playing a central role in the organization of rhodolith communities: disturbance, spatial heterogeneity and stress. In addition, effects of disturbance may depend on precisely how diversity is quantified, since all diversity and richness measures are influenced by sample size, and statistical procedures [20]. However it is evident that disturbance is a strong factor determining the resulting assemblages associated to rhodolith beds.

The Intermediate Disturbance Hypothesis and Stabilization Experiments

The results obtained show that, in general, the assemblages associated with rhodolith did not follow the IDH prediction. In general, the associated assemblages response to increased stability not necessarily follows the predictions the dynamic equilibrium model [21], the IDH or even a facilitation model, which is an indicator that assemblages reflect site variations, including species present in the surrounding habitats, hydrodynamics, and some other physical or biological factors. The experiment started on March 2003 at Sketrick Island was the only one that apparently supports the Intermediate Disturbance Hypothesis. In this experiment, approximately 20% of the rhodolith thalli were lost from the three artificially stabilized samples by physical disturbance resulting in a decrease of the abundance of most species. According to the ANOSIM for the presence/absence data, there were no significant changes in the composition of the assemblages, only in the abundance of species. The lack of a monotonic pattern of species richness in the experiments suggests that individuals of mobile species arrived and left randomly, which may be a function of biological interactions or selection of rhodolith substrata on the basis of some physical or chemical cue. In stabilized rhodolith, evenness was affected by the degrees and the duration of stability reinforcing the idea of shifts in the distribution of the abundance within assemblages. Whereas the IDH predicts that diversity will peak at some intermediate intensity or frequency of disturbance, Huston's dynamic equilibrium model predicts that the location of the diversity peak depends on the rates of population growth and competitive displacement in the community (Huston, 1979). The mixed results observed for the associated assemblages reflect the dynamic environment that rhodoliths are. Stability plays an important role and may be even more relevant during summer times, when storms and wind are milder. That artificially stabilized samples had similar species richness to natural rhodolith may mean that it takes more than six months after stabilization to reach the intermediate disturbance levels and assemblages are still in the first stages of succession after disturbance. However, the fact that differences were observed in some cases as early as after two months, suggest that there was enough time for colonization, both by adults and larvae, to produce at least part of an IDH curve. Artificial stabilization proved difficult due to the nature of rhodolith thalli, which were not robust enough to undergo periods of stability longer than six months without been smashed or broken apart. Turn over of rhodoliths, together with cleaning by herbivore organism, has been mentioned as important to prevent fouling of thalli [42, 14]. During our stability experiments we observed the detrimental effect of bacterial and epifloral growth; after two months of stability rhodoliths were covered by a fine film of organisms that can not be efficiently removed by herbivores, preventing photosynthesis and calcification. Thalli loose pigmentation and became very fragile and easy to break resulting in most case on the complete lost of thally on experiments by the end of the 6th month.

Assemblage Composition

Most of the species recorded on artificially stabilized rhodolith occur naturally in rhodolith beds and some other habitats, such as *Laminaria* kelp grounds [59] and boulders in Strangford Lough (personal observations). In this case, the sources of species for recovery of the rhodolith environments after disturbance are non-disturbed rhodolith areas or some other habitat such as kelp environments. Key components of benthic ecosystems, including amphipods, isopods, and other small crustaceans do not have planktonic larvae, but have direct development [15, 53]. Nevertheless, active swimming or passive drift through the water column are efficient modes of colonisation for them [37], and defaunated sites often may be colonised within a few hours, days or weeks after disturbance [39, 25, 48]. The presence of adult organisms of different species on stabilized rhodolith shortly after experiments deployment indicates that migration of at least some of the mobile fauna took place. As observed in many other marine environments [34, 10, 37, 52, 18, 36, 31], mobile organisms, such as the amphipods *Microdeutopus versiculatus*, *Corophium sextonae*, *Aora gracilis*, *Lysianassa ceratina* and *Jassa falcata*, rapidly colonised the stabilized rhodolith samples following deployment and presented high abundances by the second month. In the case of molluscs, it is more likely that they represent new recruitment, since sizes were relatively small (5 to 10 mm), and juveniles may migrate just before they are too heavy to be carried by currents or waves [58]. Dominant species were consistently present in both artificially stabilized and natural rhodolith. *Corophium sextonae* and *Aora gracilis* were the most important epifaunal species at both rhodolith beds. Their abundance fluctuated in the different experiments, most likely related to natural fluctuations of the populations, interspecific competition, the reduction in disturbance, and to the increase of habitat complexity by biogenic structures, such as their own tubes [57, 19] and plant biomass [24], although the effect of each factor is difficult to estimate. *C. sextonae* colonised experiments and built tubes all over the stabilized rhodoliths, which resulted on abundances even close to 5,000 individuals after 4 months of stability. This density of tubes was not observed on natural rhodoliths highlighting the importance of rotation to regulate biodiversity on rhodolith beds by preventing dominance of a single species. The analyses suggest that the assemblages were restricted in abundances and richness due to movement disturbance, although not all effects of disturbance are negative. In accordance with Glasby [16], the observations of the present study support the general idea that the differences in marine assemblages are highly related to substratum movement, which can influence the settlement of different species. Nevertheless the starting date and the time elapsed since stabilization influenced the successional patterns followed by rhodolith assemblages. For organisms that inhabit disturbed habitats, such as loose rocks, the combination of all these factors will be important in structuring the community [43, 44, 26, 2]. Natural ecosystems are a reflection of the interactions between disturbance effects and the organism responses [29, 45, 22, 51, 53, 2] resulting in a particular species composition. The results of stabilization experiments reflect the vulnerability of rhodolith assemblages to changes to hydrodynamic regimes, this should be taken into consideration to prevent impacts by coastal development is proposed. Reduction of currents and wave action could result on increased sedimentation and thalli fouling which in turn may end on the complete lost of entire rhodolith beds.

Conclusion

The assemblages associated with rhodolith are highly affected by physical disturbance although biological interactions may play a major role in determining assemblage composition that was not covered in the present study. The general predictions of the Intermediate Disturbance Hypothesis for community structure after disturbance had only limited support, which highlights the uniqueness of each of the beds and their associated assemblages. Changes to rhodolith assemblages in response to altered hydrological conditions, for example through climate change or coastal modification, will be difficult to predict and will depend on the interactions of seasonal factors with recruitment, migration processes and species relationships. These unpredictable processes may result in a persistently altered assemblage after two months or may result in transient fluctuations to the community that makes management of rhodolith environments difficult. Nevertheless, it is clear that stability results on a rapid increase of thalli fragility and a dominance of reduced number of specie, which in turns reduces biodiversity.

Acknowledgments

We would like to thank C. Blake, A. Davies and S. Vize for their assistance in the field. This study was funded by the Mexican Council for the Sciences and Technology (CONACYT) and the Queen's University of Belfast. G. Hinojosa-Arango thanks CONACYT for the scholarship granted for postgraduate studies (162307).

Appendix

Table I. Summary of the two-way ANOVA tests for the effects of disturbance on the assemblages associated with stabilized and loose rhodolith (Stability), time elapsed since stabilization (Date), and the interactions of these two factors (Interaction). Significant *p*-values are in bold

a) Experiment started May 2002

		df	MS	*F*	*P*
Total richness Sketrick	Date	2	36.2	1.75	0.216
	Stability	1	10.9	0.53	0.482
	Interaction	2	37.6	1.81	0.205
	Residual	12	20.7		
Total richness Zara Shoal	Date	2	260.4	5.86	0.017
	Stability	1	12.5	0.28	0.606
	Interaction	2	197.2	4.44	0.036
	Residual	12	44.4		

b) Experiment started March 2003

		df	MS	*F*	*P*
Total richness Sketrick	Date	2	708.22	354.11	0.000
	Stability	1	76.06	38.03	0.000
	Interaction	2	228.22	114.11	0.000
	Residual	12	2.00		
Total richness Zara Shoal	Date	2	369	3.19	0.078
	Stability	1	53	0.46	0.510
	Interaction	2	207	1.79	0.209
	Residual	12	116		

c) Experiment started September 2003

		df	MS	*F*	*P*
Total richness Sketrick	Date	2	12.67	2.02	0.176
	Stability	1	589.39	93.88	0.000
	Interaction	2	0.22	0.04	0.965
	Residual	12	6.28		

c) Experiment started September 2003					
		df	MS	*F*	*P*
Total	Date	2	144.5	*6.19*	*0.014*
richness	Stability	1	24.5	*1.05*	*0.326*
Zara Shoal	Interaction	2	99.5	*4.26*	*0.040*
	Residual	12	23.3		

Table III. Summary of the two-way ANOVA analysis for effects of starting date on the assemblages associated with the three artificial stabilized experiments. Time elapsed since stabilization (Period since stabilization), and the interactions of these two factors (Interaction) were the sources of variation in the richness of the assemblages. Significant *p*-values are in bold

Sketrick		df	MS	*F*	*P*
Richness	Starting Date	2	254.5	9.56	0.001
	Period since stabilization	2	544.1	20.43	0.000
	Interaction	4	344.8	12.95	0.000
	Residual	18	26.6		
	Total	26			
Zara Shoal					
Richness	Starting Date	2	284.7	4.25	0.031
	Period since stabilization	2	212.5	3.41	0.066
	Interaction	4	218.8	3.27	0.035
	Residual	18	66.9		
	Total	26			

Table IV. Summary table for the patterns followed by the assemblages associated with rhodolith beds in Strangford Lough in response to disturbance and succession. Experiments are listed by rhodolith bed with starting date of experiments.

Possible scenarios according to Figure 5.1			
a IDH	b positive/negative	c no-effect	d idiosyncratic
Sketrick March-2003	Sketrick Sep-2003	Sketrick May-2002	-
-	-	-	Zara May-2002
-	-	Zara March-2003	-
-	-	-	Zara Sep-2003

References

[1] W. E. Arntz and H. Rumohr, *J. Exp. Mar. Biol. Ecol.* 64, 17-45 (1982).

[2] J. J. Bell and K. A. Barnes, *Biol. Bull.* 205, 144-159 (2003).

[3] C. Blake and C. A. Maggs, *Phycol.* 42, 606-612 (2003).

[4] D. L. Breitburg, *Oecol.* 65, 173-184 (1985).

[5] M. G. Chapman, *J. Exp. Mar. Biol. Ecol.* 275, 95-116 (2002).

[6] K. R. Clarke and R. N. Gorley, *PRIMER v5. Plymouth Marine Laboratory*. England, 2001
[7] J. H. Connell, *Sci.* 199, 1302-1310 (1978).
[8] P. K. Dayton, *Ecol. Monogr.* 41, 351-389 (1971).
[9] P. K. Dayton and R. R. Hessler, *Deep-Sea Res.* 19, 199-208 (1972).
[10] R. L. Dean and J. H. Connell, *J. Exp. Mar. Biol. Ecol.* 109, 217-247 (1987).
[11] M. N. Dethier, *Ecol. Monogr.* 54, 99-118 (1984).
[12] R. Dial and J. Roughgarden, *Ecol.* 79, 1412-1424 (1998).
[13] B. C. Emerson and N. Kolm, *Nat.* 434, 1015-1017 (2005).
[14] M. S. Foster, *J. Phycol.* 37, 659-657 (2001).
[15] D. R. Franz and Y. Mohamed, *J. Exp. Mar. Biol. Ecol.* 133, 1-13 (1989).
[16] T. M. Glasby, *Mar. Ecol. Prog. Ser.* 215, 37-47 (2001).
[17] O. Hamerlynk and J. Mees, *Oceanol. Acta. Spec. Issue*. 11, 205-212 (1991).
[18] D. D. Hardin, J. Toal, T. Parr, P. Wilde and K. Dorsey, *Mar. Environ. Res.* 37, 165-193 (1994).
[19] P. J. Hayward and J. S. *Ryland, Handbook of the Marine Fauna of North-West Europe*. Oxford University Press. Oxford, UK (2002).
[20] R. J. Hobbs and L. F. Huenneke, *Conser. Biol.* 6, 324-337 (1992).
[21] G. Hinojosa-Arango, C. A. Maggs and M.P. Johnson, *Ecol.* 90(2), 517-528 (2009).
[22] M. A. Huston, *Am. Nat.* 113, 81-101 (1979).
[23] K. Johst and A. Huth, *Diversity Distrib.* 11, 111-120 (2005).
[24] J. H. Lawton, *Oikos.* 71, 367-374 (1994).
[25] K. M. Leber, *Ecol.* 66, 1951-1964 (1985).
[26] S. A. Levin, J. Dushoff and J. A. Keymer, *Scientia Marina.* 62, 171-179 (2001).
[27] M. Liiri, H. Setälä, J. Haimi, T. Pennamenand H. Fritze, *Oikos.* 96, 137-149 (2002).
[28] K. A. McGuinness, *Oecologia.* 71, 409-419 (1987).
[29] D. Maurer, R. L. Keck, J. C. Tinsman and W. A. Leathem, *Mar. Environ. Res.* 4, 299-319 (1981).
[30] R. W. Osman and R. B. Whitlatch, *J. Exp. Mar. Biol. Ecol.* 311, 117-145 (2004).
[31] H. Page, J. Dugan and J. Bram, *Spatial and temporal variability in early successional patterns of an invertebrate assemblage at an offshore oil platform. MMS OCS Study, 2005-003.* Coastal Research Centre, Marine Science Institute, UCSB. Santa Barbara (2005).
[32] R. T. Paine and S. A. Levin, *Ecol. Monogr.* 51, 145-178 (1981).
[33] S. Perkol-Finkel and Y. Benayahu, *Mar. Environ. Res.* 59, 79-99 (2005).
[34] C. H. Peterson, *Ecol.* 56, 958-965 (1975).
[35] M. Rejmánek, E. Rejmánková and W. Holzner, *Preslia, Praha.* 76, 207-222 (2004).
[36] C. S. Reynolds, *New Zealand J. Ecol.* 19, 219-225 (1995).
[37] W. B. Savidge and G. L. Taghon, *J. Exp. Mar. Biol. Ecol.* 115, 137-155 (1988).
[38] K. Shea, S. H. Roxburgh and E. S. J. Rauschert, *Ecol. Letters.* 7, 491-508 (2004).
[39] K. M. Sherman and B. C. Coull, *J. Exp. Mar. Biol. Ecol.* 46, 59-71 (1980).
[40] S. D. Smith and M. J. Rule, *J. Exp. Mar. Biol. Ecol.* 277, 25-41 (2002).
[41] W. P. Sousa, *Ecol. Monogr.* 49, 227-254 (1979).
[42] D. L. Steller and M. S. Foster, *J. Exp. Mar. Biol. Ecol.* 194, 201-212 (1995).
[43] D. Thistle, *Mar. Ecol. Prog. Ser.* 6, 223-228 (1981).

[44] C.R. Townsend, R. M. Thompson, A. R. McIntosh, C. Kilroy, E. Edwards and M. R. Scarsbrook, *Ecol. Letters*. 1, 200-209 (1998).

[45] S. J. Turner and C. D. Todd, *J. Exp. Mar. Biol. Ecol.* 166, 251-272 (1993).

[46] A. J. Underwood, *Experiments in ecology: their logical design and interpretation using analysis of variance*. Cambridge University Press. Cambridge (1997).

[47] I. Valiela, *Marine Ecological Processes*. Springer. New York (1995).

[48] R. W. Virnstein and M. C. Curran, *Mar. Ecol. Prog. Ser.* 29, 279-288 (1986).

[49] D. A. Wardle, L. R. Walkerand R. D. Bardgett, *Sciencexpress*. 10, 1126/Science. 1098778 (2004).

[50] R. M. Warwick and K. R. Clarke, *Mar. Ecol. Prog. Ser.* 92, 221-231 (1993).

[51] R. M. Warwick and K. R. Clarke, *Mar. Biol.* 118, 739-744 (1994).

[52] R. M. Warwick, H. M. Platt, K. R. Clarke, J. Agard and J. Gobin, *J. Exp. Mar. Biol. Ecol.* 138, 119-142 (1990).

[53] L. Watling and E. A. Norse, *Conserv. Biol.* 12, 1180-1197 (1998).

[54] S. Widdicombe, M. C. Austen, M. A. Kendall, R. M. Warwick and M. J. Jones, *Hydrobiologia*. 440, 369-377 (2000).

[55] D. M. Wilkinson, *J. Biogeography*. 29, 1109-1115 (2002).

[56] J. B. Wilson, *New Zealand J. Ecol.* 18, 176-181 (1994).

[57] D. Witman, *Ecol. Monogr. 55*, 421-445 (1985).

[58] R. S. S. Wu and P. K. Shin, *Mar. Biol.* 128, 475-487 (1997).

[59] R. N. Zajac and R. B. Whitlatch, *Mar. Ecol. Prog. Ser.* 10, 15-27 (1982).

In: Invertebrates: Classification, Evolution and Biodiversity
Editor: Rafael Riosmena-Rodriguez
ISBN: 978-1-62618-789-4

Chapter 6

Some Affinities between Mollusks (Gastropoda, Bivalvia, Polyplacophora and Cephalopoda) and the Environment in the Gulf of California and the Western Mediterranean Sea

Oscar Efraín Holguín Quiñones[1*], José Luis Sánchez Lisazo[2] and Humberto Wright López[3]
[1,3]Centro Interdisciplinario de Ciencias Marinas,
IPN. Departamento de Pesquerías and Biología Marina. México
[2]Departamento de Ciencias del Mar and Biología Aplicada,
Edificio de Ciencias V, Universidad de Alicante. Alicante, España

Abstract

The Gulf of California, also known as Sea of Cortez, and the Mediterranean Sea are ecosystems considered as spots of high marine biodiversity. Both regions are comparable from various physiographical and climatic perspectives. Both are enclosed seas characterized by a negative hydrological balance derived from high evaporation rates that are partially offset by the inflow of surface oceanic currents and, to a lesser extent, by precipitation and freshwater discharges from rivers; as a result, these seas are saltier than oceans. The Gulf of California, however, has a broad communication with the Pacific Ocean and is bordered by the coasts of a single country (Mexico), in contrast with the Mediterranean Sea, which connects with the Atlantic Ocean through the Gibraltar Strait and involves littorals of European, Asian and African countries, many of which share a complex centuries-long history of coastal settlements and exploitation of marine resources. These two basins are regarded as having a high biological diversity and share a similar number of marine mollusk species (2,113 in the Mediterranean Sea and 2,193 in the Gulf of California). It is worth noting that, when the surfaces covered by these basins

* Corresponding author: Email: oholguin@ipn.mx.

are compared, the Gulf concentrates this high biodiversity within a rather small area. Furthermore, considering that studies focused on the Gulf of California are still scarce, it is estimated that the number of faunal records in the area will double as investigation works progress. Of the 293 families of marine mollusks reported in the literature reviewed for both basins, at least 120 share between one and four genera. This chapter aims to compare the Mediterranean Sea and the Gulf of California in terms of physiographic features, the diversity of mollusks inhabiting both basins by family, genera and number of species represented, and the role of these organisms as objects of commercial exploitation in fisheries.

Introduction

The geological history of the Mediterranean Sea and the Gulf of California has its origins in the primary displacements of the Earth's crust tectonic plates and the continental drift. The Gulf of California is one of the youngest marginal bodies and was formed by the separation of the North American plates and the Pacific Plate at least some 12 million years ago [1]. The Baja California Peninsula is being affected by a gravitational slide due to the fracture of plates made of cortical material and experiences a drift that heads northwest towards the Pacific Plate [2], [3]. The Mediterranean Sea has experienced a more ancient and complex geological evolution involving several continental convergence stages and expansion of the ocean, which gave origin to its basic configuration some 40 million years ago [4].

The inhabitants of the Ebro valley in the late Palaeolithic (10 000 – 11 000 years ago) used mollusks from the Mediterranean coast as food source and shells as ornaments [5]; likewise, accumulations of mollusk shells used as food by Palaeolith peninsular natives have been found in the Gulf the California's coasts and islands, along with cave paintings that date back more than 10,000 years, so that vestiges and traces of marine mollusks represent an important source of potential information on the relationship of hunters-gatherers with coasts in prehistoric times. [6].

The Mediterranean is a structurally complex sea characterized by some physical, environmental and biological features that resemble those in the Gulf of California. Both seas, however, display considerable differences, since the Mediterranean is a nearly closed sea connected with the Atlantic Ocean by a narrow strait measuring 12 kilometers deep by 300 meters long that separates Europe from Africa, and with the Read Sea by a narrow artificial navigation canal that separates Africa from Asia; it receives considerable freshwater discharges from the north of the basin, and low-salinity water through the Bosforo strait, being oligotrophic and located in more northern lattitudes (34°-44° N) compared to the Gulf (21°-32° N). Both basins experience a high evaporation and, in general, display low-energy waves.

The Gulf of California and the Mediterranean Sea are ecosystems regarded as spots of high marine biodiversity [7], [8]. The earliest records of marine mollusks in the Gulf of California used by science came from collections by a Belgian citizen (Frederick Reigen) between 1848 and 1850 and by a U.S. naturalist (John Xantus) between 1859 and 1861; it is also acknowledged that modern oceanography and biology in the Gulf started with the expeditions by E. W. Scripps in 1939 and the remarkable expedition of Ricketts and J. Steinbeck onboard the Western Flyer ship in 1940 [9].

The Mediterranean is one of the best studied seas worldwide. Knowledge on its biota dates back to Aristotle's times (384-322 B.C.); afterwards, since the early fourteenth century to the first third of the twentieth century, many naturalists and biologists became interested in the knowledge of life in its waters, but it was until after the Second World War that the marine science focusing on this basin started to undergo an unprecedented boost [10]. Both basins have been intensively investigated since the decade of 1950 to date by universities, museums, funds, institutes, research centers, FAO and NGOs, with the participation and sponsorship of the international scientific community, especially from countries bordering these basins.

Mollusks are deemed the largest Phylum after arthropods, and are among the best known biological groups; however, no agreement has been reached on the exact number of living species described, which range from 50 000 to 120 000 taxa, depending on the criteria of several authors [7], [11], [12], [13], [14], [15]. Gastropods, bivalves and cephalopods group the most important marine species in the Phylum in terms of economic use for multiple purposes; those large groups prosper in a great diversity of habitats and display a broad latitudinal distribution in seas and oceans, as well as from supralittoral fringes to the deep ocean [13], [16], [17].

For year 2000, 1,376 species belonging to the Phylum Mollusca were estimated to inhabit the Mediterranean Sea [18]; for 2010, there are references [8] that this figure reaches 2,113 species recorded, of which more than 1,200 species in 621 genera and 235 families are described and illustrated only for the marine waters of Andalucia, Spain, most of which are distributed across the western Mediterranean [8], [17]. For its part, the Gulf of California, although still little studied, included 2,193 species reported in 2005, belonging to 672 genera and 208 families [19]. This number of species is similar to the one in the Mediterranean; however, taking into account that the Gulf of California is nearly ten times smaller, this evidences the higher malacological abundance per unit of area in the Gulf basin.

The current variety of habitats populated by organisms from the most diverse groups has played a key role in contributing to the high biodiversity levels observed in the two marine basins, similar to other megadiverse regions, although in the Mediterranean this large amount of species occurs in an area having one of the lowest productivity worldwide, similar to the Sargasso Sea, and has a low population density. Despite the above, fishing activities are intense and have been practiced for centuries [20].

The Western Mediterranean Sea

The name *Mediterranean Sea* derives from latin (*Medi Terraneum*), meaning "a sea in between land". Ancient peoples also gave other names to it; its current features and biocenosis originated a little over 5 million years ago due to the reopening of the Gibraltar Strait connection during the Pliocene, which reestablished seawater circulation, flooding of the basin and passage of a great diversity of life forms; this is the largest enclosed sea worldwide, with 3,860 km long in an east-to-west direction, a surface area of a little over 2,5 million km^2 and a total water volume of 3.7 million km^3 [10, 20, 21, 22].

The Mediterranean Sea is located in the collision zone of the African and Eurasian plates [22], and is the place of origin not only of civilization in Europe and some Asian and African countries, but also of the first marine biologist (Aristotle), who proposed concrete ideas on many Mediterranean organisms, their physiology and even on certain interrelations between them [23]. The Mediterranean is considered as one of the most investigated and inventoried seas worldwide in all fields of oceanography, since the geological, physical, chemical, biological and climatic conditions and components, among others, have been systematically recorded in coasts, waters and deep zones. However, the Mediterranean Sea is still cause for research in many fields, including biogeography, since despite having been studied for centuries, it is believed that it shelters some species still unknown to science, along with many others for which their distribution has not been established accurately, still others that have migrated through the Suez canal or that have been either accidentally or intentionally introduced for commercial purposes, and others whose distribution range has broadened outside the Mediterranean Sea. One of the reasons behind the current lack of accurate information is that some elements of the Mediterranean flora and fauna have experienced rather radical changes as a result of the ongoing human intervention [21].

The Western Mediterranean Sea is bordered by six countries: Spain, France, Italy, Tunisia, Algeria and Morocco, with total territorial borders of 8,500 km, not considering islands except for Sardinia in the northern coast [44]. The Sicilia strait separates Sicilia from the Tunisian coast and divides the sea into two major regions: the western Mediterranean Sea, with an area of 860,000 km^2, and the eastern Mediterranean, with 1.65 million km^2 [10], [21], [23], [8]. The western Mediterranean basin is influenced by the Atlantic Ocean and can be divided according to ecological factors into a north subzone comprising the coasts of Spain, France and Italy, and a south subzone comprising the North African coast.

The Mediterranean Sea is essentially oligotrophic, all its littoral waters are affected by eutrophication, although to varying degrees; thus, it has a low nutrient content, since the organic matter that settles in deep waters is transported by sea currents running from the Mediterranean Sea to the Atlantic Ocean [24]. Seawater in the western basin is characterized by low nutrient levels and high temperatures most of the year, compared to the Atlantic Ocean at the same latitude, with variations in surface water ranging from 13 to 27 °C throughout the year [10].

The current flowing from the Atlantic to the Mediterranean Sea measured recently is 0.88 ± 0.06 Sv (Sv= I millón de m^3 s^{-1}), with water masses having a salinity of 36 ppm, which flow to the Sicilia canal, where salinity increases to nearly 38 ppm as a result of mixing with more saline water ascending from deeper layers, coupled with heavy evaporation processes [25]. Water from the Atlantic that flows superficially through the Gibraltar Strait partially compensates for the output of more saline deep water (38.4 ppm) through this same strait; the salinity of outgoing water ranges between 36.15 and 37.9 ppm in the summer, and from 36.5 to 38.45 ppm in the winter [26].

Evaporation in the basin is about 10 times higher than the magnitude that would correspond to the actual Mediterranean water deficit, leading to a negative hydric balance (nearly 1.0 m per year), being of 140 cm/year in the western Mediterranean basin; this is compensated by a water inflow of 95.6 cm/year from the Atlantic plus 33.5 cm/year from precipitations and 19.8 cm/year from rivers [25]. Precipitation in the western basin fluctuates from year to year and from north to south, but water flowing from rivers in the north is estimated to account for 92% of the total incoming volume, which contrasts with only 8% from the south [20]. Long-term changes in the freshwater regime are foreseen throughout the Mediterranean as a result of the decline in flow from all rivers and, mostly, from the dramatic contention of the Nile River due to the giant Assuan dam in Egypt from 1964. Likewise, evaporation has increased because surface temperatures tend to be higher; these processes have led to the steady rise in salinity observed in the Mediterranean Sea [25].

The negative hydrological balance in the western Mediterranean results in an active vertical exchange of water masses characterized by an intensive mesoscale eddy activity (eddies ranging from 50-150 km) superimposed to the surface movement across the whole basin, and which is crucial for the diffusion of all water masses [25].

The Mediterranean Sea includes a large variety environments and habitats across its long littorals (40,000 km), in its nearly 5,000 islands and islets, coastal lagoons, wetlands, marshlands, estuaries, seabeds various types, phanerogam seagrass beds, clusters of algae, rocky reefs, ship remains and even artificial reefs.

The Mediterranean Sea is amongst the most diverse seas worldwide in terms of biological richness. Up to 17,000 species have been recorded in the Mediterranean biota to date (from bacteria to mammals), and nearly 8,000 of these are invertebrate macrofauna, with crustaceans and mollusks being those with the highest diversity (13.2 and 12.4%, respectively) [8].

A peculiarity worth noting, which has a marginal effect on the development of supra- and mesolittoral habitats, is the narrow tidal range, from 1.8 cm in the western Mediterranean´s Spanish coast to 10.4 cm in Italian coasts [26, 27].

The Mediterranean marine biota derives primarily from the Atlantic fauna, but the wide climatic and hydrographic range has contributed to the co-occurrence and survival of organisms of temperate and subtropical affinity. The opening of the Gibraltar Strait some 5.3 million years ago led to the so-called "Gibraltar cascade" and the Atlantic biota invaded the Mediterranean basin, which was undergoing a desiccation process and was hypersaline [21]. Many species underwent speciation processes and a high percentage of them ended up being endemic to the Mediterranean zone [11]; according to J. Templado (pers. Comm, as per the review by Coll *et al.* 2010), molecular techniques are revealing increasing numbers of cryptic species and endemism; as a result, the percentage of endemic Mediterranean mollusks is likely around 22% (468 of the 2,113 species currently known).

Human pressure on the Mediterranean Sea has been going on for centuries, but the opening of the Suez Canal in 1869 has had a marked impact on the Mediterranean biota by facilitating the passage of marine species from the Indo-Pacific region [10]. The lessepsian migration from the Red Sea started in the decade of 1940 and led to a rise in the number of benthic and planktonic species, although some recent introductions to the Mediterranean Sea result from the rapid increase in maritime traffic from the second half of the twentieth century [47].

Waters from many rivers that drain into the sea carry a high load not only of nutrients but also of pesticides, fungicides and a toxic mixture of diverse chemicals used in industry and agriculture, in addition to pollution by viruses and microorganisms derived from sewage. The urban population living in coastal areas doubled between 1960 and 1980 to reach 80 million inhabitants, and is expected to surpass 170 million persons from 1980 to 2025 [24].

The tourism boom since more than a century ago along the Mediterranean coast is changing the physiognomy of coastal areas; urbanization and industrial facilities are leading to an ever-increasing reduction of natural areas [10, 24] and have resulted in a high occupation level of the western Mediterranean coast along virtually the whole extension of the western European littoral, including facilities and constructions of all sorts.

Most Mediterranean fishery resources are deemed overexploited. According to information published by FAO, Italy, France, Spain and Greece have the biggest fishing fleets of the Mediterranean Sea, with the highest number of vessels operating in the basin; only in 1998 these caught 960,000 tons of fish and shellfish, representing more than 50% of the 1.7 million tons fished in this area that year, which was the highest volume recorded ever since, given that in subsequent years the fishery fell below 1,500,000 tons [24].

Fisheries influence marine biodiversity due to the local decline of some species because fishing activities lead to habitat destruction. Virtually all fish and shellfish populations exploited in the Mediterranean basin are fished well beyond the sustainable yield, which, added to habitat degradation, constitutes an ever-increasing threat to their sustainable exploitation [24].

Gulf of California

The Gulf of California is a Mexican partially enclosed or marginal sea, bordered on the west by the Baja California peninsula, one of the largest worldwide, and by the coasts of the Sonora, Sinaloa and Nayarit states on the east; hence, it is located between two arid environments (the peninsula and the Sonoran desert) and opens on the south to the Pacific Ocean through a wide mouth. This water body is over 1,100 km long (from the Colorado river mouth to Cabo Corrientes, Jalisco), with a width ranging from 100 to 222 km. The marine surface covers 258,593 km^2 approximately and reaches a maximum depth of 3,700 m [7], [28].

The Gulf was originally named "Sea of Cortez" in 1539 by the Spanish sailor Francisco de Ulloa after Mexico's conqueror Hernan Cortes, who had discovered the Californias in 1524. The movement of the tectonic plates involved in this part of the world triggered the steady separation of the Baja California peninsula that has occurred since the end of the mid-Miocene (12 million years ago, approximately), and a steady marine invasion into the newly formed basin took place over the past 10 million years [1]. It is considered that the Gulf of California maintains the same configuration since 4.5 million years ago, and moves in a NW direction at 5 cm/year [7].

The Gulf is a scarcely investigated basin; however, the interest in studying it has been on the rise since the mid-twentieth century, not only because of its biological richness but also for comprising a large variety of environments and habitats across its long continental littorals, about 900 islands and islets, extensive coastal lagoons, estuaries, marshlands, wetlands, natural seabed diversity, phanerogam seagrasses and marine algae, mangrove forests, rocky and coral reefs which are used by many marine and terrestrial species as reproduction, nesting, feeding and shelter grounds.

The Gulf of California is a subtropical system that has been divided into four provinces: Upper Gulf, Ballenas Canal-Salsipuedes trough, Lower Gulf and Gulf Mouth [29]. Productivity rates are exceptionally high, similar to those reported for the Bengala Bay and upwelling areas in the western coast of the Baja California peninsula [30]; additionally, these productivity rates are two to three times those observed in the Pacific and Atlantic oceans at similar latitudes [7]. The inflow of cold deep water carries nutrients into the gulf and leads to a rise in productivity, mainly in the southern and central regions; nutrient levels in the northern Gulf are high all the year round [28].

The Gulf of California biota comes from tropical Central and South America; the Caribbean Sea, when there was interoceanic communication before the Panama Isthmus formed; the cold-temperate California coasts during the glaciar periods; and some species may have arrived from the tropical western Pacific Ocean [9].

The Gulf's biological-ecological richness justifies its being considered as one with the highest diversity worldwide; the invertebrate macrofauna is diverse, comprising at least 4,854 known species and subspecies, including 2,193 species of mollusks, and with 20% (460 taxa) of the total number of mollusk species being endemic [19]. Given the presence of numerous invertebrate species and that new species belonging to groups scarcely studied are described each year, mostly from unexplored deep zones, it is believed that the current figure may even double in the region [9].

The mean evaporation rate in the Upper Gulf is 1.1 m/year, whereas precipitations in the upper and medium Gulf range between 4 and 8 cm per year [30], [31], and exceed 100 cm per year in the south along the Sinaloa and Nayarit coasts, contrasting with the peninsular east-south coastline, where precipitation is lower than 20 cm per year. Since evaporation exceeds precipitation, there is a net inflow of surface water from the Pacific, estimated at 1.7×10^4 m^3/s; for this reason, the Gulf is considered as a unique evaporation basin in America [32].

The basin's hydrological balance is characteristic of adjacent seas with little precipitation and freshwater inflow drained by rivers. Salinity has increased not only because of the high evaporation rate but also as a result of the drastic decline in freshwater discharges that started over 80 years ago from the construction of dams along rivers flowing in the basin, mostly because of the virtually total contention of fluvial water by multiple dams along the Colorado river built since the decade of 1930. Nearly all the remaining rivers discharging to the Gulf have been dammed [7].

The circulation of surface waters in the Gulf results from three main causes: cold water from the subartic region derived from the California Current characterized by low salinity and temperature (22 °C and 34.6 ppm); warm water from the Tropical Eastern Pacific, of intermediate salinity and temperature (25 °C and 34.8 ppm); and water from the Gulf, of high salinity and temperature (22-25 °C and 34.9 ppm) [32], [33]. In the summer, surface salinity may be as high as 39 ppm in shallow coastal areas at the north, whereas in the central Gulf salinities resemble those in the ocean (35.0 – 35.8 ppm). The occurrence of El Niño and La Niña phenomena in the eastern Pacific Ocean causes fluctuations in surface temperature in the Gulf of up to 3 °C above and 4 °C below the climatological mean [31], with marked effects on the biota. Incoming warm waters increase the sea surface temperature at the southern and central Gulf when El Niño occurs; additionally, there are local seasonal currents that affect the surface water layers [9].

Tides in the Gulf of California are semidiurnal, diurnal and mixed, depending on the region. At the northern end of the Gulf a large tidal range is observed, between 6 and 10 m, which gives rise to marked tidal currents in the straits formed between Angel de la Guarda island and the Baja California peninsula, on the one hand, and between Tiburón island and the Sonora coast. Intermediate tidal ranges occur in central and southern areas, with 2.7 m in Mazatlán, Sinaloa, and 2.3 m in La Paz, Baja California Sur [7].

Table 1. General characteristics of the Mediterranean Sea and Gulf of California and literature sources reviewed

	Mediterranean Sea	Gulf of California	Source
Location	37°N (Algerian coast) to 44°N (NW of Italy)	21°N (southern end) to 32°N (northern end)	[41], [42]
Origin	40 million years	12 million years	[4], [1]
Maximum length	3,860 km East-West (1500 km East-West, at Western Medit.)	1100 km NO-SE	[8], [7]
Width	800 km North-South (550 km N-S, at Western Medit.)	100 km min to 222 km max (Mean. 150 km East-West)	[42], 28]
Area	2.5 million km^2 (0.85 million km^2 Western Medit)	258, 593 km^2	[21], [7]
Costline length	40,000 km (8,500 km approx. Western Medit.)	3,000 km	[44], [43]
Maximum depth	5,121 m (3,700 m Western Medit.)	3700 m	[10], [28]
Water volume	3.7 million km^3 (860,000 km^3 Western Medit.)	145 000 km^3	[10], [28]
Mean evaporation rate	100 cm/yr (up to 140 cm/yr Western Medit.)	110 cm/yr	[25], [30]
Inflow volume	$0.88 \pm 0.06 \times 10^6 m^3$ water/s	$1{,}7 \times 10^4 m^3$ water/s	[25], [32]
Precipitation	<100 a 335 cm/yr (Western Medit.)	4-8 cm/yr to the north and over 100 cm/yr at the southern end	[25], [9]
Salinity	38.0 ppm (Western Medit.)	35.0 ppm in the central Gulf, to 37.5 ppm in the northern Gulf	[25], [9]
Surfce seawater temperature	13 °C Min., 26 °C Max. (Western Medit.).	10 °C Min., 32 °C Max.	[10], [9]
Tides	1.8 cm at Alicante, Spanish coast; 10.4 cm at the Tyrrhenian Sea, Italian coast	From 2.3 m in the south (Mazatlán) to 10 m in the north (Colorado river delta)	[26], [9]
Climate	Temperate and rainy winter; dry and warm summer (Mediterranean climate)	Mild winter with scarce precipitation; very warm summer and scarce rainfall	[23], [7]
No. of macroinvertebrate species	7,803	4,854	[8], [7]
No. of mollusk species described	2,113	2,193	[8], [19]

	Mediterranean Sea	**Gulf of California**	**Source**
No. of endemic mollusk species	468 approx (22%)	460 (21%)	[8, 11], [19]
No. of families in the literature reviewed	226 (of 293)	209 (of 293)	This chapter
No of genera in the literature reviewed	623	666	This chapter
No. of genera shared	227	227	This chapter
No. of species in the literature revieweed	1,094	2,119	This chapter
Fisheries production	1,500 000 t	950,000 t	[24], [45]
Total catch x km^2	0,3 a 0,4 t/km^2 (Western Medit.) 0.66 t/km^2 (total Medit.)	3,67 t/km^2	[10], [45] This chapter

In the Gulf, dominant winds come from the northwest in the winter-spring and from the southeast in the summer-autumn, which contribute to produce upwelling events in both coasts and foster the formation of mesoscale eddies at the south; tropical storms or hurricanes are formed during this time of the year in tropical northeastern Pacific waters, which generally reach the Gulf with variable intensities in categories 1 – 5 (119 – 250 km/h) on the Saffir-Simpson scale, along with rainfall.

Oxygen concentrations tend to be high throughout the Gulf, dropping from 5 – 6 ml/l at the surface to 1 ml/l at depths of 300– 500 m in the northern Gulf, whereas in the central and southern portions low concentrations are recorded at intermediate depths of 150 m [30].

According to the latest fishery production records (2010) [44], 950,000 tons of marine products were caught in the Gulf of California, it is worth noting that the statistical figures for Baja California and Baja California Sur are reduced by 50% and 25%, respectively, since these states have coasts facing both the Gulf and the Pacific.

Conclusion

For their extension, littorals in the western Mediterranean Sea (8,500 km) and the Gulf of California (3,000 km) comprise a great variety of environments and habitats derived from the interaction of geological, physical, chemical, biological and atmospheric processes, all of which have favored a huge biological diversity despite of the Mediterranean being an oligotrophic sea. In both seas, the heavy exploitation of marine resources has led to a decline in catches of the major species, although to a lesser extent in the Gulf of California. Species richness in both basins is evident in any animal phyla, with mollusks ranking second after arthropods; the former is a group of considerable economic importance, as well as in relation to the livelihood of coastal inhabitants.

Some shared characteristics that enable to establish similarities between the Mediterranean Sea and the Gulf of California include inner basins with similar evaporation rates, benign climates, moderate water temperatures in winter and high in the summer, high salinity relative to oceans, low-energy waves, except under strong winds, exceptional mollusk diversity and high levels of endemism (22 – 23%). The differences include the area occupied (km^2) and water volume contained (Km^3), which are considerable in the Mediterranean and relatively modest in the Gulf; productivity and population density with high figures in the Gulf *versus* low in the Mediterranean; low precipitations in nearly all the Gulf and low to intermediate in the Mediterranean; much longer coastline in the Mediterranean; narrow tidal ranges in the western Mediterranean and much wider in the Gulf, especially in the upper Gulf; great diversity of commercially exploited mollusk species in the Gulf of California, contrasting with a relatively low number of species exploited in the Mediterranean Sea, even including cultivated and introduced species. In general, fishing volumes per unit of area are high in the Gulf and relatively low in the Mediterranean (Table I).

For the present chapter, we quantified 1,094 mollusk species recorded for the Mediterranean Sea in the literature reviewed [11], [15], [17], [19], [34], [35], [36], [37], [38], [39], [40], grouped into 623 genera and 236 families; gastropods are the dominant group, accounting for 71.20% of the total, followed by bivalves with 23.58%, cephalopods with 3.58% and polyplacophorans with 1.46%; for the Gulf of California we quantified 2,119 species grouped into 666 genera and 211 families; gastropods were again the dominant group, with 69.70% of the total number of species, followed by bivalves with 26.38%, polyplacophorans with 2.97% and cephalopods with 0.77% (Tables 2; 3, 4 and 5).

Following the biogeographical concept of vicariant species, we found at least 182 genera containing species in separated biogeographical areas (Gulf of California and Mediterranean Sea) inhabiting similar habitats. Many species stand out in both basins for their economic importance, belonging to the following families. Class Bivalvia: Arcidae, Glycymeridae, Pinnidae, Mytilidae, Gryphaeidae, Ostreidae, Pectinidae, Spondylidae, Carditiidae, Lucinidae, Chamidae, Cardiidae, Veneridae, Tellinidae, Donacidae, Semelidae, Hiatellidae and Solenidae; Class Gastropoda: Patellidae, Turbinidae, Neritidae, Cerithidae, Strombidae, Turritellidae, Cassidae, Ranellidae, Muricidae, Buccinidae, Olividae, Olivellidae, Conidae; Class Polyplacophora, the family Chitonidae; and Class Cephalopoda: Octopodidae, Loliginidae, Omastrephidae, Sepiidae and Sepiolidae.

The Mediterranean Sea currently faces a number of issues, namely overpopulated coasts, pollution and overexploitation of marine resources; thus, efforts between countries and international organizations (UNEP, AEMA, MEDPOL, FAO, NGOs and research centers) shall be undertaken to avoid further deterioration and restore terrestrial and marine ecosystems. The Gulf of California is still a well-preserved sea, despite the fact that since the past three or four decades it has shown signs of issues, mostly overexploitation of the highly demanded marine resources; urban settlements have risen considerably; tourism service facilities spread across 3,000 km of; furthermore, in the more developed regions rivers have been dammed, industries coastline have been established, farming and mining activities have increased in areas draining to the sea, with the subsequent population growth, which in some locations has led to pollution issues in coastal waters. Although these issues are not exclusive of both seas and also occur elsewhere in the world, the high biodiversity in both basins and the threats currently faced warrant the adoption of measures to preserve the ecosystems in them as a top priority.

The primary problems and threats affecting a large number of taxonomic groups in both seas are common to other areas of the world, such as overfishing, transport of sediments derived from deforestation, river damming, catastrophic events (hurricanes and earthquakes), unplanned urbanization, habitat alteration and loss, climate change, settlement of alien species, marine pollution resulting from sewage drainage, channel dredging and alteration of the dynamics in littoral waters resulting from the construction of ports, marinas and other facilities.

The information in the following tables (II, III, IV, V) was obtained from: Salvini-Plawen [11], Kerstitch and Bertsch [15], Gofas, Moreno and Salas [17], Hendrickx and Brusca [19], Skoglund [34], [35], Tarruela and López [36], Soppelsa *et al.* [37], Keen [38], Houston [39], Reyes-Gómez [40].

Table 2. Polyplacophora families and genera reported in the Mediterranean Sea and the Gulf of California; shared genera and total number of species reported in the literature reviewed for this chapter

FAMILIES	Genera Mediterranean Sea	Genera Gulf of California	Genera shared in both basins	Total No. of species (all genera) Medit. / Gulf	
Leptochitonidae	2	2	*Leptochiton*	5	4
Acanthhochitonidae	1	4	*Acanthochitona*	7	13
Chitonidae	1	2	*Chiton*	3	4
Ischnochitonidae	5	5	*Ischnochiton, Lepidochitona*		34
Callochitonidae	1			1	
Lepidopleuridae		4			5
Mopalidae		1			3

Total number of families: 7
Mediterranean Sea: 5 families, 10 genera and 16 species
Gulf of California: 6 families, 18 genera and 63 species
Shared genera: 5

Table 3. Bivalve families and genera reported in the Mediterranean Sea and Gulf of California; shared genera and total number of species reported in the literature reviewed for the present chapter

FAMILIES	Genera Mediterranean Sea	Genera Gulf of California	Genera shared in both basins	Total No. of species (all genera) Medit. / Gulf	
Solemydae	1	1	*Solemya*	1	2
Acharacidae		1			1
Nucinellidae		1			1
Nuculanidae	2	3	Nuculana, Sacella	2	15
Nuculidae	2	2	*Nucula, Ennucula*	7	6
Sareptidae	1	1	Yoldiella	1	1
Neilonellidae		1			1
Arcidae	5	7	*Arca, Barbatia, Anadara*	10	31
Noetiidae	2	1		3	3
Glycymerididae	1	2	*Glycymeris*	3	8
Limopsidae	1			3	
Philobridae		1			1
Mytilidae	13	16	*Crenella, Modiolus, Gregariella, Lithophaga* *Mytillus galloprovincialis* (Especie compartida)	22	28
Pinnidae	2	2	*Pinna, Atrina*	3	4
Pteriidae	2	2	*Pteria, Pinctada*	2	2
Pectinidae	11	10	*Chlamys,Cyclopecten*	17	20
Spondylidae	1	1	*Spondylus*	2	3
Isognomonidae		1			2
Malleidae	1	1		1	1
Gryphaeidae	1	2		1	2
Ostreidae	3	7	*Crassostrea*	3	7
Dimyidae		1			1
Plicatulidae		1			4
Limidae	5	3	*Lima, Limaria, Limatula*	8	5
Anomiidae	3	2	*Anomia, Pododesmus*	3	4

Table 3. (Continued)

FAMILIES	Genera Mediterranean Sea	Genera Gulf of California	Genera shared in both basins	Total No. of species (all genera) Medit. / Gulf	
Placunidae		1			1
Crassatellidae		2			9
Cardiniidae		1			1
Carditidae	3	4	*Cardites*	3	6
Condylocardiidae		1			1
Astartidae	4			5	
Corbiculidae		2			3
Vesicomyidae		2			3
Lucinidae	7	8	*Lucina, Ctena, Myltha*	7	23
Ungulinidae	2	3	*Diplodonta*	4	8
Thyasiridae	3	2		9	2
Galeommatidae	3	4		3	4
Erycinidae	1	2	*Lasaea*	1	2
Leptonidae	2	1	Lepton	2	2
Kelliidae	1	3	*Bornia*	2	7
Montacutidae	3	5	*Mysella*	5	10
Sportellidae		3			8
Bernardinidae		2			2
Chamidae	2	3	*Chama, Pseudochama*	2	14
Cardiidae	4	6	*Laevicardium*	15	17
Veneridae	18	17	*Timoclea, Pitar, Dosinia*	28	72
Petricolidae	1	2		1	6
Myidae	1			1	
Mactridae	4	9	*Mactra, Spisula*	9	14
Pharidae	2			2	
Tellinidae	10	8	*Tellina, Macoma*	17	53
Coperellidae		1			2

FAMILIES	Genera Mediterranean Sea	Genera Gulf of California	Genera shared in both basins	Total No. of species (all genera) Medit. / Gulf	
Psammobiidae	2	5		6	15
Donacidae	3	3	*Donax*	4	11
Semelidae	3	5	*Abra*	9	28
Solenidae	3	1	*Solen*	4	4
Cultellidae		12			
Trapezidae	1			1	
Myidae	1	2	*Sphenia*	1	2
Corbulidae	1	1	*Corbula*	1	12
Hiatellidae	3	2	*Hiatella, Panopea*	2	2
Pholadidae	2	9	*Pholas chiloensis* (shared species)	2	12
Teredinidae	3	2	*Bankia subtruncata* (shared species) *Nototeredo*	4	5
Xylophagaidae	2			2	
Spheniopsidae		2			2
Gastrochaenidae	1	1	*Gastrochaena*	1	3
Pandoridae	1	1	*Pandora*	2	6
Lyonsiidae	1	2	*Lyonsia*	1	4
Periplomatidae	1	2		1	7
Thraciidae	1	4	*Thracia*	1	10
Clavagellidae	1			6	
Poromyidae	1	1	*Poromya*	1	2
Cuspidariidae	3	4	*Cardiomya, Cuspidaria*	4	8
Verticordiidae	1	4	*Haliris*	1	4

Total number of families: 74
Mediterranean Sea: 56 families, 162 genera and 258 species
Gulf of California: 67 families, 219 genera and 559 species
Shared genera: 59

Table 4. Gastropod families and genera reported in the Mediterranean Sea and the Gulf of California; shared genera and total number of species reported in the literature reviewed for the present chapter

FAMILIES	Genera Mediterranean Sea	Genera Gulf of California	Genera shared in both basins	Total No. of species (all genera) Medit.	Gulf
Patellidae	3	2	*Patella*	7	2
Lepetidae	1				
Lepetellidae	1			1	
Lotiidae	1	3	*Tectura*	1	16
Sissurellidae	3	2		3	2
Haliotidae	1			1	
Lepetodrilidae		1			1
Fissurellidae	4	8	*Diodora, Fissurella, Emarginula*	12	25
Turbinidae	3	8	*Astraea, Arene, Homalopoma*		24
Trochidae	10	4	*Calliostoma*	32	26
Skeneidae	10	2	*Parviturbo*	17	6
Lissospiridae	1			2	
Neritidae	2	2	*Theodoxus*	2	3
Phasianellidae	1			8	
Phenacolepadidae		2			2
Titiscaniidae		1			1
Pyropeltidae		1			1
Addisoniidae	1	1	*Addisonia*	1	1
Seguenziidae		1			1
Provannidae	3	1		3	3
Littorinidae	3	4		3	7
Rissoidae	17	7	*Rissoina, Alvania, Rissoa*	66	24
Aderobidae	1			1	
Anabathridae	2	1		3	1
Barleeidae	1	2		2	14

FAMILIES	Genera Mediterranean Sea	Genera Gulf of California	Genera shared in both basins	Total No. of species (all genera) Medit. / Gulf	
Assimenidae	1	1		1	1
Hydrobiidae	3			5	
Ctiloceratidae	1			1	
Skeneopsidae	1			2	
Elachisinide		1			2
Truncatellidae	1	1	*Truncatella*	1	1
Vitrinellidae		13			88
Caecidae	2	3	*Caecum*	7	28
Iravadiidae	2			2	
Tornidae	2	1		2	2
Rastodentidae		1			1
Vermetidae	4	6	*Vermetus, Serpulorbis, Dendropoma, Petaloconchus*	7	12
Turritellidae	1	2	*Turritella*	3	16
Cingulopsidae	1			2	
Modulidae	1	1		1	3
Cerithiidae	4	3	*Bittium, Cerithium*	11	15
Litiopidae		1			3
Diastomatidae		1			1
Finellidae		1			4
Planaxidae	3	2	*Fossarus*	4	7
Potamididae	1	1		1	4
Siliquaridae	3			3	
Batillaridae		1			1
Strombidae		1			4
Hipponicidae		2			4
Vanikoridae	2	1		3	1
Calyptraeidae	2	4	*Calyptraea, Crepidula*	3	28

Table 4. (Continued)

FAMILIES	Genera Mediterranean Sea	Genera Gulf of California	Genera shared in both basins	Total No. of species (all genera) Medit. / Gulf	
Capulidae	1	1	*Capulus*	1	1
Xenophoridae	1	1	*Xenophora*	1	1
Naticidae	8	5	*Natica, Sinum, Neverita*		24
Lamellaridae	1	2	*Lamellaria*		4
Triviidae	2	2	*Trivia*		13
Cypraeidae	4	1		4	7
Ovulidae	5	6		5	6
Tonnidae	1	1			1
Cassidae	1	3		1	4
Ficidae		1			1
Ranellidae	4	2	*Cymatium*	4	10
Atlantidae	1			1	
Aclididae	1			6	
Personidae		1			4
Bursidae		3			3
Pterotracheidae	2	2	*Firoloida, Pterotrachea*	2	2
Cerithiopsidae	4	2	*Cerithiopsis*	5	20
Triphoridae	9	2	*Metaxia, Triphora*	13	21
Epitoniidae	8	9	*Acirsa, Opalia, Cirsotrema*		48
Janthinidae	1	2	*Janthina*	1	3
Eulimidae	9	9	*Eulima, Melanella*	22	32
Muricidae	17	38	*Hexaplex, Muricopsis, Stramonita*		82
Coralliophilidae	1	3	*Coralliophila*	1	9
Buccinidae (Incluye Nassaridae)	7	19	*Cantharus, Fusinus, Nassarius, Opeatostoma*	16	79
Cystiscidae	1			7	
Columbellidae	3	22	*Anachis, Coumbella, Mitrella*	8	57

FAMILIES	Genera Mediterranean Sea	Genera Gulf of California	Genera shared in both basins	Total No. of species (all genera) Medit. / Gulf	
Volutidae	2	1		2	2
Harpidae		2			2
Olividae		2			21
Olivellidae		1			16
Vasidae		1			1
Marginellidae	3	5	*Granulina, Volvarina*	4	9
Mitridae	1	2	*Mitra*	3	17
Costellaridae	1	2		2	2
Cancellaridae	3	8		3	28
Conidae	1	1	*Conus*	1	30
Borsoniidae	2			3	
Clavatulidae	1			1	
Clathurellidae	1			2	
Drillidae	2			2	
Mangeliidae	2			16	
Horaiclavidae	1			1	
Mitromorphidae	1			3	
Raphitomidae	5			12	
Mathildidae	1			5	
Cornirostridae	2			1	
Omalogyridae	3			4	
Cimidae	2			5	
Terebridae		2			36
Turridae	2	53		2	179
Rissoellidae	1	1	*Rissoella*	1	1
Architectonicidae	8	4	*Discotectonica, Pseudotorini, Heliacus*	10	8
Ebalidae	1			2	
Pyramidellidae	12	14	*Chrysallida, Odostomia, Turbonilla*	87	145
Amathinidae	1	1		4	3

Table 4. (Continued)

FAMILIES	Genera Mediterranean Sea	Genera Gulf of California	Genera shared in both basins	Total No. of species (all genera) Medit. / Gulf	
Murchisonellidae	1			1	
Onchididae		2			3
Melampidae		4			7
Siphonaridae	2	2	*Siphonaria, Williamia*	2	5
Trimusculidae	1	1	*Trimusculus*	1	2
Acteonidae	2		*Acteon*	2	5
Cylichnidae	2	3	*Cylichna*	2	12
Scaphandridae	2			2	
Aglajidae	3	2	*Aglaja*	3	3
Philinidae	3			10	
Diaphanidae	3	1		3	1
Bullidae	1	1	*Bulla*	1	2
Ringiculidae	1			2	
Hyalogyrinidae	2			2	
Haminoeidae	3	3	Atys, *Haminoea*	10	7
Gastropteridae	1			2	
Retusidae	3	2	*Volvulella*	5	4
Rhizoridae	1			1	
Juliidae		2			2
Cylindrobullidae		1			1
Oxynoidae	3	2	*Oxynoe*	3	2
Caliphyllidae	1	1		1	1
Limapontidae	4			9	
Stiligeridae	2			2	
Elysiidae	2	2	*Elysia*	2	3
Lobigeridae	1			1	
Volvatellidae	2			2	

FAMILIES	Genera Mediterranean Sea	Genera Gulf of California	Genera shared in both basins	Total No. of species (all genera) Medit. / Gulf	
Aplysidae	5	5	*Aplysia, Phyllaplysia*	7	8
Umbraculidae	1	1	*Umbraculum*	1	1
Peraclidae	1			3	
Cymbulidae	1			1	
Ellobiidae	5			5	
Akeridae	1			1	
Dolabriferidae	1			1	
Runcinidae	1			7	
Limacinidae	1			4	
Tylodinidae	2	1	*Tylodina*	2	1
Pleurobranchidae	4	3	*Pleurobranchus, Berthella, Berthellina*		5
Cavoliniidae	7	3	*Cavolina*	12	3
Goniodorididae	3	2	*Okenia*	12	2
Onchidorididae	3	2	*Acanthodoris*	5	2
Corambide	1			1	
Triophidae		3			5
Notodoridae		1			1
Polyceridae	7	2	*Polycera, Polycerella*	9 + 3	3
Chromodorididae	2	5	*Chromodoris, Hypselodoris*	14	17
Conualeviidae		1			3
Aldisidae		1			1
Rostangidae		1			1
Dorididae	1	1	*Doris*	5	1
Discodorididae	11	5	*Discodoris, Taringa, Peltodoris*	18	5
Cliopsidae	1			1	
Clonidae	1			2	
Hermaeidae	4			4	
Elysiidae	2	1	*Elysia*	5	1

Table 4. (Continued)

FAMILIES	Genera Mediterranean Sea	Genera Gulf of California	Genera shared in both basins	Total No. of species (all genera) Medit. / Gulf	
Bosellidae	1			1	
Gasteroptidae	1			1	
Aegiridae	1			3	
Phyllididae	1			1	
Cadlinidae	2			4	
Corambidae	1			1	
Asteronotidae		1			1
Dendrodorididae	2	2	*Dendrodoris, Doriopsilla*	5	5
Tritonidae	2	1	*Tritonia*	5	2
Hancockiidae	1			1	
Hedylopsidae	3			4	
Scyllaeidae		1			1
Bornellidae		1			
Dendronotidae		1			1
Tethyidae	2	1	*Melibe*	2	1
Bornellidae		1			1
Dotidae	1	1	*Doto*	9	2
Arminidae	1	2	*Arminia*	2	2
Zephyrinidae	1	1	*Janolus*		1
Dironidae		1			1
Flabellinidae	3	1	*Flabellina*	8	8
Eubranchidae	1	1	*Eubranchus*	5	1
Tergipedidae	4	1		11	1
Glaucidae	1	1	*Glaucus*	1	1
Microchedylidae	1				
Hedylopsidae	1				

FAMILIES	Genera Mediterranean Sea	Genera Gulf of California	Genera shared in both basins	Total No. of species (all genera) Medit. / Gulf	
Rumicinidae	1				
Babakinidae	1	1	*Babakina*	1	1
Piseinotecidae	3			3	1
Lomanotidae	1			1	
Facelinidae	8	3	*Aeolidella, Baeolidia*	10	4
Aeolidiidae	5	3		9	3
Spurillidae	1	1	*Spurilla*		2
Fionidae	1			1	
Embletoniidae	2			2	
Calmidae	2			2	
Otinidae	2				

Total number of families: 199
Mediterranean Sea: 153 families, 424 genera and 779 species
Gulf of California: 162 families, 421 genera and 1477 species
Shared genera: 113

Table 5. Cephalopod families and genera reported in the Mediterranean Sea and the Gulf of California; shared genera and total number of species reported in the literature reviewed for the present chapter

FAMILIES	Genera Mediterranean Sea	Genera Gulf of California	Genera shared in both basins	Total No. of species (all genera) Medit. / Gulf	
Spirulidae	1			6	
Sepiidae	1			3	
Sepiolidae	5			6	
Loliginidae	2	2		4	4
Enopleurethidae	1			1	
Thysanoteuthidae	1	1	*Thysanoteutis rhombus* (shared species)	1	1
Ommastrephidae	4	2		5	2
Architeuthidae	1			1	
Onychoteuthidae	1	1	*Onychoteuthis*	1	1
Octopodidae	5	1	*Octopus*	8	9
Tremoctopodidae	1			1	
Ocythoidae	1			1	
Argonautidae	3	1	*Argonauta*	1	3

Total number of families: 13
Mediterranean Sea: 13 families, 27 genera, 39 species
Gulf of California: 6 families, 8 genera and 20 species
Shared genera: 4

Clase: POLYPLACOPHORA

Chitons: Some 650 living species have been described [13]. Eighteen species have been reported in the Mediterranean [11], [17]. Sixty three species have been reported in the Gulf of California [19], [34], [35], [38], [39]. Chitons live in hard substrata, especially in low coastal water from high-tide levels.

Class: BIVALVIA

Clams and oysters. Some 15,000 species have been described, mostly from marine and brackish habitats; some families live in freshwater environments [13]. About 400 species have been described in the Mediterranean Sea [8], 258 of them reported [11], [17], [36], [37]. Bivalves are very abundant in the Gulf of California, 565 species have been reported [19], [38], [34], [35], [39]. These are exclusively benthic organisms found in nearly all types of substrates, from the intertidal zone to the deep sea, and are abundant in sedimentary (sand, clay, silt, mud, rocks) and rocky bottoms.

Class: GASTROPODA

Snails, limpets, nudibranchs and others: Some 90 000 species are known, inhabiting all habitat types (marine, brackish, freshwater and terrestrial) [13]. Up to 1530 species have been described in the Gulf of California [7] and 1564 species are found in the Mediterranean Sea [8], [17], being common in all substrate types, almost exclusively benthic from splash fringes to the deep sea.

Clase: CEPHALOPODA

Octopuses, squids, cuttlefish and argonauts: Some 650 exclusively marine species. About 53 species are known in the Mediterranean Sea 53; nine of them are endemic [11]. Cuttlefish and octopuses can be good swimmers but almost always live in association with the seabed.

Acknowledgments

Authors thank the Centro Interdisciplinario de Investigaciones Marinas at Instituto Politécnico Nacional, México and the Marine Sciences Department and Applied Biology at Universidad de Alicante, Spain, for the financial support received from August 2011 to July 2012.

References

[1] Calmus, T. (2008). *Revista Universidad de Sonora*, N° 20, enero-marzo 2008, 4-6.
[2] Rusnak, G. A., Fisher, R. L. & Shepard. F. P. (1964). *American Association of Petroleum Geologists Memoir, 3*, 59-75.
[3] Lyle, M., Ness. G. E., Dauphin, J. P. & Simoneit, B. R. T. (1991). 403–423, AAPG, Memoir 47.
[4] Maldonado, A. (1989). In: Margalef. Ed. Omega, Barcelona. 374 pp.
[5] Álvarez-Fernández, E. (2005). *Munibe (Antropologia-Arkeologia), 57*, 359-368 pp. San Sebastián, España.
[6] Fujita, H. (2002). *Fundación Mexicana para la Educación Ambiental A. C. México.* 35-55 p.
[7] Brusca, R. C., Findley, L. T., Hastings, P. A., Hendrickx, M. E., Torre-Cosio, J. & van der Heiden. A. M. (2005). In: Cartron, J.L.E., G. Ceballos & R.S. Felger. Oxford University Press. 179-203.
[8] Coll, M., Piroddi, C., Steenbeek, J., Kaschner, K., Ben Rais Lasram, F. (2010). *et al. PLoS ONE, 5*(8), e11842. doi:10.1371/journal. pone.0011842.
[9] Brusca, R. C. y Findley, L.T. (2005). *Arizona-Sonora Desert Museum*, 429 pp.
[10] Margalef, R. (1989). Omega, Barcelona. 374 pp.
[11] Salvini-Plawen, L. V. (1986). In: Riedl, R. Pp 257-400. Omega, Barcelona España. 858 p.
[12] Purchon, R. D. (1977). *2nd Edition. Pergamon Press.*
[13] Pearse, V., Pearse, J., Buchsbaum, M. & Buchsbaum, R. (1987). *The Boxwood Press*, Pacific Grove, CA. 848 pp.
[14] Brusca, R. C. & Brusca, G. J. (2003). *2nd Edition. Sinauer Associates, Inc*. USA, 936 p.
[15] Kerstitch, A. &. Bertsch, H. (2007). *2nd. Ed. Sea Challengers*, Monterey, Calif. USA. 124 pp.
[16] Brusca, R. C. & Brusca, G. J. (1990). *Sinauer, Sunderland*, Massachussetts, USA. 922 p.
[17] Gofas, S., Moreno y, D., Salas. C. (2011). Vol.I, 1-342; Vol. II 343-798. *Servicio de publicaciones e Intercambio Científico*, Universidad de Málaga.
[18] Bianchi, C. N. & Morri, C. (2000). *Mar Poll Bull, 40/5*, 367-376.
[19] Hendrickx, M. E. & Brusca, R.C. (2005). 195-310. In: Hendrickx, Brusca y Findley. *Museo Desierto Arizona-Sonora*, Tucson, USA. 429 pp.
[20] Cruzado, A. (1989). 128-149. In: Margalef, R. Omega, Barcelona. 374 pp. (1989).
[21] Schmidt, Glaubrecht y Golani. (2004). 464-499. In: Hofrichter R. Vol I. Omega, S.A. Barcelona. 592 pp.
[22] Tichy, G., Tomek, C., Hsü, K. J. & Hofrichter, R. (2004). 56-101 Pp. in Hofrichter, R. Vol I. Omega, S.A. Barcelona.
[23] Hofrichter, R. (2004). Vol I. Omega, S.A. Barcelona.
[24] Maack, T. y Rademaker-Wolff, V. (2004). 500-523. In: Hofrichter, R. Vol I. Omega, S.A. Barcelona.
[25] Klein, B. y Roether, W. (2004). 258-287. In: Hofrichter, R. Vol I. Omega, S.A. Barcelona.
[26] Hopkins, T. S. (1989). 102-127. In. Margalef R. Omega, Barcelona. 374 pp. (1989).

[27] Pfannschmidt, S., M. Heb, M., Melzer, R., Hofrichter, R., Wilke, M., Türkay, M. y Herzer, K. (2004). 288-423 In: Hofrichter, R. Vol *I*. Omega, S.A. Barcelona.
[28] Bray, N. A y Robles, J. M. (1991). In: J.P. Dauphin and B.C. Simoneit. *Am. Assoc. Pet. Geol. Mem.*, *47*, 511-553.
[29] Roden, G. I e Emilson, I. (1980). Centro de Ciencias del Mar y Limnología. *UNAM*. México. *90*, 67.
[30] Álvarez-Borrego, S. (1983). 427-449 in: B. H. Ketchum. Elsevier, Amsterdam.
[31] Lavín, M. F., Godínez, V. M. & Álvarez, L. G. (1998). *Estuar. Coast. Shelf Sci.*, *47*, 769-795.
[32] Roden, G. I. (1964). 30-58. In: Tj. H. Van Andel & G.G. Shor. Mem. *Am. Petrol. Geol.*, *3*, 30-58.
[33] Roden, G. I. & Groves, G. W. (1959). *Marine Research Journal*, *181*, 10-35.
[34] Skoglund, C. (2001). *The Festivus*, *32*, 1-119.
[35] Skoglund, C. (2002). *The Festivus*, *33*, 1-286.
[36] Tarruella-Ruestes, A. & López-Soria, J. (2006). SPIRA, Vol. *2* (1). *Associació Catalana de Malacologia*.
[37] Soppelsa, O., Crocetta, F. & Fasulo, G. (2007). *Boll. Malacol.*, *43* (1-8): 21-32.
[38] Keen, M. A. (1971). *Standford Univ. Press*. Stanford, California. 1064 p.
[39] Houston, R. S. (1980). 130-204 pp. In: Brusca, R.C. 2a Ed., *The University of Arizona Press*, Tucson, USA.
[40] Reyes-Gómez, Q. (2004). *Bullettino Malacologico*, Roma, Spl. 5:69-82. Societá Italiana di Malacologia.
[41] http://go.hrw.com/atlas/span_htm/namerica.htm.
[42] http://go.hrw.com/atlas/norm_htm/europe.htm.
[43] http://www.biodiversidad.gob.mx/pais/pdf/CapNatMex/Vol%20I/I04_Losecosistemasc os
[44] Hofrichter, R., Kern, W., Beckel, L., Domnig, I., Zankl, A., y Hein, A. (2004). 102-195 pp, In: Hofrichter R. Vol I. Omega. 592 pp.
[45] http://www.conapesca.sagarpa.gob.mx/wb/cona/anuario2010
[46] Zander, C. D. (2004). 424-463 In: Hofrichter, R. Vol I. Omega, S.A. Barcelona.
[47] Pérés, J. M. (1989). 200-234 pp. In Margalef R. Omega, Barcelona. 374 pp.

In: Invertebrates: Classification, Evolution and Biodiversity ISBN: 978-1-62618-789-4
Editor: Rafael Riosmena-Rodriguez

Chapter 7

Ascidians Biodiversity of Tropical Eastern Pacific

Betzabé Moreno-Dávila*
Laboratorio de Botánica Marina,
Universidad Autónoma de Baja California Sur (UABCS) C.P., La Paz, BCS, Mexico

Abstract

The Tropical Eastern Pacific islands ranging from Revillagigedos to the Galapagos Islands are very poorly studied. Throughout this region, there is a lack of studies about the biodiversity of sea squirts. During adulthood, ascidians are sessile organisms and have a variety of shapes and colours, but have unique characteristics to identify them in the field, such as a tunic that covers the body and two siphons, one oral and one atrial. Research in the southern Gulf of California, Southern Mexico, Costa Rica and Panama project the presence of three orders, 11 families, 16 genera and 35 species. Of these, four are the most represented families and eight are the most abundant species. Among the species recorded are five invasive species.

Introduction

Tropical Eastern Pacific (TEP) is a biogeographic realm which is formed by various provinces and ecoregions. The provinces are *Tropical Eastern Pacific* and *Galapagos.* Ecoregions of TEP are Revillagigedos, Clipperton, Mexican Tropical Pacific, Chiapas-Nicaragua, Nicoya, Cocos Islands, Panama Bright and Guayaquii. Ecoregions of *Galapagos* are Northern Galapagos Islands, Eastern Galapagos Islands and Western Galapagos Islands [1]. The average sea temperature in this region is usually 29.5°C in the summer and from 26°C in the winter. This region also presents a wide range of coastal and marine environments, such as rocky shores, sandy, coastal lagoons, estuaries, mangroves and coral

*E-mail address: bb.morenodavila@gmail.com.

reefs [2]. These environments tend to be suitable for the adult sea squirt, because it requires a hard substrate to attach to, whether natural environments such as rocky, mangrove root and coral reefs, or artificial environments such as docks, pilings, buoys, and boat hulls, among others. Ascidians form a diverse yet poorly studied group in these coastal Pacific waters. The knowledge of ascidian species of the TEP was based on the following studies: in the Gulf of California [3, 4, 5, 6, 7], Mexican Southeast (Oaxaca) [8], Costa Rica [9, 10, 11] and Panama [12]. The class Ascidiacea is included in the subphylum Urochordata (Tunicata). The name Urochordata refers to the notochord, a flexible rod-like structure that forms the main support of the larval tail. As the widely used alternative name Tunicata implies, the body is encased in an external tunic, or vest [13]. The ascidians or sea squirts lose their notochord and nerve cord when they metamorphose from non-feeding, briefly swimming larvae to benthic adults.

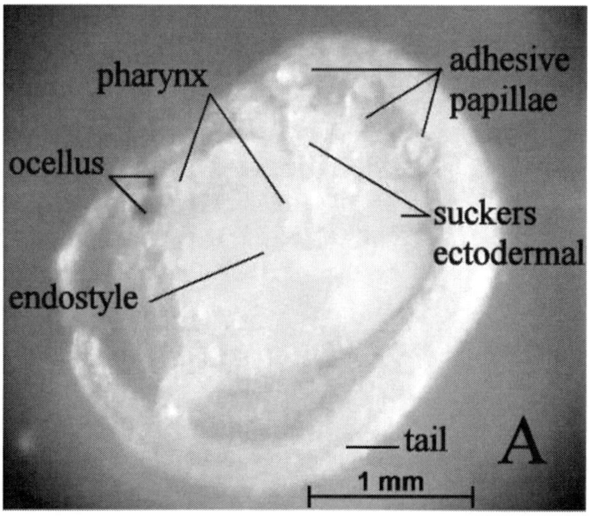

(a)

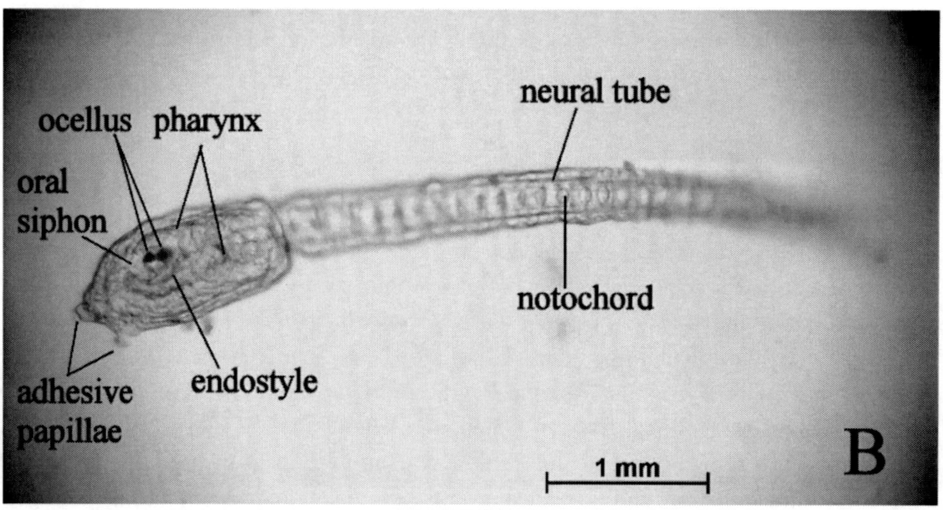

(b)

Figure 1. A) Primary morphological characteristics of an early larva of a *Trididemnum*. B) Morphological characteristics of total developing larva which are called tadpole larva.

Metamorphosis of the "tadpole" larvae includes the swift destruction of the tail and reabsorption of this musculature and of the notochord and dorsal nerve cord, transformation of the rest of the larval central nervous system, development of numerous pharyngeal gill slits, postural adjustment of the newly attached larvae so that it will be an effective benthic filter-feeder and opening of the oral and atrial siphons [13] (Figure 1A, B). Some ascidians grow to solitary adulthood.

In other species, starting with the oozooid that emerges from metamorphosis of the larva, asexual reproduction produces colonies of strobilated, generally identical individuals with blastozooids. Depending on the species, solitary ascidians may be merely a few millimetres in diameter even as adults; others may grow to several centimetres high or across [13]. The body is divided into three regions: the thorax, including the pharynx, the abdomen, which is the stomach and intestine, and the post-abdomen, including the gonads (Figure 2). The shape depends heavily on the consistency of its tunic.

If its tunic is thick and hard, the animal will often have a globular, erect shape and may carry a fairly dense coating of epibiotic organisms. This appearance is especially common among Stolidobranch. The tunic of small solitary Phlebobranch ascidians is generally clean and often transparent and soft, but it may become opaque and rather cartilaginous in large species. Many of these animals prefer less-exposed habitats with crevices and gaps among coral branches, which may provide some protection from predators. In some solitary ascidians, the tunic is encrusted with sediment [14]. These forms live in silty habitats, where they can hide in the sediment; their tunics may hold onto sand or mud, and their camouflage can be so perfect that one might fail to notice that there is a sea squirt, until it is touched [14]. The zooids of colonial ascidians are usually much smaller than the bodies of solitary ascidians. However, the colonies may become encrusted or lobate masses that are several or even many centimetres across [13].

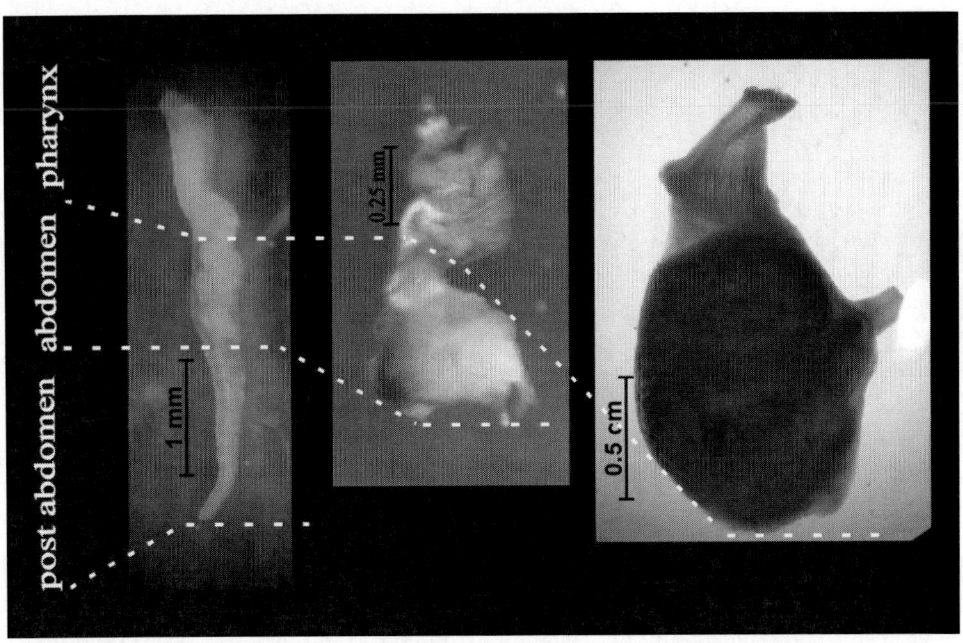

Figure 2. Regionalisation of the body according the position of: the pharynx, intestine and gonads.

Table 1. Ascidians species recorded in the Eastern Tropical Pacific; Status: N= native, I= introduce, U= unspecified. Locality: CR= Costa Rica, GC= Gulf of California, P= Panama, MSP= Mexican South Pacific Substrate: b =bivalves, c= crabs, d= docks, f =floats, mr= mangrove roots r= rocks, p= piles. References: [3, 4, 5, 7, 8, 9, 10, 11, 12, 15, 1 6]

Species	Status	Locality	Substrate	References
Ascidia. ceratodes (Huntsman, 1912)	N	GC, MSP	r, p, f,	3,8
A. incrassate Heller, 1878	I	P	r, b	12
A. sydneiensis Stimpson, 1855	N	GC, P	r, b, mr, d	3, 8,12,
Aplidium arenatum (Van Name, 1945)	N	GC	r	3
A. propinquum (Van Name, 1945)	N	GC	r	3
A. solidum (Ritter and Forsyth, 1917)	N	GC	d	5
Botrylloides diegensis Ritter and Forsyth, 1917	N	GC	d, r	5
B. nigrum Herdman, 1886	U	P, MSP	r, b, mr, d	12
B. planus (Van Name, 1902)	I	P	r, b, mr, d	12
Clavelina fasciculata Van Name, 1945	N	GC	r	3
Cystodytes dellechiajei (Della Valle, 1877)	U	GC	r	3
Didemnum carnulentum Ritter and Forsyth, 1917	N	GC, P	r, c	3,12, 16
D. perlucidum F. Monniot, 1983	I	P	r, mr, d	12
D. santaelenae Van Name, 1945	N	E	r	3
D. vanderhorsti Van Name, 1924	U	GC, P	r	5, 3
Diplosoma listerianum Milne-Edwards, 1841	U	P	r, mr, d	12
Eudistoma diaphanes Ritter and Forsyth, 1917	N	GC	r	5
E. mexicanum Van Name, 1945	N	GC	r	3
E. pachecae Van Name, 1945	N	P	r	3
Lissoclinum caulleryi (Ritter and Forsyth, 1917)	U	CG	r	3
Microcosmus exasperatus Heller, 1878	U	P	r, b, mr, d	3, 12
Polyandrocarpa zorritensis (Van Name, 1931)	U	P, PR	b, mr, d	3, 5,12
Polyclinum constellatum Savigny, 1816	U	GC, P	r, b, mr, d	3, 7,12, 16
P. laxum Van Name, 1945	N	GC, MSP	r	3,5,8
P. planum (Ritter and Forsyth, 1917)	N	GC, MSP	r	5,8
Pyura stolonifera (Heller, 1878)	N	GC, MSP	r	3,5,8
P. haustor (Stimpson, 1864)	N	GC, MSP	r	3,5,8
P. lignosa Michaelsen, 1908	N	GC, MSP	r	3,5,8
P. vannamei Monniot, 1994	N	CR	r	11
Rhopalaea birkelandi Tokioka, 1971	N	MSP, CR, P	r	3,5, 8, 9
Styela canopus Savigny, 1816	U	GC	mr	4
Symplegma brakenhielmi (Michaelsen, 1904)	U	P	r, b, mr, d	8, 12
S. reptans Oka 1927	U	GC	r	5
S. rubra Monniot1972	U	P	r, b, mr, d	12
Trididemnum orbiculatum (Van Name 1904)	I	P	r, b, mr, d	12

Other colonial ascidians are partially embedded in the common tunic; there are many pedunculate in which the abdomens of the individual zooids are enclosed within a dense mass of tunic, itself tightly attached to the substrate, while the zooid thoraxes project more or less independently from this common mass. Finally, colonies with zooids entirely embedded in a common tunic are massive colonies; the surface may be smooth and even, or contorted into folds, or even in the form of little projections [14]. Some colonies have calcareous spicules in the tunic, which give strength to the tunic and protection against some predators. Ascidians are filtering animals, which are generally not selective in the size of particles and small oral tentacles, inside the oral siphon, block large objects. Their main diet is based on particles of 0.1-0.3μM.

All ascidians reproduce sexually, but many also replicate their bodies by strobilation or budding (reproduce asexually) to form colonies [14]. The traditional classification of ascidians includes three orders: Aploustobranchia Lahille, 1887, Phlebobranchia Lahille, 1890 and Stolidobranchia Lahille, 1887; this classification is based on the internal characteristics of the pharynx, from Aploustobranchia where the pharynx is simple, lacking in longitudinal vessels and branchial folds, to Phlebobranchia, where the pharynx has longitudinal vessels and branchial papillae, and finally to Stolidobranchia, where the pharynx has longitudinal vessels and branchial folds [18].

So far, the community of ascidians in the TEP is formed of 35 species, which are mainly attached to rocks, piles, docks, and mangrove roots, among others (Table 1). Some invasive species have been registered; it is likely that their introduction was by activities such as shrimp aquaculture taking place in the estuary in Mazatlan Uriah [4] and maritime traffic [12]. This group of organisms has not been evaluated in The Red List of Threatened Species (IUCN), but it is considered important to begin assessing and determining the conservation status of species of sea squirts in the TEP. Three orders are represented in this region: Stolidobranchia, which presents fewer families (Styelidae and Pyuridae) in the area; Phlebobranchia, which is represented by four families (Ascidiidae, Cionidae, Diazonidae, Perophoridae); and Aploustobranchia, which is represented by five families (Clavelinidae, Didemnidae, Holozoidae, Polycitoridae, Polyclinidae).

Most Abundant Species of Each Order

Order Aplousobranchia Lahille, 1887

Family Didemnidae Giard, 1872

(Modified 14): colonies are usually thin, fouling; the consistency of the tunica varies with the presence or absence of spicules. Cloacal cavity has wide or narrow channels. Zooids have six lobes on the oral siphon. Muscles are restricted to the chest. A muscular process is present and there is a flat stomach wall. Larva generally well developed, with one or more blastozooid. Statocyte and ocellus always present.

Genus Didemnum Savigny, 1816

(Modified 3, 14): spicules star-shaped or spherical tiny projections, typically on the entire surface of the tunica, zooids smaller than 1 mm in diameter. Atrial aperture can be from a wide or single opening. Presence or absence of gill flap. Pharynx has four rows of stigmata. May or may not present a muscular process. Coiled sperm duct. Testes divided into one or two glands.

Didemnum carnulentum Ritter and Forsyth, 1917

(Modified 3, 14): colony encrusting of 1 mm or less in thickness, its pure white colour is due to the large number of spicules present in the tunic. The spicules measure from 0.015 to 0.065 mm in diameter and are found in the innermost layer of the colony and in deeper. Four present zooid stigma series. Sperm duct with 7 laps.

Didemnum vanderhorsti Van Name, 1924

(Modified 3, 14): fouling smooth colony type, its average thickness varies between 3 to 5 mm. Colonies are brown, due to the pigments that are along the tunic. Despite its colour, the tunic is somewhat transparent and the zooids can be appreciated from the surface. The zooids are numerous and are very close to each other, which can form rather complex systems. The spicules are few in number, are distributed very unevenly in small groups and their average diameter is less than 0.01 mm; its shape is nearly spherical, and their tips have the appearance of being just developing. The zooids are small, usually less than 1 mm long. The four sets of stigmas have at least 10 stigmata on each side in the first, second or third series.

Family Polyclinidae

Colonial Ascidians with globular or flat shape. Zooids completely embedded in the tunic, forming small groups around a common cloaca. Zooids divided into chest, abdomen and post-abdomen. The oral siphon has six or eight lobes, while the siphon has no atrial lobes and may or may not have a tongue. Small ovaries located before the set of male follicles.

Genus *Polyclinum* Savigny, 1816

Gelatinous colony tunic, sometimes with sand grains. Zooids with a long atrial languet; post-abdomen abdomen and connected by a marked neck. Longitudinal muscle restricted to the thorax. Pharynx with 10 to 20 rows of stigmata: upper limit for vessels with transverse corrugations. Part of the bowel is twisted.

Female gonads, small follicles surrounded by males. Outbreaks of embryos or larvae in the atrial cavity, or in small bags projected on the wall of the pharynx to the height of the anus.

Polyclinum laxum Van Name, 1945

Colonial ascidian shaped pad, which is extensively bound to the substrate by one of its sides (Figure 3).

This species, when free of particles, is smooth and bright, but often shows the shape and extension of the systems when it is covered by a thin layer of sand grains. The tunic has a fairly tough outer layer, but inside is soft and gelatinous, allowing the colony to form its easily bent and wrinkled surface. Its colour ranges from purple to red in live specimens. The zooids have 12 to 13 series of stigmas.

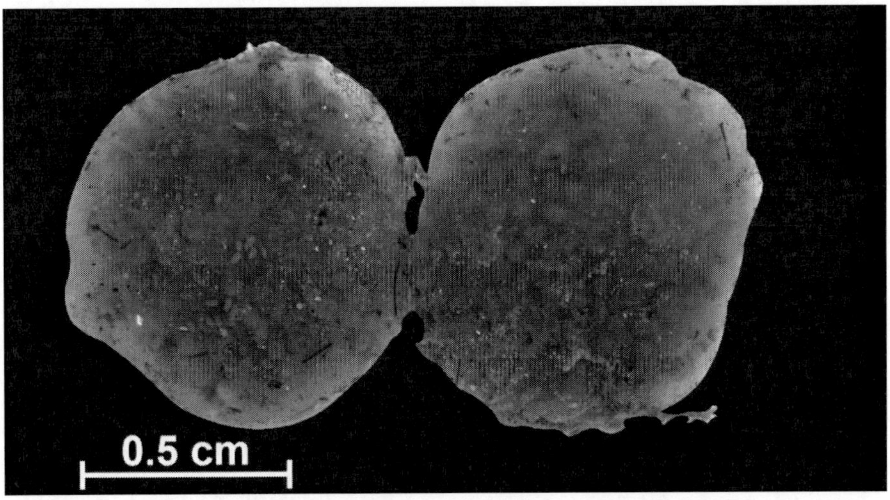

Figure 3. Colonies of Polyclinum constellatum Savigny, 1816, in preservative, collected in Oaxaca, Mexico.

Order Phlebobranchia Lahille, 1890

Family Diazonidae Garstand, 1891

Solitary or colonial Tunicates; in the case colonial forms, only the abdomen is embedded in the tunic.

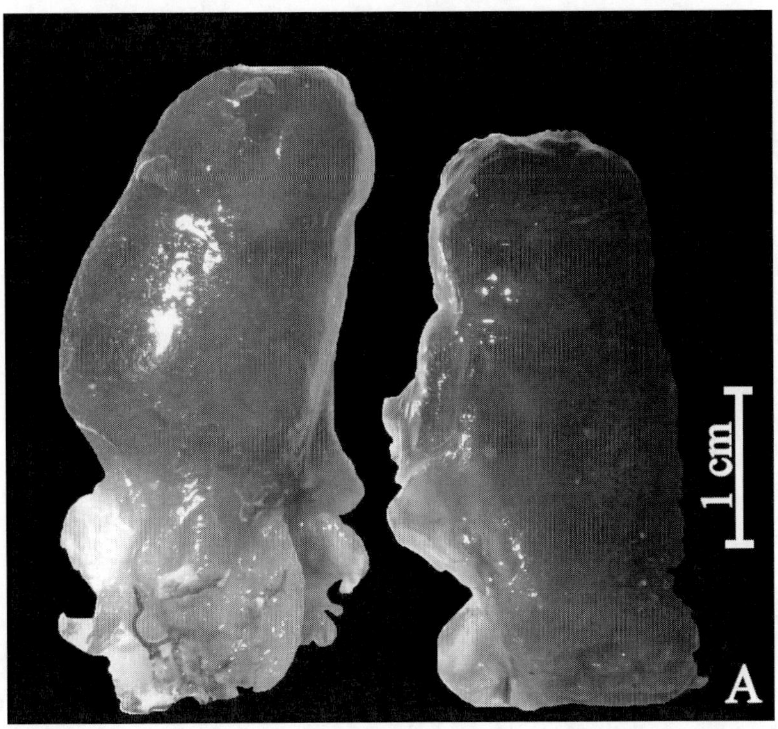

Figure 4. A) Samples of *Rhopalaea birkelandi* Tokioka, 1971 in preservative.

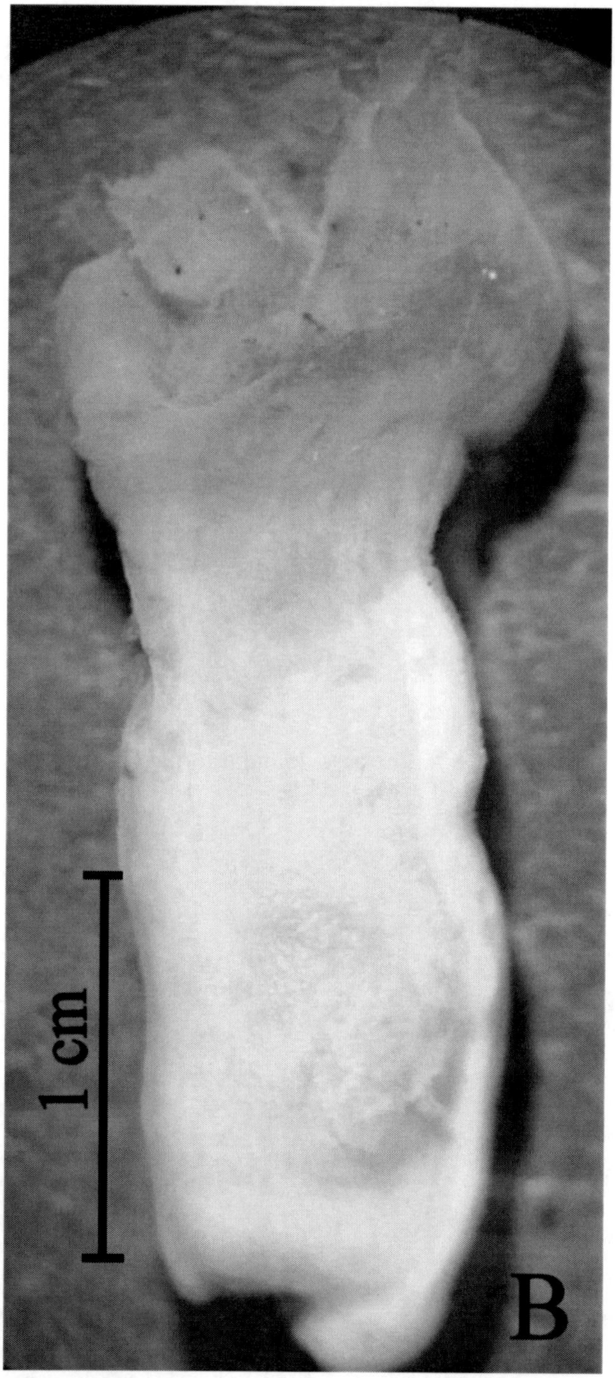

Figure 4B. Body of the ascidian, showing the small size of the abdomen; collected in Oaxaca, Mexico. Abbreviations: a= anus, ab= abdomen, as= atrial siphon, i= intestine, os= oral siphon, ph= pharynx.

Both siphons have six lobes; oblique musculature, forming two strips along the midline the abdomen. Endostilo branched, forming two smaller bands along the line half of the abdomen. Smooth inner wall of the stomach, but outside has folds.

Genus *Rhopalaea* Philippi, 1843

Solitary ascidians sometimes with two individuals connected by a thin piece of tunic. Tunic cartilage is purple or blue, without scaling. Body divided into thorax and abdomen. Often more than 20 branchial tentacles. Branchial papillae and vessels present longitudinally.

Rhopalaea birkelandi Tokioka, 1971

Purple tunic is live, smooth, firm and cartilaginous (Figure 4A). Thorax region is much larger than the abdomen. Muscle fibres are found throughout the mantle, the arrangement of these is the margin to the centre of the body, where they converge and are directed to the siphons, the muscles of the siphons are circular. Both distal siphons have six lobes and a yellow spot between each lobe. Simple branchial tentacles, 97 in number, alternate between long and short.

Dorsal tubercle is an oval hole. The abdomen region contains part of the stomach; the intestine is located on the side of the pharynx (Figure 4B). The anus has five well-defined lobes.

Family Ascidiidae Herdman, 1880

Solitary ascidians with a cartilaginous and translucent tunic. Siphons have six or more lobes, and small dots of colour on each lobe. Muscles are mainly developed on the right. Intestine is usually long, in some species with a lobed anus and dilated rectum. Ovary is generally surrounded by the intestine, with male follicles on the wall thereof.

Genus *Ascidia* Linnaeus, 1767

Tunic is cartilaginous, and transparent or semi-transparent. Body is elongated, oval, laterally compressed and adhered to the substrate on the rear or left side of the body. Oral siphon is terminal, usually with six lobes, but some species may have more or less than ten. The atrial siphon has five to eight lobes is located next to the body. The musculature of the body takes place mainly on the right side wall of the body. Branchial papillae project completely over the longitudinal vessels.

Ascidia ceratodes (Huntsman, 1912)

Solitary ascidian body oval, elongated and laterally compressed, with tunic clear or bright yellow to brown (Figure 5A). The atrial siphon is normally quite separate from the oral siphon; sometimes comprised of half of the body and directed more or less at right-angles along the same.

The right side of the body is composed of muscle fibres crossing in different directions (Figure 5B); the left side only has muscle fibres around the siphons.

The oral tentacles are slender and very numerous (150 or 200), with longitudinal vessels ranging from 30 to 45, separated by 3-6 stigmas. Papillae are present in the intersection between the vessels.

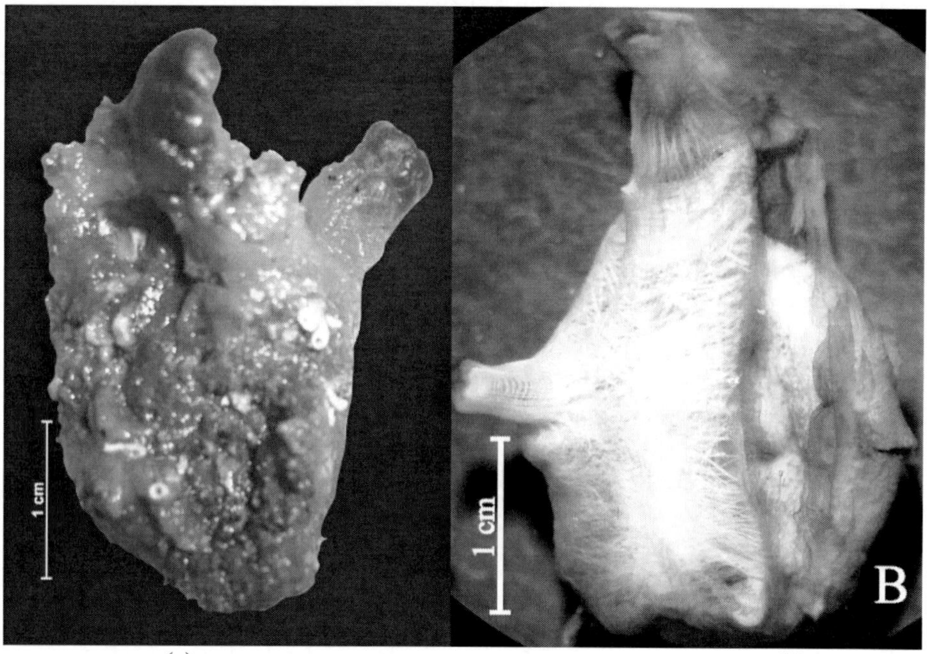

(a) (b)

Figure 5. A) Sample of *Ascidia ceratodes* (Huntsman, 1912), in preservative, collected in Oaxaca, Mexico. B) Body of the ascidian, showing the muscle arrangement in various directions over the mantle of the right side.

Ascidia sydneiensis Stimpson, 1855

The body is moderately elongated and compressed, and adheres to the substrate by a considerable part of the left side. Oral siphon has seven or eight lobes; atrial siphon has six. Transparent tunic, almost all were free of epibiontes, the coat is beige. Both siphons have numerous bands and a few longitudinal circular muscles. Between 60 and 70 branchial tentacles of various sizes are present. Digestive tract is long, covering most of the rear half of the left side. Branched gonads are seen.

Order Stolidobranchia Lahille, 1887

Family *Styelidae* Sluiter, 1895

(Modified from 14): colonial or solitary Ascidians of varied sizes. Tapeworms have a slightly fleshy tunic and body sack, with the bowel completely folded on the left side of the pharynx. The Colonial has a generally more delicate tunic, with zooids united by stolons or embedded in a common tunic, body sac or divided into two parts, with the bowel and stomach following the pharynx. Simple tentacles are present, along with four folds on each side of the pharynx (exclusive of solitary ascidians) and folds on the external wall of the stomach. Endocarps are often present. In the species descriptions, the distribution of longitudinal vessels is represented by numbers, where the numbers outside the parentheses are vessels that are located outside the fold, while that the numbers are inside the parentheses are the vessels that are located on the branchial fold.

Genus *Botrylloides* Milne-Edwards, 1841

Ascidia colonial, smooth. Zooid body divided into regions. Only the oral siphon protrudes from the robe, usually forming small groups within the colony, sharing a common cloaca. Pharynx has only three longitudinal vessels. Stigmata are straight and longitudinal. Male and female gonads are separate, located on either side of the pharynx; there is one ovary oocyte on either side, following the testicle, which has numerous glands.

Botrylloides nigrum Herdman, 1886

Purple tunic, with the front of the oral siphons having an orange tone. You can view the zooids through the tunic, which are completely embedded therein and arranged in a row or circle; the number of zooids sharing the sewer varies from 8 to 11 (Figure 6). Only the oral siphon protrudes from the tunic. Zooids are orange and semi-transparent, elongated to 1.5 mm in length. Both siphons are on the front. The oral siphon is short, while the atrial siphon is a thin transverse opening. The pharynx has three longitudinal vessels on each side, and 11 rows of stigmas. The final part of the pharynx is shifted to the left. The blind pouch comes immediately after the stomach, and the stomach is oval, with nine folds. Two female gonads are present with an oocyte on each side of the pharynx, located immediately after the male gonads, which have seven glands that are grouped in a flower shape.

Family Pyuridae Hartmeyer, 1908

Solitary ascidians in the tunic, that is generally tough, hard and covered with epibiontes. Usually have spines and scales around the siphons; they can form a small stalk. Strong muscles, with fibres longitudinally and a circular body in the siphons. Endocarps are present. Usually one gonad on each side of the pharynx, which may or may not be divided into small lobes or sacs.

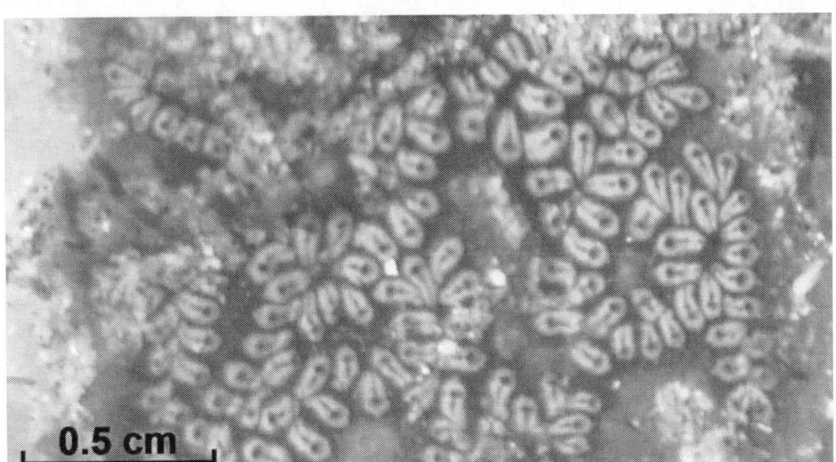

Figure 6. Colony of *Botrylloides nigrum* Herdman, 1886, in preservative, collected in Oaxaca, Mexico.

Genus *Pyura* Molina, 1782

Tunic is generally fleshy, rough and very hard, and covered by epibiontes. There are spines around the siphons and branched tentacles. Dorsal sheet tabs. Pharynx has 6-8 folds per side. The stigmata is straight, curved or spiral. Usually, there is a gonad on each side, and the gonads have many small lobes. The left gonad is surrounded by the intestine.

Pyura stolonifera (Heller, 1878)

Solitary ascidian laterally compressed body, fixed to the substrate by a small end. The outer surface is densely covered with sand par *Pyura stolonifera* (Heller, 1878) ticles (Figure 7), which, when removed from the tunic, allow you to discover a soft and supple body. In the uppermost region, siphons are observed very close together, both with four lobes and with well-developed circular bands.

Figure 7. Sample of *Pyura stolonifera* (Heller, 1878), in preservative, collected in Oaxaca, Mexico.

The coat is soft beige. All tissues of the mantle and internal organs contain spicules, varying in number, which are very abundant in some places. The branchial sac has 6 folds that are well-developed on each side.

Conclusion

Tropical Eastern Pacific has been little studied in terms of fauna of ascidians; however, we know that is represented by 11 families, with the most representative being Ascidiidae, Didemnidae, Styelidae and Pyuridae. Eight species are the most abundant recorded so far: *Ascidia ceratodes* (Huntsman, 1912), *A. sydneiensis* Stimpson, 1855, *Botrylloides nigrum* Herdman, 1886, *Didemnum carnulentum* Ritter and Forsyth, 1917, *D. vanderhorsti* Van Name, 1924, *Rhopalaea birkelandi* Tokioka, 1971, *Polyclinum laxum* Van Name, 1945 and *Pyura stolonifera* (Heller, 1878). These species have been found more commonly on substrates such as rocks, roots of mangrove and bivalves. Five invasive species have been reported in Panama (*A. incrassate* Heller, 1878, *Botrylloides planus* (Van Name, 1902), *Didemnum perlucidum* F. Monniot, 1983 and *Trididemnum orbiculatum* (Van Name 1904)), with their introduction attributed to maritime traffic. There are 12 species associated with docks and mangrove roots that are classified as "unspecified".

References

[1] M.D. Spalding, H.E. Fox, G.R. Allen, N. Davidson, Z.A. Fardaña, M. Finlayson, B.S. Halpern, A. Lombana, S.A. Lourie, K.D. Martin, E. McManus, J. Molinar, C.A. Recchia and J. Robertson. 2007. BioScience, 57(7), 573-583.

[2] T. Wilkinson, E. Wiken, J. Bezaury Creel, T. Hourigan, T. Agardy, H. Herrmann, L. Janishevski, C. Madden, L. Morgan and M. Padilla. 2009. Comisión para la Cooperación Ambiental, Montreal, pp. 200.

[3] W.G. Van Name, 1945. *Bull. Amer. Mus. Nat. Hist.* 84: 1-476.

[4] J. Salgado-Barragán, N. Méndez and A. Toledano-Granados. 2004. *Cah. Biol. Mar.* 45: 167-173.

[5] J.L. Carballo, 2006. Informe final SNIBCONABIO proyecto No. BC005, México, D.F. 51 pp.

[6] Y.B. Okolodkov, R. Bastida-Zavala, A.I. Ibáñez, J.W. Chapman, E. Suárez-Morales, F. Pedroche and F.J. Gutiérrez-Mendieta. 2007. *Ciencia y Mar* II (32): 29-67.

[7] M. A. Tovar-Hernández, E. Suárez-Morales and B. Yáñez-Rivera. 2010. *Bull. Mar. Sc.* 86(3): 637- 648.

[8] B.B. Moreno-Davila (2010). Unpublished Bachelor thesis. Universidad Del Mar, Oaxaca, 116 pp.

[9] T. Tokioka, 1971. *Publ. Seto Mar. Biol. Lab.* 19(2/3):119-122.

[10] T. Tokioka, 1972. *Publ. Seto Mar. Biol. Lab.* 19(6):383-408.

[11] C. Monniot, 1994. *Zool. J. Linn. Soc.* 110: 41-51.

[12] M.R. Carman, S.G. Bullard, R.M. Rocha, G. Lambert, J.A. Dijkstra, Roper J.J., Goodwin A.M., M.M. Carman, E.M. Vail. (2011). *Aqu. Inv.* 6(4): 371-380

[13] D.P. Abbott, A.T. Newberry and K.M. Morris. 1997. *Bishop Museum Press. Honolulu, Hawai'i,* 64pp.
[14] C. Monniot, F. Monniot and P. Laboute. 1991. *Mem. Mus. Natl. Hist. Nat*. Éditions de l'ORSTOM, Collection Faune tropicale (30): 1-247.
[15] R. L. Ritter and E. Forsyth 1917. *University of California Publishes. Zoologist,* (16): 439-512.
[16] T. Tokioka 1967. *Bulletin of the American Museum of Natural History*, (251): 1-247.
[17] X. Lahille 1890. *Faculté des Sciences de Paris*, 330 p.

In: Invertebrates: Classification, Evolution and Biodiversity
Editor: Rafael Riosmena-Rodriguez
ISBN: 978-1-62618-789-4

Chapter 8

Macro Invertebrates Associated to *Zostera marina* Meadows at Pacific Baja California Sur Coastal Lagoons

Rafael Riosmena-Rodríguez[1], María del Carmen Méndez-Trejo[1], Jorge Manuel López-Calderón[1], Juan Manuel López-Vivas[1], María Mónica Lara-Uc[1], Gerardo González-Barba[2], Carlos Sánchez-Ortiz[2] and Gustavo Hinojosa-Arango[3]
[1]Programa de Investigación en Botánica Marina,
[2]Departamento de Biología Marina,
Universidad Autónoma de Baja California Sur, México
[3]Centro para la Biodiversidad Marina y la Conservación A.C. (CBMC), México

Abstract

Meiofaunal assemblages in seagrasses from Northwest Mexico were studied using NaGISA protocol. We evaluated spatiotemporal changes by sampling in the subtropical lagoon Laguna San Ignacio in December 2008 and April 2009. We took triplicate sediment core samples (10x15 cm dimensions) to quantify the abundance and taxa composition of invertebrates. We found 8 invertebrates taxa in the seagrass meadows of Zostera marina (ascidians, anomurans, amphipods, isopods, diplostraca, mollusks, ophiuroids, and polychaetes) with high spatiotemporal variability in their abundances. Mollusks, amphipods and polychaetes were the most frequent taxa observed. There were significant differences ($p<0.05$) between the abundances of amphipods found in the Mouth and Pitahaya zones. One of the ecological services provided by meiofaunal assemblages is that gray whales (Eschrichtius robustus) that visit this lagoon in the winter season, feed on the amphipods that live within the rhizomes of Zostera marina. Seagrass meadows in Northwest Mexico provide essential ecosystem functions; they form a community on which many species rely on whether as a habitat, food source or as a fishery ground for men. The ongoing monitoring of these meadows will serve as a baseline time series from which the effect of climate change on this community will be assessed. Natural Geography in Shore Areas (NaGISA) protocol represented a good method for our biological estimations in relation to other seagrass monitoring protocols reviewed during this study.

Introduction

Coastal lagoons are well known systems in where key biological processes (high primary production, fisheries, recruitment, feeding grounds) occur [1]. In Mexico, the Pacific side of the Baja California Peninsula has five coastal lagoons: Bahia San Quintín, Punta Banda, Laguna Ojo de Liebre, Laguna San Ignacio and Bahía Magdalena which are critical areas for marine and coastal biological interactions. Those lagoons have been a relevant site as feeding and nursery grounds for black turtles (*Chelonia mydas agassizii*), loggerhead turtles (*Caretta caretta*), olive ridley turtles (*Lepidochelys olivacea*) and hawksbill turtles (*Eretmochelys imbricata*) [2, 3, 4] It is also a well known nursery and feeding site for the grey whale (*Eschrichtius robustus*) on winter months [5, 6, 7], a feeding site for migratory birds [8, 9] like brant geese (*Branta bernicla*). The diet of the black turtle is composed mainly of red seaweeds belonging to the family Gracilariaceae in Bahía de Los Angeles [10], a combination of several species of green, red seaweed and seagrasses in Bahia Magdalena [11, 12] and mainly of seagrass in Laguna San Ignacio [13]. All the species mentioned above are included in the Mexican endangered species list [14] and in the IUCN endangered species red list [15]; while seagrasses are mentioned in the NOM-022-SEMARNAT-2003 for wetlands.

Seagrasses and seaweeds are important key species in terms of biomass and their contribution in the food web [16]; they also are known because of their significant carbon fixation [17]. Seagrass meadows and seaweed beds are good substrate for a wide variety of fish and invertebrates [18], providing the physical structure for a wide range of organisms that will increment local biodiversity [19, 20, 21]. Seagrass meadows and seaweed beds function as refuges and food source for many larvae of local invertebrates and fish species [22, 23, 24]. There are few reports on the spatial and temporal variability of meiofaunal assemblages in seagrass meadows [25]. Studies performed in meadows of Africa show that meiofaunal assemblages are stratified according to seagrass species and the depth-distribution of the seagrass meadows [26, 27]. The presence of these assemblages is a fundamental aspect for the species-richness that distinguishes bare-sand bottoms from seagrass-dominated bottoms [28].

In Mexico the study of spatiotemporal variability of meiofaunal assemblages is less than scarce [25]. Here we analyze the spatiotemporal changes of the meiofaunal community that dwells within the rhizomes of the eelgrass *Zostera marina* in Laguna San Ignacio, by means of the seagrass protocol outlined by the NaGISA project.

Material and Methods

We made two surveys in Laguna San Ignacio in December 2008 and in April 2009 (Figure 1). We choose sampling sites far from human influence, selecting both tidal and subtidal meadows. Eelgrass populations found in Laguna San Ignacio have both annual and perennial life cycles. SCUBA gear was used for the subtidal sampling (up to 10 m deep). In each sampled meadow a 30 m transect was set, along this transect we estimated the shoot density (25x25 cm quadrant) and the canopy height in a total of four replicates. Three replicates of sediment cores were randomly extracted in the same transect to quantify the invertebrates found within the eelgrass rhizomes. Core size was 15x10 cm, and they were sieved in the field using three different grid sizes (3000 μm, 840 μm, and 420 μm). The large-size grid was used to collect all the plant material, the medium-size grid collected all the eelgrass seeds and meiofauna, and the small-size grid was used to collect microfauna. All the samples were preserved in alcohol (70%) for later analysis in the laboratory. Invertebrates were identified to the lowest taxonomic level possible using available identification keys [29, 30]. A Trimble GPS unit was used to mark all the sampling sites as well as the perimeter of each sampled meadow, in order to estimate the coverage (in hectares) of each meadow. Seawater temperature and salinity were recorded at each sampling site with a bucket thermometer and a refractometer. During the sampling design two more seagrass sampling protocols were compared with the NaGISA sampling protocol in order to evaluate advantages and disadvantages and their cost-effectiveness.

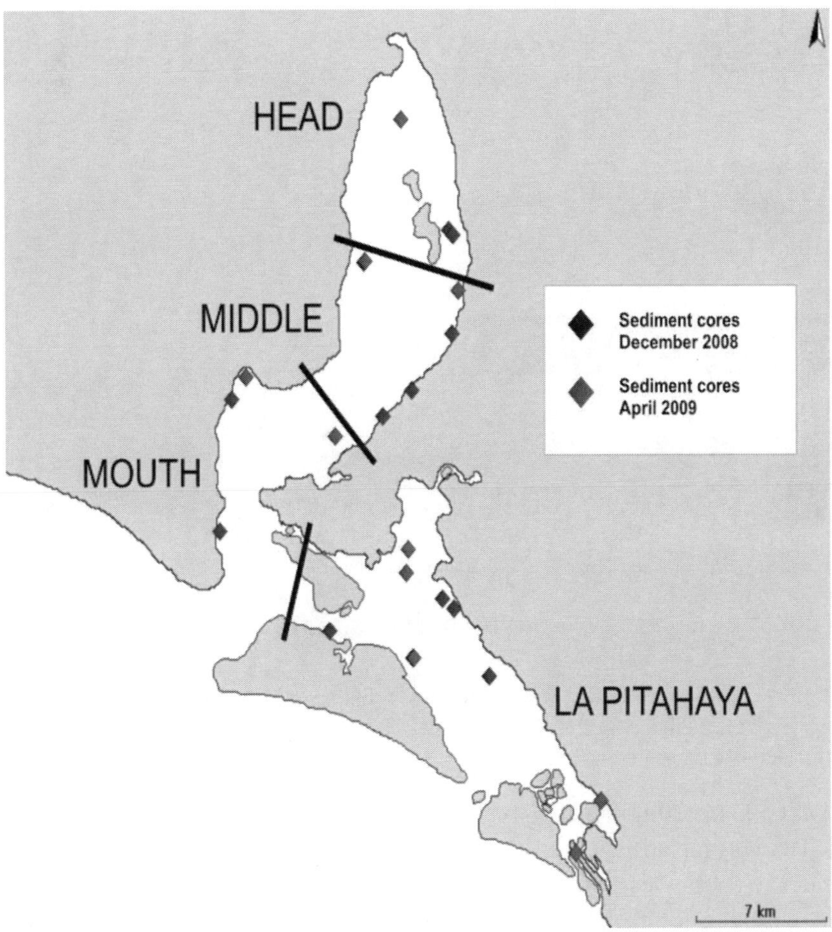

Figure 1. Meiofauna sampling locations in Laguna San Ignacio in December 2008 (blue) and April 2009 (red). The lagoon was divided in four sections, from north to south: Head, Middle, Mouth, and La Pitahaya.

Results

After performing an intensive sampling in Laguna San Ignacio (Figure. 1) we found very high variability in the abundance of invertebrates in the entire lagoon. During the sampling of December 2008 the dominant group was that of mollusks, therefore, we present that data in a separate graph (Figure 2), focusing on the different number of classes found in the four sections of the lagoon. The greatest richness of mollusk as well as the largest variability was found in the Middle section of Laguna San Ignacio. The Head section was the lowest with only one genus present (*Nassarius*). La Pitahaya and the Mouth were the second and third most important sections for mollusk diversity. We performed a non-parametric test (Kruskal-Wallis one way analysis of variance) and found no significant differences (p=0.49) in the richness of classes between the four sections.

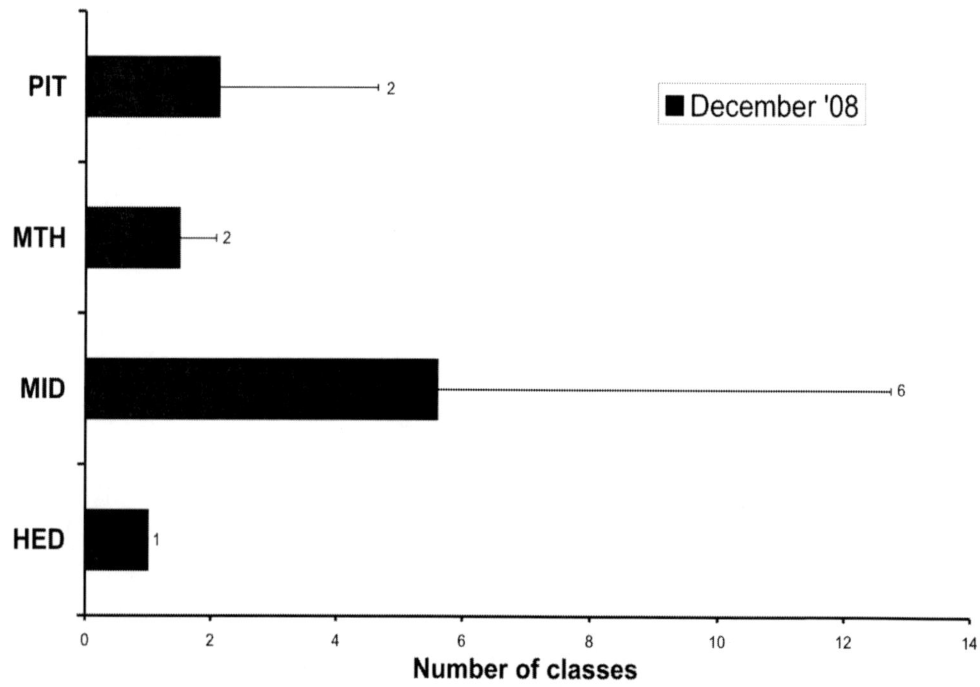

Figure 2. Mollusk classes found in the four sections of Laguna San Ignacio (Head= HED, Middle= MID, Mouth= MTH, La Pitahaya= PIT) in December 2008. Correspondence of each section of the lagoon is indicated in Figure 1.

There was a marked difference in the sampling of April 2009, the abundance of mollusks decreased, data was included with the rest of the taxonomic groups in Figure 3. Ophiuroids and anomurans were the least common groups in the meiofaunal assemblages of the lagoon, only found during December 2008 in the Mouth and La Pitahaya sections, respectively (Figure 3C, D). Something similar happened with the ascidians group, however they were more abundant, and they are present in all the sections of the lagoon with the highest abundance in the Head and Middle sections (>1000 organisms/m^3).

The dominant groups in the assemblages of Laguna San Ignacio were three: polychaetes, amphipods, and mollusks (Figure 3) in December 2008 and April 2009. In April, as well as in December, mollusks had the greatest abundances, except at the Mouth where amphipods were slightly greater (Figure 3C). On the other hand, polychaetes were present in the assemblages of the four sections in both seasons, except in the Head were they occurred only in December (Figure 3A). Meiofaunal assemblages had their greatest densities in La Pitahaya and it was the only section with the occurrence of anomurans (Figure 3D).

We looked for statistical spatial and temporal differences in the abundance of the different taxonomic groups found in Laguna San Ignacio. For the differences between seasons a Mann-Whitney test was applied for each taxon; for the differences between sections a Kruskal-Wallis analysis of variance was applied. The only significant difference ($p<0.05$) found in the abundance of invertebrates between December 2008 and April 2009 was that of amphipods (Figure 3). For the spatial comparisons, the amphipod group was also the only one with significant differences, specifically between the Mouth and La Pitahaya sections (Figure 3C, D).

Assemblages at the Head and the Middle consisted of six groups: amphipods, isopods, diplostraca, polychaetes, ascidians, and mollusks; while at the Mouth and La Pitahaya meiofaunal assemblages consisted of seven groups: amphipods, isopods, diplostraca, polychaetes, ascidians, ophiuroids, anomurans, and mollusks. Ophiuroids appeared only at the Mouth and anomurans at La Pitahaya (Figure 3C, D).

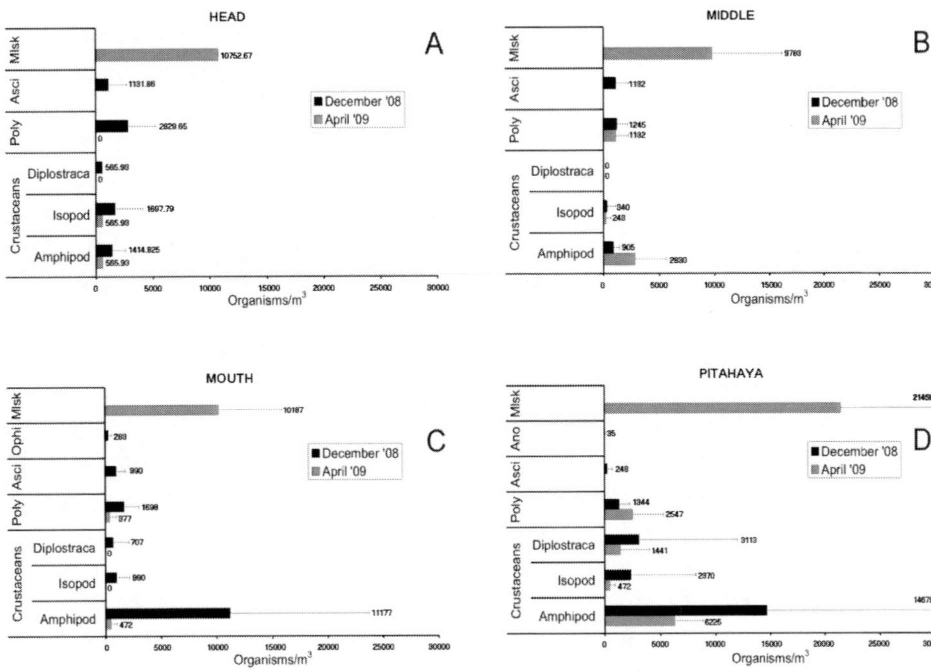

Figure 3. Invertebrate groups (Polychaetes= Poly, Ascidians= Asci, Mollusks= Mlsk, Ophiuroids= Ophi, Anomurans= Ano) found in the rhizomes of *Zostera marina* in Laguna San Ignacio in December 2008 (black bars) and in April 2009 (gray bars).

Discussion

We investigated for the first time the meiofaunal assemblages found within the rhizomes of eelgrass (*Zostera marina*) growing in the Mexican wetland Laguna San Ignacio. Although there is high spatial and temporal variability three of the eight invertebrate groups found in the lagoon were dominant: mollusks, amphipods, and polychaetes. Meiofaunal assemblages had greater mollusks densities in winter than in spring. Assemblages at the northern portion of Laguna San Ignacio (Head and Middle sections) are dominated by mollusks of the genus *Nassarius*. Only in the southern portion of Laguna San Ignacio (Mouth and La Pitahaya) there were significant differences ($p<0.05$) in the abundance of invertebrates (amphipods). Lower abundances of amphipods in the Mouth section could have important implications for the documented decrease in the number of gray whales entering Laguna San Ignacio in the last ten years [5]. It has been hypothesized that gray whales (*Eschrichtius robustus*) occasionally feed on the amphipods within *Z. marina* rhizomes [31], thus gray whales could be moving to nearby wetlands (e.g. Laguna Ojo de Liebre) to give birth and teach their calves feeding techniques. We have found that the number and extension of eelgrass meadows in Laguna San Ignacio is also diminishing and that the permanence of these meadows in the lagoon is jeopardized [32](Riosmena-Rodriguez unpublished data) due to a deficient management plan for the lagoon. A fewer number of gray whales in Laguna San Ignacio could have serious negative consequences to the economy of the residents of the lagoon (less ecotourism) and to local biodiversity.

Meiofaunal assemblages together with the *Z. marina* meadows provide ecosystem services fundamental for coastal environments, in Mexico the seagrass community has been overlooked until recently. However, in other parts of the world, the higher biodiversity and the role as a nursery ground of the seagrass meadows compared to bare-sand areas has been documented [24, 22]. A primordial service that these invertebrate assemblages provide is the recycling of nutrients in the sediment column. By consuming a fraction of the detritus from dead leaves, algae and dead animals, these nutrients become available for the sustaining of the seagrass meadow. Polychaetes are also fundamental in the process of oxygenation of the upper sediment column by constantly digging tunnels in the sediment. As we already mention, amphipods in these meiofaunal assemblages constitute a food source for bottom feeders like gray whales [6].

Meiofaunal assemblages in Laguna San Ignacio were highly variable in space and time, we found no significant differences ($p<0.05$) for any taxonomic group except for amphipods in southern Laguna San Ignacio (Figure 3C, D, Mouth and Pitahaya). This is opposite to what has been observed in other water bodies [33, 34 26]. It is urgent to expand the number of sediment cores in the lagoon together with a long-term monitoring of meiofaunal assemblages that will permit the identification of spatial and temporal trends. The current loss of eelgrass meadows in Laguna San Ignacio [32 Riosmena-Rodriguez unpublished data) and their accompanying meiofaunal assemblages will have a negative impact in the fishery resources, coastal resilience to disturbances, water quality, productivity and the stability of the entire ecosystem, this phenomena is already happening worldwide [35, 36]. Global warming and coastal development as touristy, industrial or fishing activities might damage the meadows down to disappear/substitute them from extensive areas [16, 37, 38] ranging from local to regional scales. The impact of this modification is intense and active restoration techniques have been developed to recover the areas [39]. Techniques range from regular transplants of adult plants using typical gardening techniques to the active plant of seeds in new areas for 'natural' development; they can also be planted in small containers and transplant the young plants into the field. Recently, the Mexican government has changed the legislation NOM-022-SEMARNAT-2003 for wetlands suggesting that restoration and mitigation activities should be applied to recover lost areas as a consequence of coastal development, but the regulation does not specify a course of action or recommend any restoration or mitigation politics, thus, leaving an critical gap.

Besides NaGISA seagrass sampling protocol we review two more protocols, one from Texas Parks and Wildlife [40] and another from SeagrassNet [41] to determine advantages of each one of them for our research needs. Protocol from Texas Parks and Wildlife (TPW) was the most extensive, including hydrographic (nutrient, chlorophyll a, dissolved oxygen) and sediment (redox condition, grain size, total organic carbon) surveys combined with seagrass monitoring (below-ground biomass, leaf area index, chlorophyll concentration). Therefore logistics and cost are too high because it gives the most comprehensive view of environment health, but the human resources and equipment needed make it difficult to implement in a worldwide scale. On the other hand, SeagrassNet protocol does not require such a detailed survey of the environment because it was designed to be implemented worldwide. However it requires above-ground and below-ground biomass measurements and 12 replicates along 50m transects for shoot density estimations. When only two or three people are working with the monitoring (as is our case) those measurements are cumbersome, nevertheless they give relevant information about the meadows but they are not fundamental to assess the general conditions of the meadow and its biodiversity. The NaGISA protocol is the most suited of all for us, its cost and logistics are low, the monitoring can be performed with only two or three persons, and it gives us the opportunity to survey three meadows in about five hours, with a replicate effort. There is only one modification we made to the protocol, we used the 25 cm instead of the 50cm quadrant, because most of our surveys were performed using SCUBA diving and the air in our tanks was a limiting factor for the shoot count, fighting tidal currents was also oxygen-consuming.

There are a huge number of species that rely on seagrass meadows in different stages of their lives or permanently [42, 24, 22, 5, 43, 20]; humans also rely on these meadows, not only for commercial fishery [16], but for medicinal, nutritional purposes, and climate regulation too [44, 18]. Meiofaunal invertebrates are one of the key aspects of seagrass meadows and also a biological indicator of the health condition of these communities given that they are sesile and sensitive organisms that cannot displace from a site in the event of environmental disturbances. We must keep track on them in the long term, in order to promptly detect effects caused by global warming or human disturbances, which will give us the opportunity to take proper actions (mitigation/restoration). The final objective being the conservation of seagrasses and the ecosystem services they provide.

Acknowledgments

Authors wish to thank to the 'Laguna San Ignacio Ecosystem Science Program' (Dr. J. Urban and Dr. S. Swartz) for their support. RRR thanks to the 'Project CONACYT-

SEMARNAT' for the support throughout this research. JLC was supported by CONACYT grant 41393.

References

[1] Valiela, and J. M. Teal, *Nature,* 280, 652-656. (1979).
[2] J. A. Seminoff, A. Resendiz, W.J. Nichols, and T. T. Jones, *Copeia,* 610-617. (2002b).
[3] S. C.Gardner and W. J. Nichols, *Chelon. Conservat. and Biol.,* 4(1), 197-199. (2001).
[4] V. Koch, W. J. Nichols, H. Peckham and V. Toba, *Biol Conservat,* 128(3), 327-334. (2006).
[5] J. Urban, L. Rojas-Bracho, H. Perez-Cortes, A. Gomez-Gallardo, S. L.Swartz, S. Ludwig and R. L. Jr .Brownell, *J. Cetacean Resear.* 5(3), 281-295. (2003).
[6] J. Caraveo-Patiño, and L. A. *Soto, Hydrobiol,* 539, 99-107. (2005).
[7] J. Caraveo-Patiño Y. Wang, Luis A. Soto, K. Ghebremeskel, C Lehane, M. A. Crawford (2009), Eco-physiological repercussions of dietary arachidonic acid in cell membranes of active tissues of the Gray whale. *Marine Ecology,* 30: 437–447
[8] J. S. Sedinger, M. P.Herzog and D. H .Ward, Auk, 121(1), 68-73. (2004).
[9] D. H. Ward, A. Reed, J. S. Sedinger, J. M. Blacks, D. V. Derksen and P. M. Castelli, *Glob Chan. Biol.* 11, 869-880. (2005).
[10] J.A. Seminoff, A. Resendiz, W.J. Nichols, *Journal of Herpetology* 36, 447–453. (2002a).
[11] M. López-Mendilaharsu, S.C. Gardner, R. Riosmena-Rodríguez, and J. Seminoff, Aquatic conservation: *Marine and freshwater ecosystems*, 15, 259-269. (2005).
[12] M. López-Mendilaharsu, S.C. Gardner, R. Riosmena-Rodriguez, and J. A. Seminoff, *J. Mar. Biology Association of the UK.* (2008).
[13] L. Santos-Baca, M. S. Thesis (2008). Evaluación de los hábitos de alimentación de la tortuga verde Chelonia mydas, en Bahía Magdalena, B. C. S., México, utilizando la técnica de isótopos estables (delta 13 C y delta 15 N). *Centro de Investigacioens Biololgias* 45 p.
[14] NOM-059-ECOL-2001 (http://dof.gob.mx/nota_detalle.php?codigo=5173091& fecha= 30/12/2010
[15] International Union for Conservation of Nature. 2009. IUCN Red list of threatened species. Version 2009.1 http://www.iucnredlist.org
[16] F.C. Coleman and S.L. Williams, *Trends in Ecology and Evolution,* 17, 40-44. (2002).
[17] C.M. Duarte, and C.L. *Chiscano, Aquatic Botany,* 65, 159-174. (1999).
[18] M. De la Torre-Castro and P. Rönnbäck, *Ocean and Coastal Man.,* 47,361-387. (2004)
[19] Schneider, F. I. and Mann, K. H. *Biol. Ecol.,* 145, 101-117. (1991).
[20] F. A. Casares and J.C. Creed, *J. Coastal Res.,* 24, 790-797.(2008).
[21] Conservation International. 2008. Economic Values of Coral Reefs, Mangroves, and Seagrasses: A Global Compilation. Center for Applied Biodiversity Science, Conservation International, Arlington, VA, USA.
[22] A. Arrivillaga and D. M. Baltz, *Bull Mar. Sci.,* 65, 301-319. (1999).

[23] H.Kirkman, 1997. Seagrasses of Australia, Australia: State of the Environment Technical Paper Series (Estuaries and the Sea), Department of the Environment, Canberra.
[24] E. Perkins-Visser, T.G. Wolcott and D. L Wolcott. *J. Exp. Mar. Biol. Ecol.,* 198, 155-173. (1996).
[25] S. E. Gómez-Noguera, and M.E. Hendrickx, *Mar. Poll. Bull.,* 34, 582-587. (1997).
[26] M. De Troch, S. Gurdebeke, F. Fiers and M. Vincx, *J. Sea Res.* 45, 45-61. (2001).
[27] C. Nozais, R. Perissinotto and G. Tita, *Est. Coast. Shelf Sci.,* 62, 325-338. (2005).
[28] C. Boström and E. Bonsdorff, *J. Sea Res.,* 37, 153-166. (1997).
[29] J. L. Barnard and G. S. Karaman, The Families and Genera of Marine Gammaridean Amphipoda (Except Marine Gammaroids), Part I y II. Records of the Australian Museum Supplement, 13. 1991.
[30] R. C. Brusca and M.E.Hendricks, (2005). Crustacea 4. Peracarida: Lophogastrida, Mysida, Amphipoda, Tanaidacea and Cumacea. In: Hendricks, M. E., Brusca R. C. and Findley L.T. (eds) Listado y Distribución de la Macrofauna del Golfo de California, México. Parte 1. Invertebrados. Arizona-Sonora Desert Museum, pp.429
[31] B. J. Le Boeuf, M. H. Pérez-Cortés, R. J. Urbán, B. R. Mate and U. F. Ollervides, *J. Cetacean Res. Man.,* 2(2), 85-99. (20009.
[32] Riosmena-Rodríguez UNPUBLISHED Data about San Ignacio Lagoon.
[33] R. Novak, *Netherlands J. Sea Res.,* 16, 380-388. (1982).
[34] K. A. Willems, Y. Sharma, C. Heip and A. J. J. Sandee. *Netherlands J. Sea Res.* 18, 418-433. (1984).
[35] B. Worm, , E. B. Barbier, N. Beaumont, J. E. Duffy, C..Folke, B. S. Halpern, J. B. C Jackson,. H. K.. Lotze, F. Micheli, S. R. Palumbi, E. Sala, K. Selkoe, J. J. Stachowicz, R. Watson, *Sci.* 314, 787-790. (2006).
[36] A. R. Hughes, S. L. Williams, C. M. Duarte, K. L. Jr. Heck and M. Waycott, *Frontiers Ecol. Environm.,* 7, 242-246. (2009).
[37] M. R. Johnson, S. L. Williams, C. H. Lieberman, A. Solbak, *Estuaries,* 26, 106-115. (2003).
[38] S. Wyllie-Echeverria, T. E. Jr. Mumford, J. K.Gaydos, and S. Buffum, (2003). Zostera marina declines in San Juan County, WA Wescott Bay taskforce mini-workshop. pp. 1-18.
[39] P. Rodríguez-Salinas, R. Riosmena-Rodriguez, G. Hinojosa-Arango, R. Muñiz-Salazar, *Ecol. Engin.doi:10.1016/j.ecoleng.*2009.09.004. (2009).
[40] (www.tpwd.state.tx.us)
[41] F. T. Short, L. J. McKenzie, R. G.Coles, K. P. Vidler, J. L. Gaeckle, SeagrassNet Manual for Scientific Monitoring of Seagrass Habitat, Spanish edition. University of New Hampshire Publication. pp 75. (2008).
[42] Z. Garcia-Esquivel and V. M. Bricelj, *Biological Bulletin,* 185, 42-55. (1993).
[43] M. S. Lindberg, , D. H. Ward, , T. L. Tibbitts and J. Roser, *J. Wildl. Man.,* 71, 534-540. (2007).
[44] R. S. Felger, M. B. Moser and E. W. Moser, 1980. Seagrasses in Seri Indian culture. In: Handbook of Seagrass Biology: An Ecosystem Perspective. Phillips, R. C., and McRoy, C. P. (eds). *Garland STPM Press,* New York. pp. 260-276

In: Invertebrates: Classification, Evolution and Biodiversity ISBN: 978-1-62618-789-4
Editor: Rafael Riosmena-Rodriguez

Chapter 9

Biodiversity Associated to Sargassum Forest at the Gulf of California

Alvin Noé Suarez-Castillo[1,2], Rafael Riosmena-Rodriguez[2], Gustavo Hernandez-Carmona[1], María del Carmen Méndez-Trejo[1], Juan Manuel López-Vivas[1], Carlos Sánchez-Ortiz[3], María Mónica Lara-Uc[1] and Jorge Torre-Cosio[4]
[1]CICIMAR-IPN, Playa de Santa Rita s/n, La Paz, México
[2]Programa de Investigación en Botánica Marina,
[3]Dept. Biol. Mar., Universidad Autónoma de Baja California Sur,
Carretera al Sur, La Paz, México.
[4]Comunidad y Biodiversidad Calle Isla del Peruano,
Colonia Lomas de Miramar, Guaymas, Sonora, México

Abstract

The *Sargassum* forests are one of the most characteristic and critical habitats for conservation in the rocky shores at the Gulf of California. Therefore, it is necessary to know more about the associated populations. The study was carried out in 2008 in five localities around La Paz Bay, México. The biodiversity associated with sub-tidal *Sargassum* forest was monitored by applying a modification of the NaGISA rocky shore protocol. The richness of fish species was estimated for long of transects of 50 m. The cover percent of the biotic and abiotic component of substrate, the composition of epibenthic invertebrates and the richness of macroalgae species were estimated inside of 1 m^2 quadrants, and the density (thalli number) and size of *Sargassum* were estimated inside 625 cm^2 quadrants. The substrate was composed principally of rock, *Sargassum* and other seaweeds. In total, 27 species of fishes, 23 species of macroalgae, 38 species of macro epibenthic and 17 micro epibenthic invertebrates were registered. Differences in species composition of macroalgae associated with *Sargassum* forests were observed between the locality El Saladito and The Sauzoso. During the study, the size of *Sargassum* showed temporal variation suggesting growth differences. The total number of epibenthic invertebrates found was 633 (mean = 3.7 individuals/m^2), which were classified into six taxa; the Cnidaria was the most abundant. The cover percent of substrate components, the community structure of fauna and morphometric variables of the algae showed a spatial-temporal variation. The results show the baseline that can be obtained, which in turn will provide the information necessary to understand the ecological role of these brown algae for the maintenance of biodiversity and the community-associated changes. *Sargassum* forest is a critical habitat for conservation in the Gulf of California because it is sensitive to environmental and human-induced impacts. Their role in coastal areas is related to species recruitment, the fixation of CO_2 and drift material, which are extremely important to sandy beach stability and food sources for insects from terrestrial environments. *Sargassum* forest comprise a relevant near shore ecosystem where species related to fisheries, protected species and those with high biodiversity treat it as a refuge There are several fisheries related who are close to this ecosystem and legal protection does not exist to protect this critical habitat for conservation in the Gulf of California.

Introduction

The presence of biological substrates in marine rocky areas, including communities of macrophytes have great influence on the organisation and development of the biological community [1]. This is because that diversifies the range of space resources that can be used, increases the area available for colonisation by sessile species and provides a refuge from predators and unfavourable environmental conditions [2]. In the temperate and subtropical waters, the species of the genus *Sargassum* dominates in coverage and biomass over other algae, forming large forests ranging from a few meters to several hectares [3, 4]. These forests form a secondary substrate for surrounding communities, such as epiphytes [5], epifauna [6], epibentic invertebrates [7] and fish [8]. Even *Sargassum* has been recognised as a critical habitat for the establishment and development of some of their associated species (e.g. Shrimps and fish; [9, 10, 30]).

In the rocky sub-tidal Gulf of California, species of the genus *Sargassum* have been established as the most important and conspicuous components of the phycoflora [7]. In the 1990s, about 1,519 ha of the west coast of the Gulf was estimated to be covered by *Sargassum*, making a total biomass of 180,711 wet tons [3, 11, 12]. Despite that, in the Gulf of California there is a gap of information on the faunal groups associated with *Sargassum* forests and their ecological importance in coastal areas, because most of the research has been directed toward understanding *Sargassum* as an organism (autoecology; [13,14,15]) and its potential uses [17,18,19,20]. Therefore, the objective of this research was to provide information on the biodiversity associated with these coastal communities of *Sargassum* forests from the Gulf of California, using the sampling protocol proposed by NaGISA (Census of Marine Life) [24].

Materials and Methods

Bahia de La Paz is a wide body of water in the south-western part of the Gulf of California [21, 22], which is 90 km long and 60 km wide with an approximate area of 4,500 km^2 [22]; it has a semi-elliptic form with the main axis orientated north-northeast-south-

southeast and geographical limits to the west and south by land, to the east by Espiritu Santo Island and in the north by the Gulf of California [21]. The bay has 2 mouths, where the water flows with a semidiurnal tidal change [21, 22]. In the south-eastern part of the bay there is a coastal lagoon protected by a sandy bar [21, 22]. Into this lagoon most of the freshwater runoff flows, and is also communicated to the ocean with a channel between the sandy bar and the continent [22]. Upwellings are present in the communication channels that form the mouths in the bay at the north and south [22], but nutrients appear in pulses in general [23].Temperature and salinity fluctuate seasonally in relation to the major water masses present [41]. The study was conducted in sub-tidal *Sargassum* forests located in five locations in the Bay of La Paz: The Quelele, The Sauzoso, The Saladito, Punta Coyote and Botafuego (Figure 1).

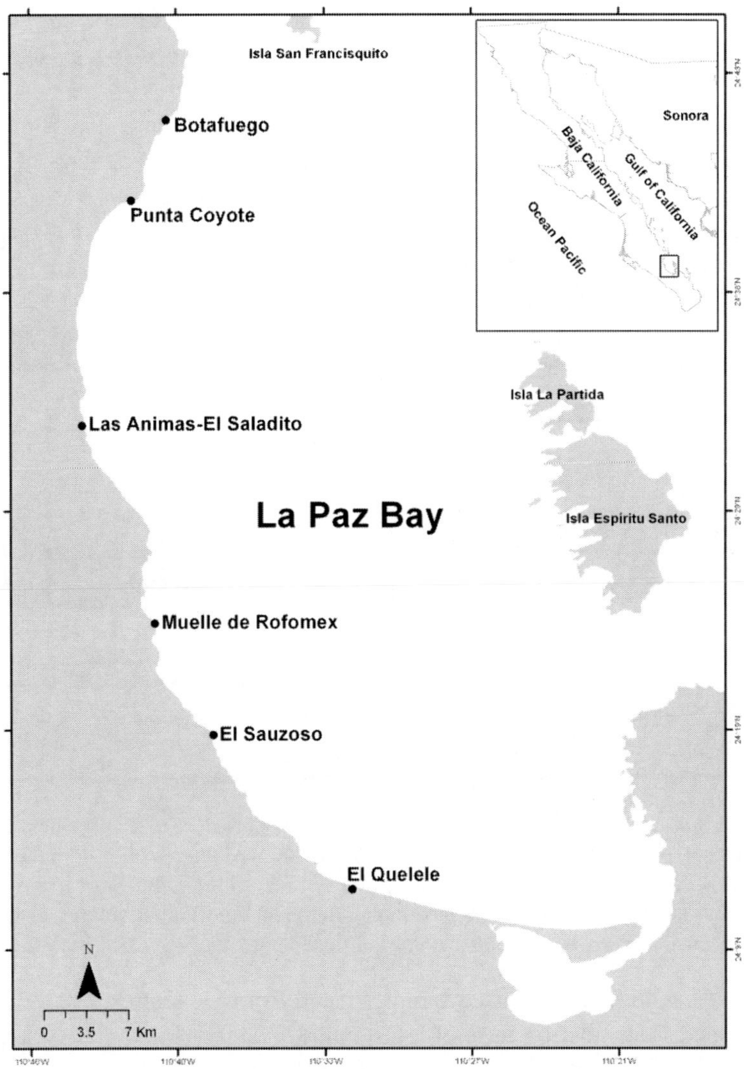

Figure 1. Location of study sites for the spatio-temporal variation of epifauna associated with Sargassum forests.

A database since 2000 has been developed for this area for monitoring purposes. The sites were chosen based on a similarity in the type of substrate (boulders) and the depth at which *Sargassum* forests were (1.5-3.5 m deep). The sampling was carried out on a monthly basis from March to June 2009 by SCUBA diving, except for the month of May due to logistical problems. The proposed methodology for estimating the biodiversity associated with *Sargassum* forests is based on the sampling protocol NaGISA for marine rocky areas [24].

The basic protocol was adjusted in order to meet the goal of implementation in this habitat, as described below. We recorded *in situ* the composition of fish swimming 50 cm from the bottom and along two transects of 50 m (three transects in April), distributed in the middle and along (parallel to the coastline) the forest. Along the transect we conducted a total of 5 groups of quadrants, one placed every 10 m. This consisted of a quadrant of 1 x 1 m and another of 25 x 25 cm (Figure 2). The percent cover of biotic and biotic components on the substrate and the composition of epibenthic invertebrates (number of species and abundances) visible (> 2 cm) were recorded within the 1 m^2 quadrant.

In the quadrant of 625 cm^2, the density (number of branches) and size of *Sargassum* was recorded (a total of 30 thalli, 3 thalli per quadrant). The different components of the substrate, the composition of epibenthic invertebrates and the size of *Sargassum* were recorded from March to June. The composition of fish and macroalgae was recorded in June, and *Sargassum* density was recorded from April to June. In the case of epibenthic invertebrates, colonial organisms (e.g. Corals, sea fans and sea pens) were counted as an individual.

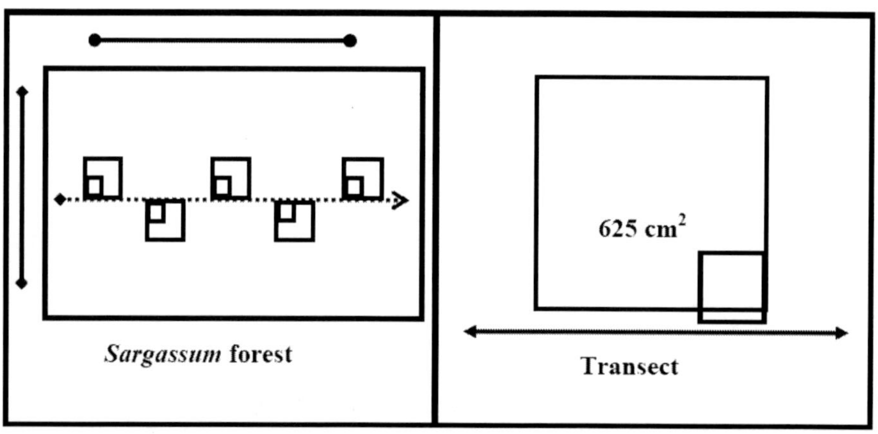

Figure 2. A. *Sargassum* forest layout: here, a 50 m transects are shown with 5 replicates (each with one complete quadrant set). The transects (second transect is not shown) were localised in the middle part of *Sargassum* forest. B. *Sargassum* forest transect/quadrant set: record the richness of fishes in the 50 m transects; record the percentage coverage of substrate components/individual number of visible (> 2cm) epibenthic invertebrates in the 1 m^2 quadrant; record of density/size of *Sargassum* in the 625 cm^2.

In the case of the epifauna, we used samples taken from September 2000 to August 2001 (except October and December because of logistical problems) using SCUBA at 8 permanent sites marked with cement blocks between 9:00 am to 12:00 am at 2–2.5 depth.

To sample the micro invertebrates, we sampled entire plants taken in a 25 cm x 25 cm quadrant, where all of the *Sargassum* plants were collected from the holdfast to ensure representative samples and, avoiding loose water, stored in plastic bags that were used to fix the material with formalin at 4% and transported to the laboratory.

Once in the laboratory, the mobile animals were segregated filtering the samples with the help of freshwater into a mesh of 0.5 mm; the organisms fixed to the thalli were picked up with pincers. All of the animals were placed in individual containers and transferred to 70% alcohol for identification using the keys of [25, 26, 27, 28, 29].

The organisms observed were photographed and identified *in situ*. In some cases, it was necessary to collect the organism for identification at the lowest possible taxonomic level.

The identification was made using specialised texts for each of the groups in the region: fish [31, 32], epibenthic invertebrates [25, 33, 34, 35, 36, 37, 38] and macroalgae [39].

Results

In the *Sargassum* forests, the percentage of coverage of different biotic and abiotic components that make up the substrate changed during the study period (Figure 3).

At baseline (March), *Sargassum* was the biotic component with the lowest percentage of coverage compared with other components, as were the bare rock (rock) and other algae. In the following months (April-June), *Sargassum* had the highest percentage of coverage compared to other biotic and abiotic components, except for El Sauzoso in June. Spatially, there was also a variation in the percentage of recovery of biotic and abiotic components of the substrate (Figure 3). *Sargassum* was the biological component that showed the greatest percentage of coverage, while abiotic substrates, like rocks and sand, had a higher percentage of coverage in comparison with other biological substrates (e.g. *Porites panamensis, Phragmatopoma californica*, among others).

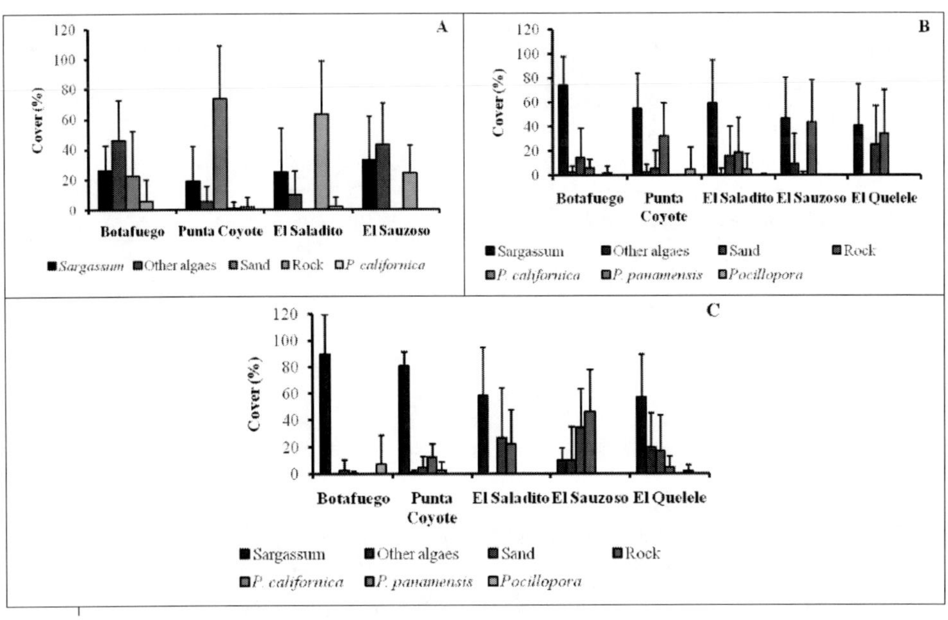

Figure 3. Spatial-temporal variation of the percent coverage of substrate components inside *Sargassum* forests at La Paz Bay. A: March, B: April, C: June.

Biodiversity Associated to Sargassum Forest

Fish. A total of 27 fish species belonging to six orders and 17 families were observed among *Sargassum* forests in June 2009 (Table 1). It was common to see juveniles of *Mycteroperca rosacea* and *Paralabrax maculatofasciatus*, adult *Balistes polylepsis* and *Abudefduf troschelii* and individuals of endemic species in the Gulf of California, such as *Stegastes rectifraenum*. The most diverse family was the Pomacentridae with 4 species, followed by the family Labridae and Serranidae with 3 species each (Figure 4A).

Table 1. Species composition of fish associated with *Sargassum* forests at La Paz Bay. Trophic Functional Group (TFG) is presented along with their importance. Zoob: zoobentivorous, Zoop: zooplanktivorous, Herb: herbivorous, Pisci: piscivorous, Omni: omnivorous, Oppo: opportunistic, FR: national fishery resource, OE: ornamental species, E: species endemic to Gulf of California

Order/Family	Species	TFG	Importance
Anguilliformes			
Muraenidae	*Muraena lentiginosa*	Zoob	
Aulopiformes			
Synodontidae	*Synodus evermanni*	Zoob	
Mugiliformes			
Mugilidae	*Mugil* spp.	Herb	FR
Perciformes			
Chaetodontidae	*Chaetodon humeralis*	Omni	OS
Cirrithidae	*Cirrithus rivulatus*	Zoob	
Gerreidae	*Gerres cinereus*	Omni	FR
Haemulidae	*Haemulon maculicauda*	Zoob	FR
	Haemulon sexfasciatum	Zoob	FR
Lutjanidae	*Hoplopagrus guentherii*	Zoob	FR, OS
Labridae	*Halichoeres dispilus*	Oppo	OS
	Halichoeres notospilus	Zoob	
	Thalassoma lucasanum	Zoob	OS
Labrisomidae	*Labrisomus xanti*	Zoob	
Pomacanthidae	*Holacanthus passer*	Omni	OS
	Pomacanthus zonipectus	Omni	OS
Pomacentridae	*Abudefduf troschelii*	Zoop	
	Stegastes rectifraenum	Omni	OS, E
	Stegastes flavilatus	Omni	OS
	Stegastes acapulcoensis	Omni	
Serranidae	*Epinephelus labriformis*	Zoob	FR
	Mycteroperca rosacea	Pisci	FR
	Paralabrax maculatofasciatus	Zoob	FR
Rajiformes			
Urolophidae	*Urobatis concentricus*	Zoob	
Tetraodontiformes			
Balistidae	*Balistes polylepsis*	Zoob	FR
	Pseudobalistes naufragium	Zoob	
Diodontidae	*Diodon histrix*	Zoob	
Tetraodontidae	*Canthigaster punctatissima*	Zoob	FR

The relative importance of trophic functional groups (Figure 4B) indicated that the most important group is that of zoobentivores, which represented 59.3% of species (16 species), followed far behind with 25.9% omnivores (7 species), while the rest of the groups had a value of 3.7% (1 species). The study site that presented the greatest richness of fish was El Saladito (18 species). *Haemulon maculicauda* was the only species observed in all five of the studied localities, and only five species were observed between 3 and 4 locations (Table 2).

Table 2. Species composition of fish associated with *Sargassum* forests in the five localities of study at La Paz Bay: Botafuego (B), Punta Coyote (PC), El Saladito (ESl), El Sauzoso (ESz), El Quelele (EQ). x: present, blank space: absent

Species	B	PC	ESl	ESz	EQ
Muraena lentiginosa			x		
Synodus evermanni			x		
Mugil spp.			x	x	
Chaetodon humeralis			x	x	
Cirrithus rivulatus	x				
Gerres cinereus				x	x
Haemulon maculicauda	x	x	x	x	x
Haemulon sexfasciatum	x		x	x	
Hoplopagrus guentherii	x				
Halichoeres dispilus				x	x
Halichoeres notospilus			x		
Thalassoma lucasanum	x	x	x		
Labrisomus xanti			x		x
Holacanthus passer	x		x		
Pomacanthus zonipectus			x		
Abudefduf troschelii		x			x
Stegastes rectifraenum	x	x	x		
Stegastes flavilatus	x		x		
Stegastes acapulcoensis			x		
Epinephelus labriformis	x	x	x		
Mycteroperca rosacea				x	
Paralabrax maculatofasciatus				x	x
Urobatis concentricus			x		
Balistes polylepsis		x		x	
Pseudobalistes naufragium			x	x	
Diodon histrix	x	x	x		x
Canthigaster punctatissima	x				
Total fish taxa	11	7	18	10	7

Table 3. Species composition of epibenthic invertebrates associated with *Sargassum* forests at La Paz Bay. Mobility, Trophic Functional Group (TFG) and their importance is present. ? and blank space: no information published, S: sessile, M: mobile, P: pivoting, Filt: filter feeding, Oppo: opportunistic, Detr: detritivorous, Cora: corallivorous, Bioe: bioerosionator, Herb: herbivorous, Omni: omnivorous, Zoob: zoobentivorous, FR: national fishery resource, OE: ornamental species

Phylum/Class	Species	Mobility	TFG	Importance
Annelida	*Bispira rugosa monterea*	S	Filt	
	Phragmatopoma californica	S	Filt	
	Spirobranchus giganteus	S	Filt	
Crustacea	*Neaxius vivesi*	M, S	Oppo	
	Dardanus sinistrepes	M	?	
	Balanus sp.	S	Filt	
	Hermit crab	?	?	?
Cnidaria	Anemone sp. 1	S	Filt	
	Anemone sp. 2	S	Filt	
	Anemone sp. 3	S	Filt	
	Anemone sp. 4	S	Filt	
	Calliactis variegata	S, P	Filt	
	Eugorgia spp.	S	Filt	OS
	Pacifigorgia sp.	S	Filt	
	Pocillopora spp.	S	Filt	
	Porites panamensis	S	Filt	
Hydrozoa	*Aglaophenia* spp.	S	Filt	
Echinodermata	*Nidorellia armata*	M	?	OS
	Phataria unifascialis	M	Detr	OS
	Acanthaster planci	M	Cora	
	Eucidaris thouarsii	M	Bioe	OS
	Arbacia incisa	M	Omni	
	Tripneustes depressus	M	Herb	
	Echinometra vanbruntii	M	Bioe	FR, PO
	Ophiocoma aethiops	M	Omni	
	Ophiotrix spiculata	M	Omni	
	Ophionereis annulata	M	?	
	Holothuria kerfersteinii	M	?	
	Isostichopus fuscus	M	Detr	FR, PO
Mollusca	*Conus* spp.	M	?	
	Elysia diomedea	M	?	
	Serpulorbis euriciformis	S	Filt	
	Phyllonotus erythrostoma	M	Zoob	FR
	Snail sp. 1	M	?	?
	Modiolus capax	S	Filt	FR
	Pinctada mazatlanica	S	Filt	FR, PO
	Atrina maura	S	Filt	FR
Porifera	Sponge sp. 1	?	Filt	?
	Total of Taxa	**38**		

Epibenthic invertebrates. Thirty eight taxa of epibenthic invertebrate were sampled, grouped into six phyla (Table 3). In the 170 quadrants sampled (170 m^2) during the study period, 633 epibenthic invertebrates were quantified (average 3.7 individuals m^{-2}).

The most abundant phylum according to their relative abundance were the Cnidaria with 40.6% (257 individuals), followed by the rest of the taxa with a relative abundance no greater than 19.7% (Table 4). The taxa that grouped the greatest number of species were echinoderms (12 species), cnidarians (10 species) and molluscs (8 species, Table 4).

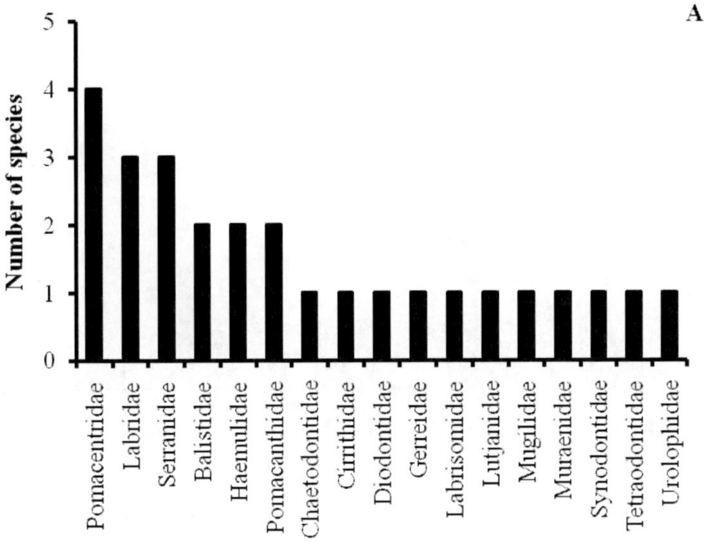

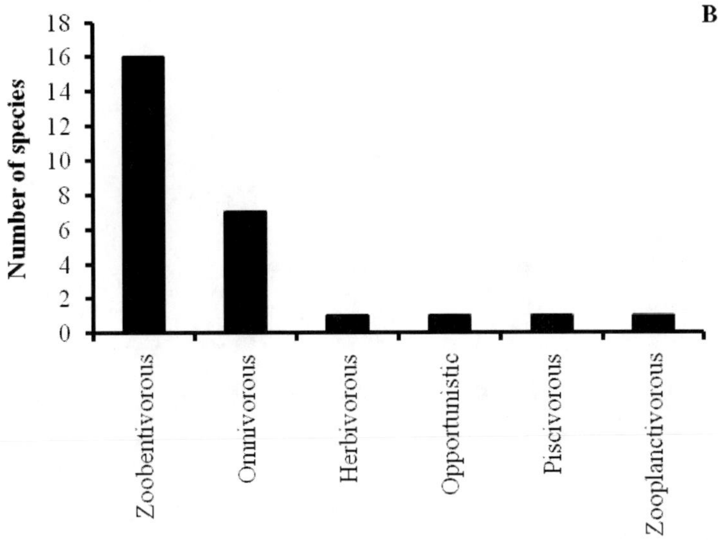

Figure 4. Distribution of the richness of fish species associated with *Sargassum* forests at La Paz Bay, in relation to the family (A) and functional trophic group (B).

Table 4. Abundance (total number of individuals), relative abundance and rank of the relative importance of the taxa of epibenthic invertebrate communities associated with *Sargassum* forests at La Paz Bay

Taxa	Abundance	Relative Abundance	Rank	Species richness
Annelida	125	19.7	2	3
Cruastacea	78	12.3	5	4
Cnidaria	257	40.6	1	10
Echinodermata	89	14.1	3	12
Mollusca	83	13.1	4	8
Porifera	1	0.2	6	1

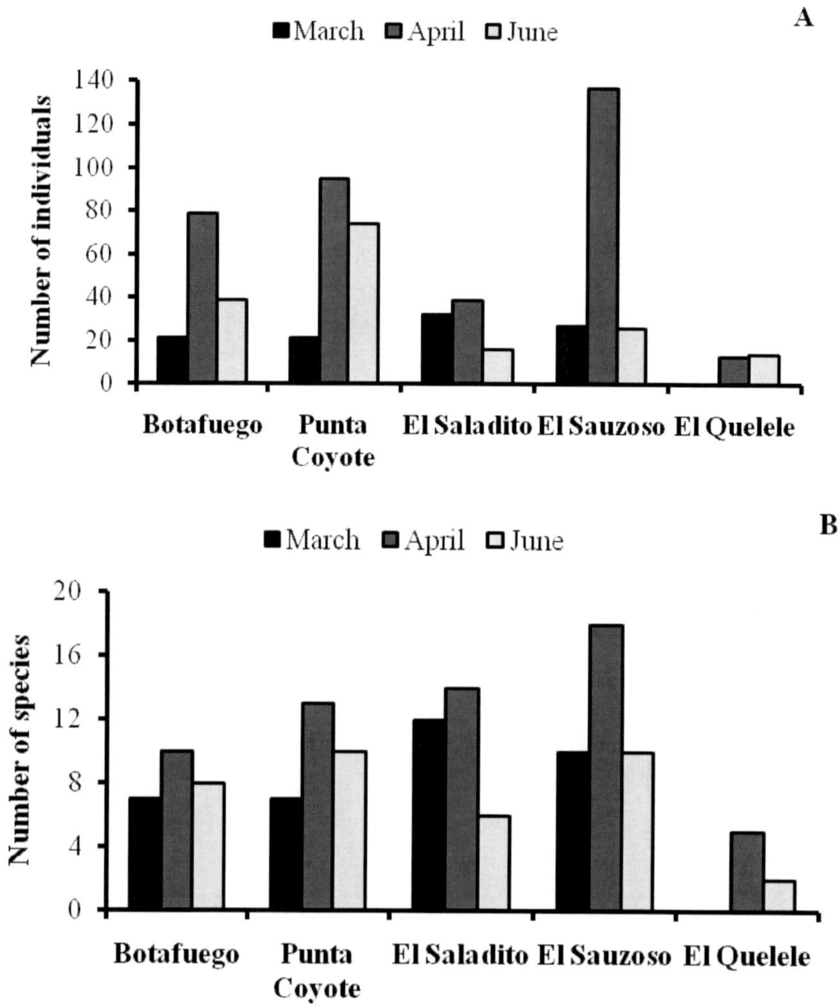

Figure 5. Spatial-temporal variation of the community structure of epibenthic invertebrates associated with *Sargassum* forests at La Paz Bay. A: abundance, B: richness of species

The epibenthic macro invertebrate community showed a pattern of temporal variation, with lower values of abundance and species richness at the beginning and the end of the study (March and June), and higher values in the middle of the study (April, Figure 5A and 5B). This pattern of variation was similar in all five studied sites, suggesting that there was no significant change spatially (Figure 5A and B). However, the community (composition and density of each species) may or may not be the same among sites (Table 5).

In The Quelele, community structure presented a pattern of maximum and minimum values at the beginning and the end of the study, respectively, because this locality was sampled from April. In the case of micro invertebrates, epibionts associated with *Sargassum* were composed of 17 taxa from 7 phyla (Table 6). In this group, the Amphipods are the most important group, followed by copepods.

Table 5. Density (number of individuals/m²) of epibenthic invertebrates associated with *Sargassum* forests at La Paz Bay. Blank space: locality not sampled, M: March, A: April; J: June

	Botafuego			Punta Coyote			El Saladito			El Sauzoso			El Quelele		
n =	10	15	10	15	15	10	10	15	10	10	15	10		15	10
Species	**M**	**A**	**J**	**M**	**A**	**J**	**M**	**A**	**J**	**M**	**A**	**J**	**M**	**A**	**J**
Bispira rugosa monterea	0.0	0.0	0.1	0.0	0.3	0.3	0.0	0.4	0.1	0.1	1.9	0.3		0.1	0
Phragmatopoma californica	0.0	0.0	0	0.5	0.9	2.8	0.2	0.5	0	0.0	0.1	0		0.0	0
Spirobranchus giganteus	0.0	0.0	0	0.0	0.4	0	0.7	0.1	0	0.0	0.0	0		0.0	0
Neaxius vivesi	0.5	0.0	0	0.5	0.3	0	0.0	0.3	0.1	0.3	1.1	0.7		0.1	0
Dardanus sinistrepes	0.0	0.1	0	0.0	0.0	0.4	0.0	0.0	0.1	0.0	0.1	0		0.0	0
Balanus sp.	0.0	0.1	0	0.1	0.0	0	0.0	0.0	0	0.1	0.2	0		0.0	0
Hermit crab	0.2	0.0	0	0.1	0.0	0.1	0.2	0.3	0	0.0	0.0	0		0.0	0
Anemone sp. 1	0.0	0.0	0	0.0	0.0	0	0.0	0.0	0	0.0	0.0	0.1		0.0	0
Anemone sp. 2	0.0	0.0	0	0.0	0.0	0	0.0	0.1	0	0.0	0.0	0		0.1	0
Anemone sp. 3	0.0	0.0	0	0.0	0.0	0	0.0	0.0	0	0.6	0.1	0		0.0	0
Anemone sp. 4	0.0	0.0	0.2	0.0	0.0	0	0.1	0.0	0	0.0	0.0	0		0.0	0.1
Calliactis variegata	0.0	0.7	0	0.0	0.0	2.5	0.0	0.0	0.3	0.0	0.3	0		0.0	0
Eugorgia spp.	0.0	0.0	0	0.0	1.3	0.8	0.0	0.0	0.8	0.0	0.0	0		0.0	0
Pacifigorgia sp.	0.0	0.0	0	0.1	0.1	0	0.0	0.0	0	0.0	0.0	0		0.0	0
Pocillopora spp.	0.0	0.1	0.1	0.0	0.4	0.1	0.0	0.0	0	0.0	0.0	0		0.0	0
Porites panamensis	0.7	2.7	2	0.1	1.9	0.2	0.1	0.0	0	0.5	2.6	0.6		0.0	0
Aglaophenia spp.	0.3	0.0	0	0.0	0.0	0	0.0	0.0	0	0.0	0.0	0		0.0	0
Nidorellia armata	0.0	0.0	0	0.0	0.0	0	0.0	0.1	0	0.0	0.0	0.1		0.0	0
Phataria unifascialis	0.0	0.0	0	0.0	0.2	0.1	0.0	0.0	0.2	0.0	0.0	0		0.0	0
Acanthaster planci	0.0	0.0	0.1	0.0	0.0	0	0.0	0.0	0	0.0	0.0	0		0.0	0
Eucidaris thouarsii	0.0	0.1	0.8	0.0	0.0	0	0.1	0.0	0	0.0	0.4	0.1		0.0	0
Arbacia incisa	0.0	0.9	0.5	0.0	0.0	0	0.0	0.0	0	0.1	0.1	0.3		0.1	0
Tripneustes depressus	0.0	0.1	0.1	0.0	0.0	0	0.1	0.0	0	0.2	0.1	0.2		0.0	0
Echinometra vanbruntii	0.0	0.0	0	0.0	0.0	0	0.0	0.0	0	0.4	0.4	0		0.0	0
Ophiocoma aethiops	0.0	0.0	0	0.0	0.0	0	0.1	0.1	0	0.0	0.4	0		0.0	0
Ophiotrix spiculata	0.0	0.0	0	0.0	0.0	0	0.0	0.0	0	0.0	0.0	0.1		0.0	0
Ophionereis annulata	0.0	0.0	0	0.0	0.0	0	0.0	0.1	0	0.0	0.3	0		0.0	0
Holothuria kerfersteinii	0.0	0.0	0	0.0	0.0	0	0.0	0.1	0	0.1	0.1	0		0.0	0
Isostichopus fuscus	0.0	0.0	0	0.0	0.1	0	0.0	0.0	0	0.0	0.0	0		0.0	0
Conus spp.	0.0	0.0	0	0.0	0.0	0	0.1	0.0	0	0.0	0.0	0		0.0	0
Elysia diomedea	0.0	0.0	0	0.0	0.0	0	1.2	0.1	0	0.3	0.9	0		0.5	1.3
Serpulorbis euriciformis	0.0	0.1	0	0.1	0.1	0	0.1	0.1	0	0.0	0.1	0		0.0	0
Phyllonotus erythrostoma	0.1	0.0	0	0.0	0.0	0	0.0	0.0	0	0.0	0.0	0		0.0	0
Snail sp. 1	0.2	0.4	0	0.0	0.4	0	0.2	0.1	0	0.0	0.0	0		0.1	0
Modiolus capax	0.0	0.0	0	0.0	0.0	0	0.0	0.1	0	0.0	0.0	0		0.0	0
Pinctada mazatlanica	0.0	0.0	0	0.0	0.1	0.1	0.0	0.0	0	0.0	0.1	0.1		0.0	0
Atrina maura	0.0	0.0	0	0.0	0.0	0	0.0	0.0	0	0.0	0.1	0		0.0	0
Sponge sp. 1	0.1	0.0	0	0.0	0.0	0	0.0	0.0	0	0.0	0.0	0		0.0	0

In the 80 sampled quadrants of 625 cm², we found 242,654 individuals throughout the entire study. The proportion per group indicated that the crustaceans were the best represented (77%), followed by molluscs (16%), and plathelmintha (4%), with other taxa under 1% (Figure 6). However, once we evaluated the dominant taxa (crustacean and molluscs) we found that Amphipods accounted for more than half of the specimens in all of the samples followed by copepods and gastropods and the rest of the taxa were below 4.3% (Figure 6) of the total abundance. No spatial differences were found among quadrants ($H_{(7,\ 1360)} = 7.4$, $p = 0.4$), and the abundance ranged from 2362 individuals to 3564 individuals, with the highest value found in quadrant 4 4 (3564 individuals ± 3370 individuals) and the lowest in quadrant 5 (2362 individuals ± 1998 individuals).

Table 6. Abundance (total number of individuals), relative abundance and rank of the relative importance of epifauna community associated with *Sargassum* forests in El Sauzoso, La Paz Bay

Taxa	Abundance	Relative abundance (%)	Rank
Arthropoda (Crustacea)			
Amphipoda	131429	54.2	1
Isopoda	3327	1.4	6
Tanaidacea	936	0.39	13
Brachyura	1146	0.47	11
Caridea	7283	3	5
Anomura	2543	1	7
Copepoda	39061	16.1	2
Cirripedia	150	0.06	17
Annelida			
Polychaeta	2320	0.96	8
Mollusca			
Gastropoda	38701	15.9	3
Bivalvia	541	0.22	15
Cnidaria			
Hydrozoa	1196	0.49	10
Echinodermata			
Ophiuroidea	558	0.23	14
Holothuroidea	1049	0.43	12
Echinoidea	271	0.1	16
Platyhelminthes			
Turbellaria	10538	4.3	4
Ectoprocta			
Gymnolaemata	1605	0.7	9

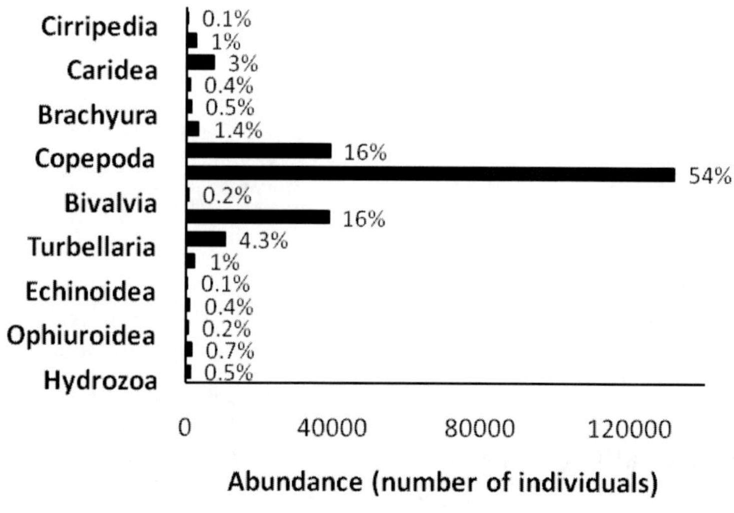

Figure 6. Number of individuals per major taxa sampled of epibionts and the proportion represented from the overall association with *Sargassum* forests in El Sauzoso, BCS.

We have found strong temporal differences in the relative abundance (Figure 7), but not in the species composition, which was consistent over time. The Kruskal-Wallis analysis showed significant differences between the months in the richness ($H_{(9,\ 1360)} = 275.7$, $p = 0.000$) with a consistent species richness trend in the sampling period. The maximal species abundance were observed in March (8025 individuals; Figure 7) and the minimum in August-September (148 individuals and 345 individuals, respectively).

Using the multiple comparison analysis of the probability, we found a consistent trend rather than abundance trends. When the abundance increased (September to April), the differences in the values became significant.

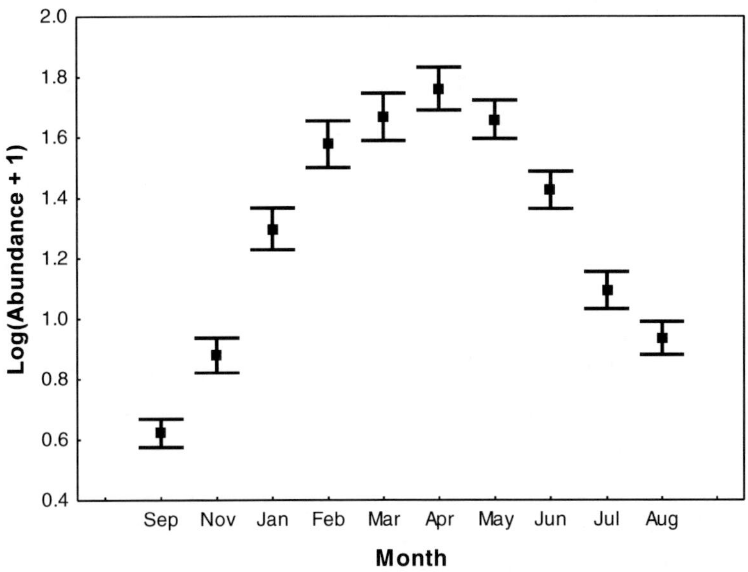

Figure 7. Monthly variation in the epifauna abundance (number of individuals log transformed) associated with *Sargassum* forests in El Sauzoso, BCS. Mean (■); standard error (vertical lines).

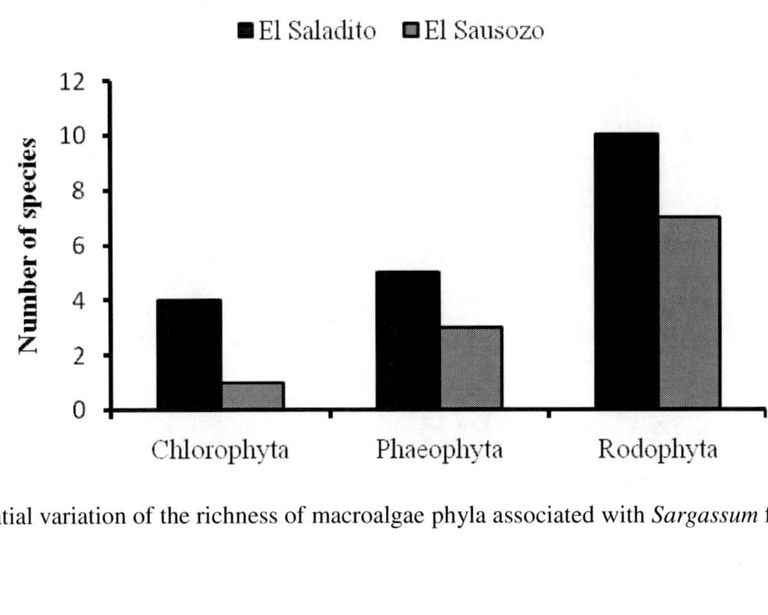

Figure 8. Spatial variation of the richness of macroalgae phyla associated with *Sargassum* forests at La Paz Bay.

Seaweeds. In June 2009, a total of 23 species of macroalgae, grouped in 3 phyla were found living among *Sargassum* forests (Table 7). The most common species observed in the cespitosa layer below the canopy were species of the genus *Sargassum*, *Padina*, *Dictyota flabellata*, *Ulva lactuca*, *Hydroclathrus clathrahrtus* and *Codium cuneatum*. The phylum Rhodophyta (red algae) was the most diverse, with 13 species, followed by the phyla Chlorophyta (green algae) and Phaeophyta (brown algae) with five species in each. Differences in species composition of macroalgae associated with *Sargassum* forests were observed between the locality El Saladito and The Sauzoso (Table 6).

Table 7. Species composition of macroalgae associated with *Sargassum* forests at La Paz Bay. x: present, blank space: absent

Phylum/Species	El Saladito	El Sausozo
Chlorophyta		
Cladohora columbiana	x	
Codium cuneatum		x
Enteromorpha sp.	x	
Halicystis ovalis	x	
Ulva lactuca	x	
Phaeophyta		
Dictyota flabellata	x	x
Hydroclathrus clathrathrus	x	
Padina mexicana	x	
Padina caulescens	x	x
Sargassum horridum	x	x
Rodophyta		
Amphiroa valonioides	x	x
Antithamnionella elegans	x	
Centroceras clavulatum	x	
Digenea simplex	x	
Gracilaria crispata	x	x
Gracilaria pachydermatica	x	x
Hypnea cervicornis		x
Laurencia spp.		x
Liagora californica	x	
Neogonolithon spp.	x	
Prionitis kinoensis		x
Pterocladia sp.	x	
Spiridia filamentosa	x	x
Total of taxa	19	11

At both sites, red algae comprised the most diverse taxon (number of species), followed by brown and green algae with fewer species, respectively (Figure 6).

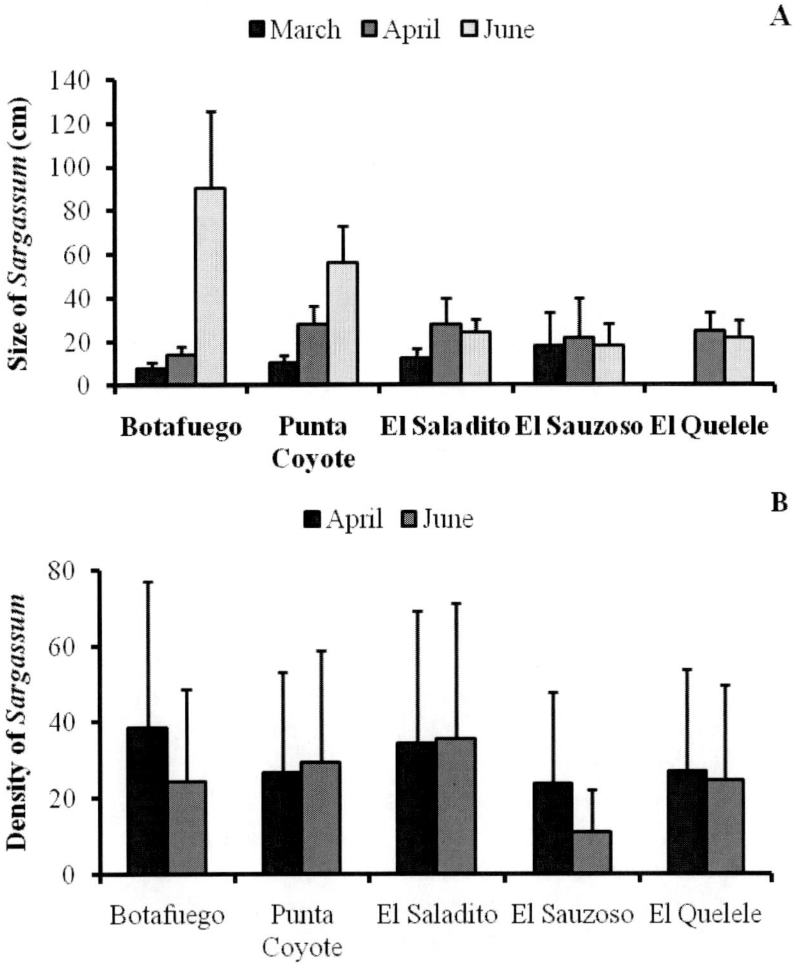

Figure 9. Spatial-temporal variation of size (A) and density (number of fronds/625 cm^2; B) of *Sargassum* forest at La Paz Bay.

During the study period, the size of *Sargassum* showed two patterns of temporal variation (Figure 9A). One was observed in the localities of Botafuego and Punta Coyote, where the size of *Sargassum* increased from lower to higher values during the course of the study.

The second pattern was observed in the remaining three sites, where lower size values of *Sargassum* were recorded at the beginning and end of the study (March and June) and the highest value was recorded during the middle of the study period (April). The difference in growth patterns of *Sargassum* between the localities allows the spatial variation of *Sargassum* size to be recognised (Figure 9A), while the density of *Sargassum* did not show significant variation in time and space (Figure 9B).

Spatial differences. In relation to the spatial comparisons we found that only 2 seaweed species (*Sargassum horridum* and *Padina caulescens*) and *Littorina* sp were significantly present in cover (Figure 10); most of the space was rocks or sand.

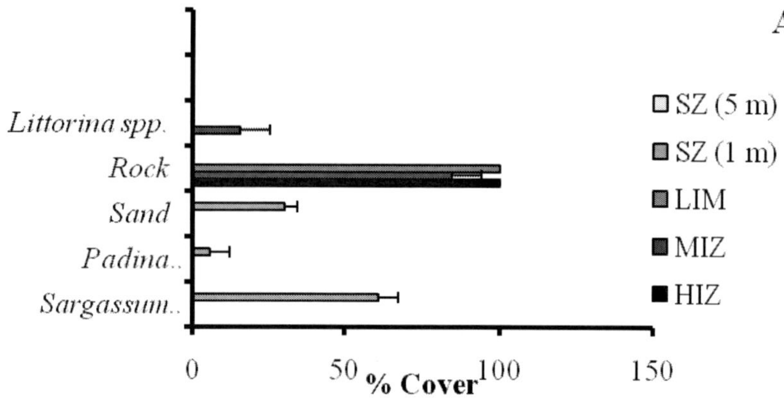

Figure 10. Percent of cover of the seaweeds at different tide levels (High: HIZ, Medium: MIZ, Low: LIM) and sub-tidal areas (1 m SZ and 5 m SZ) from el Sauzoso.

Discussion

Monitoring the marine rocky areas of the world has been performed until very recently [40]; many of the protocols are restricted to a specific area of the coast (see Reef Check methodology at www.reefcheck.org), while others include very wide areas (see PISCO methodology at www.pisco.org).

In all of the previous cases, the methods used were more related to the local conditions and facilities. However, the NAGISA methodology provided a simple method that can be applied in many areas of the world with a reasonable budget.

The results of our evaluation of the biodiversity associated with *Sargassum* forest have assessed many sites along the shore line with this methodology in a couple of days. We found a very strong temporal and spatial component that will need to be considered to refine the monitoring methods to consider the variable components in the tropics and might be considered the most consistent time of year to develop the sampling.

The temporal and spatial variation in sessile animals, seaweeds and the density of *Sargassum* plants reflects the extremely dynamic nature of the system and strongly suggests that differences in recruitment are driving the observed patterns. This is consistent with the observations of Foster *et al.* [7], who found similar trends in Bahía Concepción. In the case of invertebrates, this is one of the first studies to evaluate them all together, as presented in [7]. The species richness of invertebrates was similar to other studies [8] for a single site, but lower in the case of seaweeds [39]. The very high diversity of fish species is consistent with the observations from [1] but a higher number of species were found in our study, possibly because of the number of sites visited.

Sargassum forests harbour several hundred species (see Tables), but their role as a recruitment habitat is very well documented [42] and this is widely known for other areas of the world [2]. In addition to these in classic studies about trophic flux, in the Gulf of California *Sargassum* is considered a relevant component to keep beach sediment and provides the necessary biomass for insects during La Niña years [43].

Conclusion

The *Sargassum* forest is a critical habitat for conservation in the Gulf of California because the habitat is sensitive to environmental and human-induced impacts. Their role in coastal areas are related to species recruitment, fixation of CO_2 and drift material, which are extremely important to sandy beach stability and food sources for insects from terrestrial environments. There are several fisheries related who are close to this ecosystem, and legal protection does not exist to protect this critical habitat for conservation in the Gulf of California.

Acknowledgments

We are grateful to several Marine Biology students of the Programa de Investigación en Botánica Marina at the Universidad Autónoma de Baja California Sur. México, for helping at the field trips. This research was founded through IPN (SIP-20070752) and NaGISA "*Sargassum* forest associated biodiversity from Gulf of California". Gustavo HC thanks to "Comisión de Operación y Fomento de Actividades Académicas del IPN (COFAA)" and to the program "Estímulo al Desempeño de los Investigadores del IPN (EDI)" for the fellowship granted.

References

[1] O. Aburto-Oropeza, and E.F. Balart (2001). *Marine Ecology*, 22 (4), 283-305.

[2] F.P.P. Leite, M.O. Tanaka and R.S. Gebara (2007). *Ecology*, 67 (2), 215-221.

[3] G. Hernández-Carmona, M.M. Casas-Valdez, C. Fajardo-León, I. Sánchez-Rodríguez and E. Rodríguez-Montesinos (1990). *Investigaciones Marinas CICIMAR*, 5 (1), 11-18.

[4] T. Terawaki, K. Yoshiwaka, G. Yoshida, M. Uchimura, and K. Iseki, (2003). *Marine Pollution Bulletin*, 47, 198-201.

[5] P.O. Ang, Jr., (1986). *Marine Ecology Progress Series*, 28, 9-19.

[6] E. Ávila, N. Blancas-Gallangos, R. Riosmena-Rodríguez, and L. Paul-Chávez (2009). *Journal of the Marine Biological Association of the United Kingdom*, 1-10.

[7] M.S. Foster, L.M. McConnico, L. Lundsten, T. Wadsworth, T. Kimball, L.B. Brooks, M. Medina-López, R. Riosmena-Rodríguez, G. Hernández-Carmona, R. Vasquez-Elisando, S. Johnson, and D.K. Steller, (2007). *Ciencias Marinas*, 33 (4), 367-384.

[8] T.L. Cassaza, and S.W. Ross (2008). *Fishery Bulletin*, 106 (4), 348-363.

[9] S.D. Haecker and L.P. Madin, (1991). *Marine Ecology Progress Series*, 70, 143-155.

[10] K. Connolly, (2004). An Argument for inclusion of *Sargassum* as essential fish habitat in the dolphin and wahoo fishery management plan. Consultation Process for the south Atlantic region, 1-23.

[11] M.M. Casas-Valdez, I. Sánchez-Rodríguez, and G. Hernández-Carmona, (1993). *Investigaciones Marinas CICIMAR*, 8 (2), 61-69.

[12] Pacheco-Ruíz, J.A. Zertuche-González, A. Chee-Barragán and V. Blanco-Betancourt, (1998). *Botanica Marina*, 41, 203-208.

[13] R.M. McCourt, (1984). *Journal of Experimental Marine Biology and Ecology*, 74, 141-156.

[14] J. Espinoza-Avalos, and H. Rodríguez-Garza (1985). *Ciencias Marinas*, 11(3), 115-120.

[15] M. Muñeton-Gómez, (1989). *Investigaciones Marinas CICIMAR*, 4 (2), 257-266.

[16] R.A. Nuñez-López, and M.M. Casas-Valdez, (1996). *Revista de Biología Tropical*, 44 (2), 455-464.

[17] G. Hernández-Carmona, (1985). *Investigaciones Marinas CICIMAR*, 2 (1), 30-45.

[18] Marín, M., S. Carrillo, H. Hernández, and A. Monroy (2003). *Cuban Journal of Agricola Science*, 37 (2), 119-123.

[19] M. Casas-Valdez, H. Hernández-Contreras, A. Marín-Álvarez, R.N. Águila-Ramírez, C.J. Hernández-Guerrero, I. Sánchez-Rodríguez, and S. Carrillo-Domínguez, (2006a). *Revista de Biología Tropical, 54 (*1), 83-92.

[20] M. Casas-Valdez, G. Portillo-Clark, N. Águila-Ramírez, S. Rodríguez-Astudillo, I. Sánchez-Rodríguez, and S. Carrillo-Domínguez, (2006b). *Revista de Biología Marina y Oceanografía*, 41 (1), 97-105.

[21] A.R. Jiménez Illescas M. Obeso Nieblas, and D.A. Salas de León (1997).Oceanografía física de la Bahía de La Paz, B.C.S.. La Bahía de La Paz, Investigación y Conservación. (J. Urban-Ramírez and Mauricio Ramírez, eds.). Universidad Autónoma de Baja California Sur, Centro Interdisciplinario de Ciencias Marinas, Scripps.

[22] H.A. Alvarez, (1997). Oceanografia geológica de la Bahía de La Paz, B.C.S. La Bahía de La Paz, Investigación y Conservación. (J. Urban-Ramírez and Mauricio Ramírez, eds.). Universidad Autónoma de Baja California Sur, Centro Interdisciplinario de Ciencias Marinas, Scripps.

[23] Martínez-López, R. Cervantes-Duarte, A. Reyes-Salinas, and J.E. Valdez-Holguin. (2001). *Hidrobiologica*11: 45-52.

[24] R. Rigby, T. Kato, R. Riosmena-Rodríguez, and K. Iken, (2007). In, Rigby, R., Iken, K. and Shirayama, Y. (eds). Sampling biodiversity in coastal communities. NaGISA protocols for seagrasses and macroalgal habitats, Kyoto University Press, Kyoto, pp. 17-21.

[25] A.M. Keen, (1971*). Sea shells of tropical West America*. Marine Molluscs from Baja California to Peru. 2nd edition. Stanford University Press, U.S.A., 1065 pp.

[26] R. Smith, and J. Carlton (1975) Light's manual: Intertidal Invertebrates of the Central California Coast. University of California Press, USA.

[27] R. Brusca, (1980). *Common intertidal invertebrates of the Gulf of California*. 2nd Edition. University of Arizona Press, Arizona, 513 pp.

[28] S.I. Salazar, J.A. De León, and H. Salaices, (1988) Poliquetos (Annelida: Polychaeta) de México. Universidad Autónoma de Baja California Sur, La Paz, B.C.S., México.

[29] M. Hendrick, and J. Poutiers (1997) Guía para la identificación de especies para los fines de la pesca. Pacífico Centro-Oriental: Plantas e Invertebrados. FAO, Roma.

[30] O. Aburto-Oropeza, E. Sala, G. Paredes, A. Mendoza, and E. Ballesteros (2007). *Ecology*, 88 (9), 2220-2228.

[31] D.A. Thomson, L.T. Findley, and A.N. Kerstitch, (1979). Reef fishes of the Sea of Cortez: the rocky shore fishes of the Gulf of California. John Wiley and Sons, 302 pp.

[32] D.R. Robertson, and G.R. Allen, (2008). Shorefishes of the tropical Eastern Pacific online information system. Version 1.0. Smithsonian Tropical Research Institute. Balboa, Panamá. www.neotropicalfishes.org/sftep, www.stri.org./stefp

[33] D.W. Gotshall, (1998). Sea of Cortez marine animals. A guide to the common fishes and invertebrates. Baja California to Panama. Sea Challengers, U.S.A., 110 pp.

[34] D.W. Gotshall, (2005). Guide to marine invertebrates. Alaska to Baja California. 2nd Edition. Sea Challengers, U.S.A., 117 pp.

[35] Kerstitch, (1989). *Sea of Cortez marine invertebrates. A guide to the pacific coast*, México to Ecuador. Sea Challengers, U.S.A., 112 pp.

[36] Kerstitch, and H. Berstch, (2007). *Sea of Cortez marine invertebrates*. 2nd Edition. Sea Challengers, U.S.A., 124 pp.

[37] W. Fischer, F. Krupp, W. Schneider, C. Sommer, K.E. Carpenter, and V.H. Niem, (1995). Guía Fao para la identificación de especies para los fines de la pesca. Pacífico centro-oriental. FAO, Rome, Vol. 1: 1-646 pp.

[38] Hermosillo, D.W. Behrens, and E. Ríos-Jara, (2006). Opistobranquios de México. Guía de babosas marinas del Pacífico, Golfo de California y las islas oceánicas. CONABIO, México, 143 pp.

[39] R. Riosmena-Rodriguez and L. Paul-Chávez (1997). En: La Bahía de La Paz. Investigación y Conservación (J. Urban-Ramírez and Mauricio Ramírez, eds.). pp. 292-397 UABCS-CICIMAR-SCRIPPS.

[40] F. Wallastein, A.I. Neto, V.A. Nuno, and I. Tittley (2008). *Journal of Coastal Research* 24: 244-249.

[41] S. Alvarez Borrego, (2010) Physical, chemical, and biological oceanography of the Gulf of California. En: Richard C. Brusca, Editores, "*The Gulf of California: biodiversity and conservation*". University of Arizona Press. 24-48 p.

[42] D. Wells and J.R. Rooker, (2004). *Bulletin of Marine Science*, 74 (1), 81-99.

[43] W. B., Anderson and G. A. Polis. (1998). *Oikos* 81:75–80.

In: Invertebrates: Classification, Evolution and Biodiversity ISBN: 978-1-62618-789-4
Editor: Rafael Riosmena-Rodriguez

Chapter 10

Topological and Fractal Approaches to Systematics and Biodiversity

Valeria V. Isaeva[*]
A.N. Severtsov Institute of Ecology and Evolution of the Russian Academy of Science, Moscow, and A.V. Zhirmunsky Institute of Marine Biology, Far East Branch of Russian Academy of Sciences, Vladivostok, Russia

Abstract

This chapter presents a description and analysis of metazoan morphogenesis during evolution and development using topological and fractal concepts and methodology. The topological approach makes it possible to consider the succession of spherical surgeries, which change the topological genus of the surface. Topological singularities inevitably arising and transforming during early development destroy the preexisting pattern of symmetry. Symmetry breaking determined both genetically and epigenetically plays a key role in biological morphogenesis. Some developmental events are interpreted in terms of symmetry breakdown and related to well-known mathematical theorems. In addition to such classical symmetry of biological forms as rotating (radial), mirror (bilateral) and translational ones, the scale symmetry is widespread in fractal morphogenesis. In some animal taxa an increase of genus of the body surface up to high values is evident; these patterns evolved on the basis of fractal systems. Colonial animals with repeating modular elements are fractal organisms. For example, colonial rhizocephalan barnacles (Crustacea: Rhizocephala) have irregular fractal organization at the parasitic stage of their life cycle due to asexual reproduction by budding. Biological fractal structures as irregular forms exhibit fluctuating asymmetry. To evaluate the relationship between order and chaos in the structure of branching fractal channels of the gastro-vascular system in the jellyfish *Aurelia aurita* and the larval tracheal gills in the mayflies *Siphlonurus immanis* and *Parameletus chelifer* these patterns in symmetrical parts of the organisms were compared with deterministic fractal trees. During ontogenesis, the spatial and temporal variability of the branching systems increases and results in the breaking of the radial symmetry in each individual medusa and bilateral

[*] Corresponding author: E-mail address: vv_isaeva@mail.ru.

symmetry in mayfly larvae. The epithelial surface is an interface between the internal medium of an organism and the outside environmental medium; topological and fractal transformations during metazoan evolution and development increase this interface so ensuring better adaptation of organism to the environment. Fractal structures as well as toroid forms can be considered as a functionally optimized design in Metazoa. Topological methodology reveals a certain set of topological rules constraining and directing biological morphogenesis during evolution and development. The detailed analysis of topological and fractal form dynamics in phylo- and ontogenesis may provide additional possibilities for biological description and classification.

Introduction

Developmental and evolutionary modifications of topological patterns are discrete, qualitative steps in biological morphogenesis. Discrete character of biological morphogenesis involves topological discontinuities, topological catastrophes [1-3]. Since topology operates with the most general properties of spaces as mathematical subjects topological language and models can provide strict and adequate description and modeling of real biological objects and processes. Following R. Thom, we applied topological concepts and some theorems to describe developmental and evolutionary form dynamics involving symmetry breaking [4-10]. This chapter includes the topological description of morphogenetic patterns in onto- and phylogenesis, relationship between local and global orders, topological dependence and topological constraints of biological morphogenesis.

Topological singularities inevitably arising and transforming during early development destroy the preexisting pattern of symmetry [5, 7-10]. A topological approach makes it possible to consider a sequence of rearrangements which transform the body symmetry and alter the topological genus of the surface in the development and evolution of multicellular animals. Symmetry breaking is essential and inevitable in biological morphogenesis. Developmental symmetry transformations are determined genetically and epigenetically; some biological mechanisms of axial pattern establishment, segmentation and -right asymmetry are considered. To interpret the dynamics of symmetry transformations in metazoan morphogenesis some concepts of topology and fractal geometry are applied.

Topological Transformations in Evolution and Development

To translate traditional anatomical descriptions of metazoan morphology into topological language the external shape of a metazoan organism is modeled as a set of smooth, closed, orientable surfaces formed by epithelial layers [4, 8, 9]. Epithelial cell layers are characterized by morphological and functional connectivity, closeness of an intact surface, and apical–basal polarity, i.e., outer–inner anisotropy. Epithelial cells form cohesive cell sheets; the structural and functional connectivity of an epithelial layer is ensured by the system of specialized intercellular contacts that integrate cells into a united morphological and functional entity [11, 12]. Cell contacts maintain a closed cell layer like a whole.

The transition of cytoskeleton (cytomatrix) into integrated “histoskeleton” by means intercellular integration through cell-cell contact zones gives the possibility of coordinated cell behaviour. The establishment of apical-basal polarity of epithelial cells generates two surfaces of an epithelium with different components and functions, whereas asymmetry across the apical surface is critical for organization and pattern formation at the levels of tissues and organs [see 13, 14].

According to the theorem of elementary topology, any closed surface in three-dimensional space is homeomorphic (topologically equivalent) to the sphere with a certain number (p) of handles [see 4-10]. If there are no topological surgeries (cutting and gluing of epithelial sheets), the genus of the surface (p) is a topological invariant, and any geometrical deformations such as surface curvature, linear and angle values are not essential. The closed surfaces of the genus $p = 0$ (sphere), $p = 1$ (torus), $p = 2$ (double torus, or “pretzel”) and so on give a topological classification. The topological differences between these surfaces are fundamental and qualitative [2].

An adult metazoan organism topologically is a sphere with a set of handles. Metazoan morphogenesis may be represented as topological modification(s) of epithelial surfaces of spherical or toroid shape. Topological handles in biological objects are usually realized as channels (or canals, holes), for example, the digestive tube. Topological transitions in metazoan development occur due to topological surgeries realized by “cutting” and “gluing” of epithelial sheets [4], as “self-wounding and healing” [15]. The topological surgeries in epithelial morphogenesis occur locally and involve complex processes at the cellular level: sheet disintegration followed by cell adhesion and cytoskeletal cooperation, which give rise to a newly formed cell sheet (or sheets). Local topological surgeries inevitably lead to global topological modifications of biological forms, as the genus of the surface is a global property.

We can see two kinds of topological surgeries of epithelial surfaces in biological morphogenesis. Spherical surgeries change the genus of the surface. Another kind of topological surgeries changes the connectivity of epithelial layers by the separation of additional closed spherical surfaces from preexisting ones or by the fusion of preexisting surfaces. The developmental modifications of connectivity do not change the topological pattern of the outer surface of animal body, so the spherical surgeries changing the genus of the surface are considered here.

Topological Surgeries in Metazoan Evolution

Topological patterns in the evolution and development of Metazoa, from sponges to chordates, are considered by means of the methodology using the genus of the surface as a topological invariant. By this means metazoan morphogenesis may be represented as topological modification(s) of the epithelial surfaces of an animal body. We considered here (Figure 1) only basic topological archetypes of spatial organization for animal taxa, i.e., a plesiomorphic topological design for each taxon, neglecting variations and modifications inside the taxon. We used the interpretation of metazoan phylogeny, which consider three major clades of Bilateria: Deuterostomia, Lophotrochozoa, and Ecdysozoa [see, for example, 16. 17].

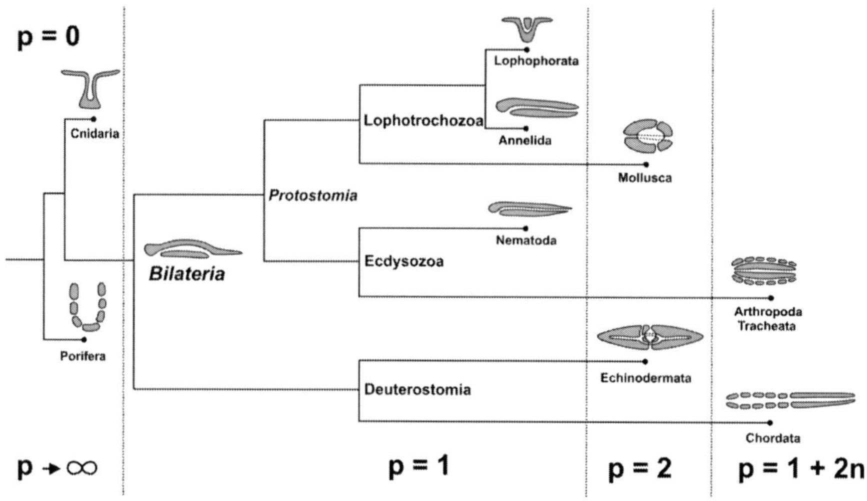

Figure 1. The scheme of main topological body plans in Metazoa using genus of surface (p).

An analysis of topological organization of the closed epithelial layer that covers the outer surface of an organism, regarding it as a continuous envelope formed in most animals by both ectodermal and entodermal (intestinal) epithelia, is applicable to the topological organization of the body surface in metazoan evolution [8-10].

In solitary cnidarians and such lower Bilateria as Acoelomorpha with a blind gut the kind of the surface is $p = 0$. The first important topological surgery in evolution of Bilateria (Triploblastica) is the appearance of a through intestinal tube instead of a blind gut. Animals of many phyla have the stable topological pattern with the through digestive tube without any other through channels, and their external surface is topologically homeomorphic to the torus, $p = 1$ (Figure 1). The appearance of a through gut is a topological modification of great evolutionary importance resulting in the differentiation of digestive tube and better utilization of nutrients from the external medium.

Subsequent evolution led to the formation of the through channels of the coelomic and respiratory systems. The surface of an organism having another additional through channel besides the digestive tube is a surface of genus 2, which is topologically equivalent to a sphere with two handles, double torus or "pretzel". For example, echinoderms exhibit a second through channel, namely the ambulacral system, besides the digestive tube. The spatial organization of the circular ambulacral system connected with the external medium by a canal is topologically equivalent to a through channel [8, 9]. So an archetype of the topological design of an adult echinoderm body can be represented by the double torus. Mollusks exhibit, besides the digestive tube, the second through channel of the coelomic system ($p = 2$). In the mollusk archetype, this through connection via coelomoducts was the result of fusion of paired coeloms.

The next evolutionary level of topological design in animals was realized through the development of the respiratory system. The system of paired tracheal tubules with lateral connections (of ectodermal origin) in higher terrestrial arthropods or paired gill slits (of entodermal origin) in a chordate archetype evolved (Figure 2). Accordingly, a parallelism of topological and functional innovations in the evolution of deuterostomes and some protostomes (higher Ecdysozoa) is obvious as a transition to the surface of the genus $p = 1 +$

$2n$, by adding an even number of channels (handles) of the respiratory system. Archetypical chordate body plan includes a through gut with several pairs of lateral gill slits ($p = 1 + 2n$; Figs. 1, 2) as well as topological design in some Hemichordates (Enteropneusta).

There are various evolutionary modifications of the archetypical body plan in many phyla. For example, the body surface in ascidians (Chordata: Urochordata) is a high-order torus due to the presence of numerous gill openings (Figure 2). Complicated disordered topological patterns evolved on the basis of through fractal-like systems (see below). In Spongia and in some Cnidaria, an increase of genus p up to very high values and the shaping of topologically complicated fractal systems (the irrigation system in sponges and gastro-vascular systems in some Cnidaria) are evident [9, 10].

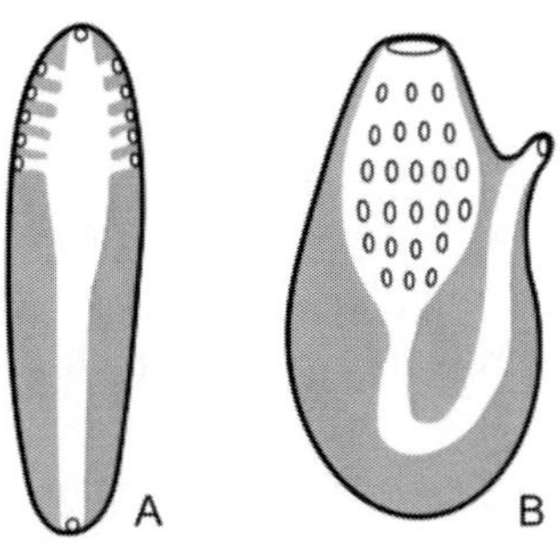

Figure 2. Topological patterns in some Hemichordata and lower Chordata:A – digestive tube with several lateral slit openings (Enteropneusta and chordate archetype),B – digestive tube with numerous pharyngeal openings (Urochordata) [after 9].

Epidermal and intestinal epithelia as well as epithelia of through coelomic and respiratory systems are boundary tissues, an interface between the internal medium of an organism and the outside environmental medium. Topological transformations of the body surface during metazoan evolution led to the increased boundary interface and resulted in better distribution of flows to and from the external medium used as the source of nutrients and oxygen and the sink of excreta, so ensuring better adaptation of organism to the environment.

Topological Modifications in Development

So, sphere, torus, double or high-order tori represent topological forms of metazoan organisms. One can observe a sequence of these forms during embryonic development: the surface of an organism (its epithelial "envelope") undergoes sequential topological surgeries, which change the topological genus of the surface.

The surface of an egg, the outer surface of blastula or early gastrula is the surface of genus 0, which is homeomorphic (topologically equivalent) to a sphere. No modification is observed in topology of the body surface during embryonic development of animals, which has the genus 0 of the surface in adult state (some solitary cnidarians and Acoelomorpha).

In most Metazoa, the development of an organism beginning from the surface of the genus 0 leads to an adult animal which genus of the surface differs from zero. A transition from blind archenteron to through digestive tube is realized during gastrulation by the formation of another opening besides the blastopore (a primary mouth); the last becomes the definitive oral opening in Protostomia, or anal opening in Deuterostomia. The epithelial surface of a postgastrulation embryo or a larva having both oral and anal openings is a surface of genus 1, topologically equivalent to a sphere with one handle or to a torus. In animals having in the adult state the genus of the body surface $p = 1$ (for instance, round and annelid worms), only the transition $p = 0 \rightarrow p = 1$, i.e. transformation of the sphere to the torus occurs during development.

Gastrulation is a topological catastrophe, transition from the sphere with field singularity (singularities) to the torus with homogeneous field; torus is a stable topological object. So a postgastrulation embryo or larva with through intestine tube is a topologically stable object; gastrulation is a way of topological stabilization in embryogenesis.

During echinoderm development, in addition to the digestive tube a second channel of the ambulacral system develops, so the topological design of an adult animal is represented by the double torus ($p = 2$; Figure 1). There are two topological transformations: $p = 0 \rightarrow p = 1 \rightarrow p = 2$ in development of echinoderms (as well as in mollusk development).

In chordate archetype, the outer surface is a surface of genus $p = 1 + 2n$ formed by intestine tube and several pairs of gill slits [8, 9]. Accordingly, during embryonic development of the higher representatives of multicellular animals, the following transformations of the genus of the body surface take place: $p = 0 \rightarrow p = 1 \rightarrow p = 1 + 2n$.

Thus, spherical surgeries during embryonic development result in the transformation of the genus of the surface. Topological surgeries modifying connectivity of germ layers and resulting in the separation of a set of additional closed epithelial surfaces from preexisting ones are another kind of topological catastrophes, for example, enterocoelic formation of mesoderm in Deuterostomia, neurulation and eye cap formation in Chordata [4, 18]. These topological surgeries of connectivity do not change the genus of the outer surface. The spatial organization of evolutionary advanced animals may be represented topologically as an outer epithelial envelope of a certain genus p embracing a number of inner closed epithelial surfaces embedded inside the outer envelope.

The detailed analysis of topological form dynamics in ontogenesis may provide additional possibilities for metazoan description and classification. For example, we have considered the dynamics of numerical characteristics of four types of channels (gut, ambulacral system, hydropore, and pores through the madrepore plate) during embryogenesis, metamorphosis and in adult animals in different classes of Echinodermata [8]. We constructed a cladogram for Echinodermata based only on these pure topological characters in the classes of recent echinoderms, and our "experimental" cladogram occurred similar to "legitimate" zoological classifications, that proved the validity of our topological approach [8].

Symmetry Transformations

Symmetry propagation and symmetry breaking are essential processes in biological morphogenesis [19], in metazoan evolution and development [20]. Symmetry breakdown is one of the fundamental processes of development [21]. Various scalar and vector fields on subcellular, cellular, and supracellular levels of biological organization are manifested in heterogeneous distribution of structural components, in biochemical gradients, in vectorized subcellular transport and other functional activities, in ionic fluxes and accompanying electric fields, in fields of mechanical tensions, in directed cell movement etc. For example, transcellular ion currents in polarized cells generate electric fields as signals and effectors of cell polarity at cellular and tissue levels [22]. In embryos, cell migration pathways during gastrulation and other morphogenetic processes visualize the morphogenetic vector fields. A relationship between local and global orders of a morphogenetic field is a kind of symmetry [7, 8]. It is possible to establish the relation between the local morphogenetic factors and integral developmental patterns and to explain symmetry breakdown in development in topological terms as well as to consider its biological mechanisms.

It has been experimentally shown that the main carrier of molecular morphogenetic information in egg cell is its cortical layer, the thin peripheral layer of ooplasm. The cortical egg field is determined and visualized by polarized distribution of membrane and cytoskeletal components, by electric field and by polarized contraction of the egg cortex during ooplasmic segregation following gamete fusion. For example, the concentration of *bicoid* protein displays an antero-posterior gradient in early Drosophila embryos and controls the zygotic activation of the segmentation gene *hunchback* [see 23]. In insects and other animals, the axial polarity of an oocyte and an egg is manifested in transcellular ion flows, which generates the extracellular electric field [22]. The egg cortical field is morphogenetic one, since it determines the axial pattern of future organism; the structurally and functionally polarized cortical layer is a carrier and indicator of the vector field on the egg surface.

In most metazoans, the animal-vegetal/anterior-posterior axis of the egg and future embryo is established during oogenesis and usually depends on the position of oocyte with respect to its cellular and non-cellular medium. Cytoskeletal systems transmit and integrate information across cells; cytoskeleton rearrangements break cell symmetry [24]. A polarized cellular architecture is fundamental to formation of embryonic axes; physical and biochemical amplification systems break cell symmetry by turning local signals [20]. Both the actin and microtubule cytoskeletal elements work together to convert this initial asymmetry into a global cell polarity [24]. For example, polarization of *Drosophila* oocytes is known to require participation of the actin and microtubule cytoskeleton as well as multiple cellular signaling systems [see 20, 24]. Egg cytoskeleton functions as a global morphogenetic determinant, which directs and maintains the anisotropy of molecular information distributed in the ooplasm determining axis polarity of an egg and a future organism [10]. Different organisms use different mechanisms to establish polarity but all rely mainly on the actin and microtubule cytoskeletal elements [24]. One of the most completely understood polarity generating mechanisms occurs in the *Caenorhabditis elegans* zygote; in this case, anterior-posterior polarity is determined by sperm entry [25].

The inevitability of egg cell polarization as the emergence of the topological singularity of the vector field on a sphere follows from the Poincaré-Hopf theorem [see 5, 7, 8, 10]. According to Poincaré-Hopf's theorem, the occurrence of a vector field singularity (or singularities) on the sphere is inevitable. The vector field on the sphere has at least one singularity (the point where the value of the field becomes zero). This singular point may be either vegetal or animal pole, or both setting the animal-vegetal differences. This singularity breaks the spherical symmetry of the egg, polarizing the egg sphere [5, 8, 10].

The second symmetry breakdown in some animals, at least in chordates, takes place following gamete contact and fusion; this symmetry breaking process involves the flow of cortical actin networks [24]. In chordates, a new singularity besides preexisting animal-vegetal egg poles emerges following sperm penetration, which location determines dorso-ventral axis of the egg and developing embryo. In ascidians, the yellow pigment moves to vegetal pole after sperm penetration and at the following stage forms a yellow crescent-like region somewhat lower the egg equator [26, 27]. So the rearrangement of the egg cytoplasm induced in some chordates by the gamete fusion results in the establishment of the dorsal-ventral polarity of future animal. Spermium penetration triggers the whole cascade of phenomena, resulting in modification of cortex organization [28, 29]. During sperm penetration in echinoderm, the fertilization cone with polymerized actin filaments arises [30]. The spermium plasma membrane is inserted into egg plasma membrane, and the sperm membrane components remains localized at the point of the penetration in sea urchin eggs during egg cleavage [31]. So the local order of sperm penetration is connected to the integral order of the egg, zygote and developing organism. The breaking of initial symmetry in development is inevitable and is programmed both genetically and epigenetically.

Ooplasmic segregation studied in eggs of ascidians [26, 27] and the annelid *Tubifex* [32, 33] as flow of cortical actin networks triggering by the sperm penetration results in the spatial-temporal vector field of the displacement of cortical egg components. The singularity of this vector field is the fixed point of ooplasmic segregation. During ooplasmatic segregation, subcellular, mainly cortical components of an egg are moving with respect to a fixed, stationary point in the course of continuous deformation. We consider ooplasmic segregation as a mapping a ball into itself [5, 8, 10]. According to Brouwer's theorem on fixed point, any continuous map of the ball into itself has a fixed point. The fixed point of continuous deformation of egg sphere is a singular point of the vector field, and presence of this fixed point is inevitable for ooplasmic segregation according to Brouwer's theorem. Thus, the local order of ooplasmic segregation is connected to integral order of the egg [5, 8, 10]. Egg symmetry transformations provide the most important morphogenetic information determining axial pattern of future animal. Thus, ooplasm of an egg and zygote possesses irreversible, long-term epigenetic memory similar to imprinting.

In a cleaving zygote, the spatial organization of blastomeres is a stable feature of taxonomic value, which is evolutionarily and genetically determined. The cleavage results in a pattern of cellular contacts on the surface of an embryo that is a discrete morphogenetic field [5-8]. Each cell of the field has a number of its edges corresponding to the number of neighboring cells. There are only five homogenous discrete fields on the sphere corresponding to five regular polyhedrons (Platonic bodies). It is significant that a cleaving embryo is considered as topological polyhedron, not geometrical one [5, 8, 10]. During synchronous cleavage divisions, only the first four blastomeres can create a homogeneous field on the embryo surface. After third division (8 blastomeres) the field pattern of cellular contacts on the embryo surface inevitably becomes topologically inhomogeneous that means the symmetry breaking of the discrete morphogenetic field. Euler's theorem states that there is a relationship between the numbers of vertices, edges and faces of a polyhedron: if a graph given on two-dimensional sphere, then $V - E + F = 2$, where V, E, F are the numbers of vertices, edges and faces. In hexagonal cell network, presence of cells with another number of sides (for example, with 5 or 7 sides) can be considered as a topological dislocation of homogeneous graph cells [34, 35]. This means that the symmetry of the discrete morphogenetic field is breaking, and the symmetry breaking is a consequence of Euler's theorem for polyhedra [5, 7, 8, 10].

Positional information for a discrete morphogenetic field on the surface of a cleaving zygote is determined by Euler's theorem and by the Gauss–Bonnet's theorem as a relationship between local and integral orders of the field [5, 10]. The number of adjacent cells for any cell of the discrete morphogenetic field is the local order of the field, and the general number of vertices, edges and faces of the given graph is the integral order of the field. So a discrete field on the spherical surface of a cleaving embryo inevitably has topological singularities, and the local singularities determine the integral pattern of a developing organism. Gauss-Bonnet theorem defines a discrete cell curvature (Gauss curvature) [see 5]. If the sum of adjacent faces becomes more than 6, the curvature of the face becomes negative. It was proposed a causal relation between cell negative curvature (which leads to invagination) and following morphogenetic cell movement during gastrulation [5, 7, 8]. At least in some animals, the localization of invagination (or immigration, ingression) during gastrulation coincides with the location of the singularities of negative curvature i.e., with the highest value of the cell field on vegetal surface of embryos, whereas cells of positive curvature are present near animal pole. Cells with a negative and positive curvature inevitably emerge on the surface of an embryo, and the negative curvature determines the localization of invagination or immigration during gastrulation [5-10]. Gastrulation is an extensive rearrangement of embryo cells. There are examples of real morphogenetic vector fields of cell movements during gastrulation with singular points.

The completion of gastrulation is topological transition from the sphere with heterogeneity and singularity of the field to the torus, which is a topologically stable object with homogeneous field [5, 8, 10]. Topological rehandlings are regular processes in ontogeny and in phylogeny and involve the emergence of discontinuities; for example, a local rehandling within the structure of cuticle and epidermis can be treated as a morphological singularity [19].

Planar cell polarity describes the coordinated orientation of cells and cellular structures along an axis within the plane of an epithelial surface; components of planar polarity are involved in oriented cell division and directional cell migration [14]. The insect integument of cotton bug, *Dysdercus intermedius*, consisting of a monolayer of epidermal cells displays planar tissue polarity in the uniform posterior orientation of cell denticles and bristles. In experimentally disturbed integument of the bug, abnormally oriented denticles tend to form small arrays with topological singularities [36, 37]. Nübler-Jung [36] proposed that the normal polarity pattern in an insect segment is the result of "global constraints" as alignment along the segment boundaries and "local constraints" as the tendency of cells to elongate along each other whereas directed cell migration after experimental influences disturbed the denticle orientation.

In a planar cell fields, self-organization of direction fields is observed and results in the emergence of a certain set of topological singularities. Only a defined set of planar topological patterns with a few topological singularities of the direction fields is possible. These patterns demonstrate a topological basis and topological constraints of biological morphogenesis in planar cell fields [10]. Thus, topological information of morphogenetic fields determines some developmental events.

Fractal Morhogenesis in Development and Evolution

The development of branching fractal structures is an important mechanism of metazoan morphogenesis. Various biological structures and processes have been shown to display fractal properties: scale invariance (self-similarity) and fractal dimension [38-40]). Fractal geometry has been proclaimed as a design principle for living organisms [39, 40]. A complex spatial organization of biological fractal structures can be quantified by the fractal dimension as a measure of the degree to which a fractal structure fill the space that provides a quantitative view of the spatial complexity of a structure.

The scale symmetry as self-similarity across scaling is typical feature of fractal morphogenesis.

Complicated and disordered to some extent topological patterns increasing the genus of the surface up to high values in sponges, cnidarians and some other groups are due to the shaping topologically complicated fractal systems. The body surface in these animals is topologically homeomorphic to a high-order torus. These patterns evolved on the basis of irregular fractal systems having both inner anastomoses and connection(s) to the external medium such as the gastro-vascular system in some Cnidaria, the branching gut and excretory system in some Plathelminthes, the tracheal system in terrestrial Arthropoda [9]. Biological evolution of the outer surface topology of the metazoan body led to the enlargement of the boundary interface.

Colonial animals with repeating modular elements are fractal organisms. Among colonial metazoans, parasitic rhizocephalan barnacles (Crustacea: Cirripedia: Rhizocephala) have irregular fractal organization at endoparasitic stage of their life cycle. There is a chaotic organization of the rhizocephalan body at this stage, including the loss of limbs, segmentation, most tissue systems specific to arthropod organization and the whole Bauplan of Arthropoda. The complication of life cycle in some rhizocephalans leads to colonial organization. Using *in vitro* culture we have visualized the chaotic fractal structure of these barnacles at their parasitic stage, with multiple modular units of reproductive and trophic systems integrated into the colonial organism. We have found asexual reproduction (blastogenesis) by budding of branching epithelial stolon of the rhizocephalans *Polyascus polygenea*, *Peltogasterella gracilis* and *Peltogaster reticulatus* [41-46]. Thus, at the parasitic stage of the life cycle in certain species of Rhizocephala, by asexual reproduction without separation of blastozooids, the clonal colonial organization develops that is a unique phenomenon among Crustacea, all Arthropoda and all Ecdysozoa. The colonial organization of the endoparasitic organism of the rhizocephalans *Polyascus polygenea*, *Peltogasterella gracilis* and *Peltogaster reticulatus* is schematically depicted (Figure 3).

Fractal-like structures are typical not only for colonial and sedentary organisms. In many phyla, branching is hidden in the internal anatomy and is not obvious from external form [48]. Most animal organisms are filled with fractal structures. The branching mechanisms in many organs and glands are similar [49]. Ramifying fractal systems, such as the blood vascular system [50-53], embedded within the outer epithelial envelope of a multicellular organism, can be represented as a high-order manifold; the genus of its surface has achieved high values because of numerous anastomoses between branches of this embedded system.

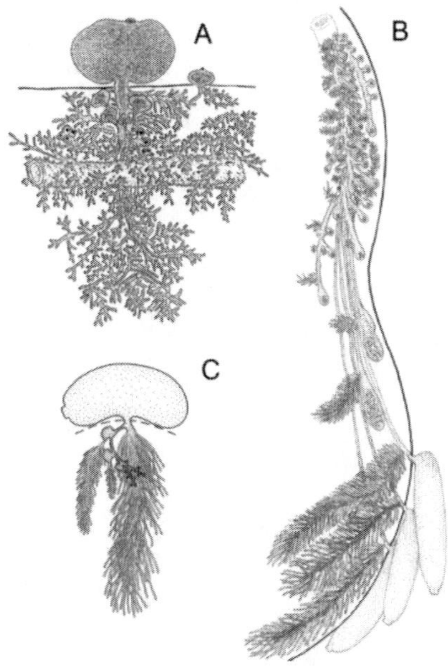

Figure 3. Schematic pictures of the fractal colonial organization of the rhizocephalan barnacles: A – Polyascus polygenea, B – Peltogasterella gracilis, C – Peltogaster reticulatus [after 43, 47].

The formation of branching respiratory trees is a classical example of fractal morphogenesis [38, 49, 54]. A conserved genetic machinery seems to control morphogenetic events such as branching and anastomosing of tubes in widely diverging animal clades [49, 55]. A biological algorithm of fractal morphogenesis in the respiratory systems of insects and chordates is similar; in particular, the same interaction of the fibroblasts growth factor with the receptor of fibroblasts growth factor is re-used in determination of many sequential steps of branching both in flies and mammals [54, 56]. Fractal algorithm with the development of a set of similar scale-invariant modules is an effective way of morphogenesis based on a relatively small genetic program. To evaluate the relationship between order and chaos in thc structure of branching epithelial channels of the gastro-vascular system in the jellyfish *Aurelia aurita* and the larval tracheal gills of the mayflies *Siphlonurus immanis* and *Parameletus chelifer* these patterns in symmetrical parts of each organism (cell clone) were compared with model graphs of deterministic fractals [10, 47]. Both systems are planar and hence are convenient for the analysis using two-dimensional models.

The larval tracheal gills of mayfly larvae are arranged on both sides of the abdomen as several pairs of metameric flat petals with a fractal tracheal pattern. Data on fractal-like organization of respiratory system in larval tracheal gills of the mayflies *Siphlonurus immanis* and *Parameletus chelifer* were collected [47]. Several pairs of the tracheal gills in a single larva provide data for quantitative analysis of tracheal tree variability and for order/chaos boundary in tracheal morphology of larval gills in these insects. A chaotic pattern of the larval gill tracheal system is quite evident; the differences in branching pattern of each pair of gill petals disrupt the bilateral symmetry of the mayfly larval body (Figure 4).

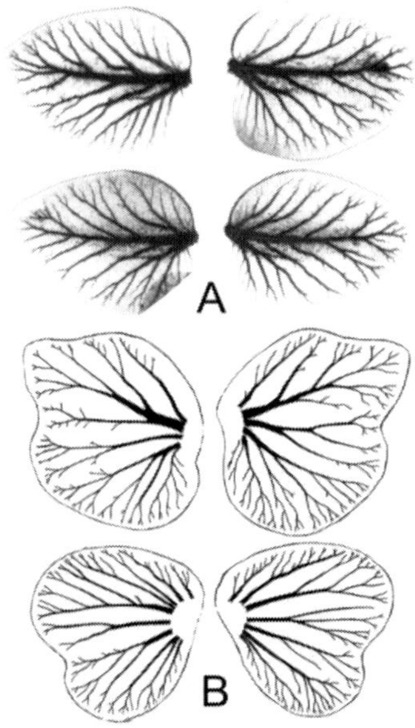

Figure 4. Variability of the branching pattern of larval tracheal gills in mayflies: A – two pairs of gill from a larva of *Parameletus chelifer* (photographs), B – upper and lower petals from a double gill pair of *Siphlonurus immanis* (sketch) [after 47].

In morphogenesis of the branching gastro-vascular channels of the jellyfish *Aurelia aurita* (Cnidaria: Scyphozoa), only two or three initial steps of branching are stereotyped and evidently rigidly determined. Presumably the most common morphological features in early development of the gastro-vascular system in *Aurelia* (4-ray symmetry, the formation of 8 branching and 8 non-branching channels, and the first 2-3 branching steps) are strictly specified genetically. The essential feature of later branching in gastro-vascular channels is its variability during ontogenesis. The branching pattern of four so named perradial channels in individual *Aurelia aurita* is represented by fractal trees with dichotomous branching (Figure 5). The irregularity of channel branching is evident. This variability in branching pattern of each perradial tree disrupts the radial symmetry of the jellyfish body. The boundary between order and chaos is revealed in the structural organization of the branching gastro-vascular channels in *A. aurita* [9, 10, 47].

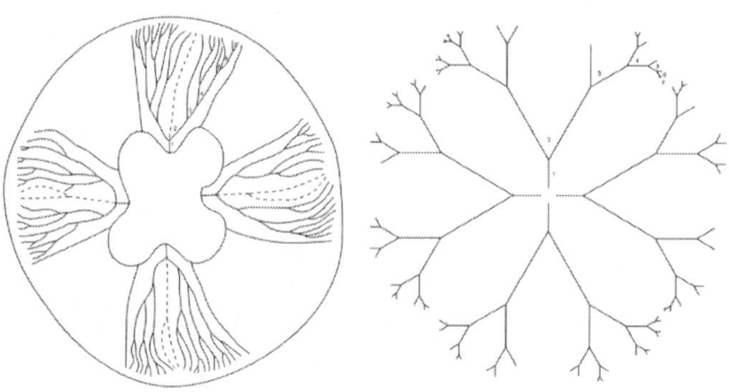

Figure 5. The branching pattern of four perradial channels in an individual *Aurelia aurita*: (left), its presentation as standard fractal trees with dichotomous branching (right) [after 47].

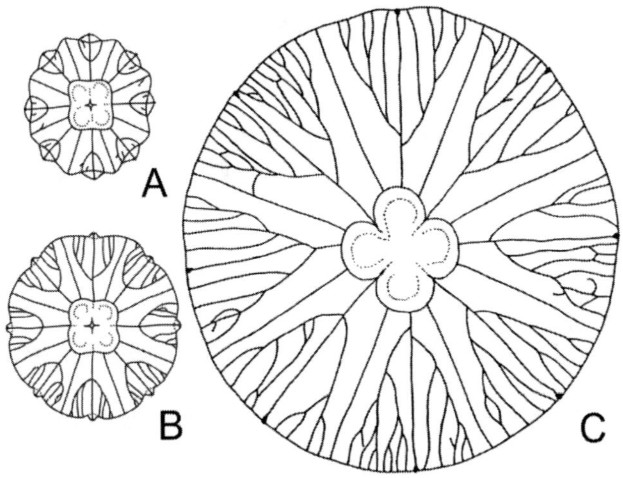

Figure 6. The formation of the gastro-vascular system in young jellyfish *Aurelia aurita*: a – a young medusa, 6 mm in diameter, b –a medusa of 8 mm, c – a later stage, 20 mm in diameter [after 57].

The formation of irregular, chaotic branching patterns of radial canals in the gastro-vascular system of *A. aurita* is a consequence of asynchronous and topographically variable development of new branches. During ontogenesis, the spatial and temporal variability of branching increases and results in the breaking of the radial symmetry in each individual medusa [57] (figure 6).

Only the first branching of the perradial and interradial channels of the gastro-vascular system of *A. aurita* in the metaephyra stage is always uniform. The ephyrae in *A. aurita* have a primitive digestive system with simple (non-branching) radial channels. In the course of their metamorphosis, differentiation of the radial channels into eight simple and eight branching channels takes place; in this process, each of the branching channels acquires the characteristic form of a trident fork [58, 59]. During further morphogenesis, the spatial and temporal variability of branching increases and results in the emergence of chaotic branching patterns as well as the breaking of the radial symmetry in each individual [57].

According to the molecular data, the *Aurelia* genus is a complex of species, and *A. aurita* includes several latent sibling species [60, 61]. Possibly, branching study of different species of the *A. aurita* complex will reveal species-specific features. At least, the differences in the structure of the gastro-vascular channels of the geographical forms described by Kramp [62] agree with the data on different species of the *A. aurita* complex.

The important feature in fractal morphogenesis is its plasticity in the course of ontogenesis providing the possibility of its adaptive responses, for example, after damage.

Fractal-like structures of an organism are a structural visualization, a record of morphogenetic dynamics including elements of chaos. During branching, a cascade of bifurcations leads to the irregular branching resulting in the breaking of radial or bilaterial symmetry of the organism.

Conclusion

Symmetry breaking plays a key role in biological morphogenesis. Topological and symmetry transformations are determined genetically and epigenetically. Segmentation along the anterior-posterior axis in Annelida, Arthropoda and Chordata is based on translation symmetry [55] including repeated units along the body axis and suggests a certain integration of a reiterated developmental program [see 63]. Annelids, arthropods and chordates all possess *Hox* gene clusters containing a number of orthologous genes expressed along the anterior-posterior axis [see 64]. The phenomenon of *Hox* colinearity is the only unambiguous example of the mapping of the spatial arrangement of *Hox* genes in their cluster (or clusters) into the spatial and temporal sequence of the gene expression along the axis of the embryo and the axial pattern of a future organism. Generally the genotype-phenotype mapping is mediated by complex epigenetic interactions of gene products, cells and cell systems in the processes of development. The relationship between genes and phenotypes is seen as a non-linear web of events, involving pleiotropy, self-organization processes, physical forces and environmental cues, all contributing to the final morphology of the organism [65]. For instance, genetically determined left-right asymmetry in Bilateria depends on the chain of events, involving morphofunctional asymmetry of components of the cell locomotor system in *Drosophila* and in chordates. It is shown a role of unconventional myosin motors in the regulation of left-right symmetry breaking in *Drosophila* [see 66].

In biological morphogenesis, in addition to such classical symmetry forms as rotating (radial), mirror (bilateral) and translational ones, the scale symmetry as self-similarity across scaling is common in fractal morphogenesis. Asexual reproduction without blastozooid separation results in irregular scale symmetry as fractal organization with multiple modules of different developmental stages. Spatial and temporal variability of branching epithelial channels increases during ontogeny, breaking, for examples, the radial symmetry of each individual jellyfish and bilateral symmetry of mayfly larvae. So the irregular scale symmetry of biological fractal structures gives rise to fluctuating asymmetry.

Iterating events in the process of segmentation (translational symmetry) and in branching of fractal channel systems (scale symmetry) are controlled by repeated expression of genes or gene network contours regulating these processes. The symmetry of the body, including the scale symmetry of the fractal structures, the translational symmetry of metamerism and other variants of morphofunctional iterations is an effective way of morphogenesis based on the relatively small genetic programs. Duplications of genes, gene clusters, gene network modules and developmental networks are based on the translational symmetry and widely used in evolution as an important mean for amplification and subsequent divergence of evolutionary innovations.

Cytoskeletal system of an egg cell is functioning as a global morphogenetic determinant of main axis of an embryo. The fundamental transformations of symmetry during oogenesis and early development determine the axial coordinates of a future organism, while in the subsequent development the scale of symmetry transformations is reducing. Chaos increases in the process of fractal morphogenesis and other manifestations of fluctuating asymmetry.

The course and outcome of biological evolution is strongly influenced by constraints, which play an important role and have both positive and negative values; evolutionary development can result in discontinuities and directionality of morphological transformations

[67]. General principles in physics, geometry and topology are applicable to many situations in biology [19]. In developmental morphogenesis, the important constraint is mechanodependence [see 68-70]. There is increasing evidence on mechanical control of tissue growth, on mechanical and geometrical constrains of supracellular patterns [69-73]. Mechanodependent modulation of gene expression was experimentally shown, for example, in the *twist* gene expression under mechanical deformation during gastrulation in *Drosophila* [74]. Tissue geometry and mechanical stress as inductive cues are entangled within the causal web of structure-function relationships that drive developmental processes [70].

To interpret the dynamics of symmetry transformations in metazoan evolution and ontogeny some concepts of topology and fractal geometry were applied. The topological approach can provide, to some degree, an explanation of some morphogenetic phenomena in biology and topological mechanisms that control morphogenesis. Topological singularities inevitably emerge and transform in biological morphogenesis. The local topological surgeries lead to the global topological modifications of biological forms. Topological methodology reveals a certain set of topological rules that constrains and directs biological morphogenesis [5, 10, 75]. The inevitable topological dependence and constraints of the symmetry transformations are integrated, fitted into genetically determined processes of biological morphogenesis, which cannot be independent of physical and topological organization of our space; the physical and topological imperative limits and directs the biological morphogenesis [10, 73].

The epithelial surface is an interface between the internal medium of an organism and the outside environmental medium; topological and fractal transformations during metazoan evolution and development increases this interface so ensuring metabolic efficiency and better adaptation of organism to the environment. It was supposed that some biological forms such as branching structures are the most functional design; these forms "are topological attractors that evolution cannot avoid" [76]. Just as branching structures, torus and multiple tori can be considered as functionally optimized biological design and topological attractors in biological morphogenesis.

References

[1] Thom, R. (1969) *Topol.*, *8*, 313.

[2] Thom, R. (1996). *Mem. Soc. Ital. Sci. Natur. Mus. Civ. Stor.Natur. Milano*, *27*, 115.

[3] Thom, R. (1997). *Structural Stability and Morphogenesis: An Outline of a General Theory of Models*, Addison-Wesley Publishing Co, New York.

[4] Presnov, E. V., Malyghin, S. N. & Isaeva, V. V. (1988). "Topological and thermodynamic structure of morphogenesis", *Thermodynamics and Pattern Formation in Biology*, Walter de Gruyter, Berlin, 337-370.

[5] Presnov, E., Isaeva, V. & Kasyanov, N. (2010). *Theor. Biosci.*, *129*, 259.

[6] Presnov, E. V. & Isaeva, V. V. (1990). *Forma*, *5*, 59.

[7] Presnov, E. V. & Isaeva, V. V. (1991). *Spec. Sci. Technol.*, *14*, 68.

[8] Presnov, E. V. & Isaeva, V. V. (1996). *Mem. Soc. Ital. Sci. Natur. Mus. Civ. Stor. Natur. Milano*, *27*, 89

[9] Isaeva, V. V., Presnov, E. V. & Chernyshev, A. V. (2006). *Bull. Math. Biol.*, *68*, 2053.

[10] Isaeva, V. V., Kasyanov, N. V. & Presnov, E. V. (2008). "Analysis situs of spatial-temporal architecture in biological morphogenesis", *Progress in Mathematical Biology Research*, Nova Science Publishers, New York, 141-189.
[11] Kolega, J. (1986). "The cellular basis of epithelial morphogenesis", *Developmental Biology*, Vol. *2*, Plenum Press, New York, 103-143.
[12] Zallen, J. A. (2007). *Cell*, *129*, 1051.
[13] Nelson, C. M. (2009). *Biochim. Biophys. Acta – Mol. Cell Res.*, *1793*, 903.
[14] Vladar, E. K., Antic, D. & Axelrod, J. D. (2009). *Cold Spring Harb Perspect Biol.* a002964, 1.
[15] Jockush, H. & Dress, A. (2003). *Bull. Math. Biol.*, *65*, 57.
[16] Peterson, K. J. & Eernisse, D. J. (2001). *Evol. Devel.*, *3*, 170
[17] Halanych, K. M. (2004). *Ann. Rev. Ecol. Evol. Syst.*, *35*, 229.
[18] Maresin, V. M. & Presnov, E. V. (1985). *J. Theor. Biol.*, *114*, 387.
[19] Bouligand, Y. (1996). *Memor. Soc. Ital. Sci. Natur. Mus. Civ. Stor. Natur. Milano*, *27*, 89.
[20] Li, R. & Bowerman, B. (2010). *Cold Spring Harb. Perspect. Biol.*, *2*, a003475
[21] Hirokawa, N., Tanaka, Y. & Okada, Y. (2009). *Cold Spring Harb Perspect Biol.*, *1*, a000802.
[22] Nuccitelli, R. (1984). "The involvement of transcellular ion currents and electric fields in pattern formation", *Pattern Formation. A Primer in Developmental Biology*, MacMillan, London, 23.
[23] Nüsslein-Volhard, C. (1991). *Development* (Suppl. 1), 1,
[24] Mullins, R. D. (2010). *Cold Spring Harbor Perspect. Biol.* cshperspect.a003392
[25] Goldstein, B. & Hird, S. N. (1996). *Development*, *122*, 1467.
[26] Sawada, T. (1988). *Zool. Sci.*, *5*, 667.
[27] Sawada, T. & Osanai, K. (1985). *Devel. Biol.*, *111*, 260.
[28] Vacquier, V. D. (1981). *Develop. Biol.*, *84*, 1.
[29] Kirschner, M. W. & Gerhart, J. C. (2005). *The Plausibility of Life*, Yale Univ. Press, Yale, New Haven.
[30] Kyozuka, K. (1993). *Bull. Mar. Biol. Stn. Asamushi*, Tohoku Univ. *19*, 1.
[31] Shapiro, B. M. (1981). *Annu. Rev. Biochem.*, *50*, 815.
[32] Shimizu, T. (1988). *Devel. Biol.*, *125*, 321.
[33] Shimizu, T. (1989). *Devel. Growth Differ.*, *31*, 283.
[34] Dubertret, B. & Rivier, N. (1997). *Compt. Rend. Acad. Sci. Fasc. B.*, *324*, 755
[35] Rivier, N., Miri, M. F. & Oguey, C. (2005). *Colloids Surf. A*, *263*, 39.
[36] Nübler-Jung, K. (1987). *Development*, *100*, 171.
[37] Nübler-Jung, K., Bonitz, R. & Sonnenschein, M. (1987). *Development 100*, 163
[38] Mandelbrot, B. B. (1983). *The Fractal Geometry of Nature*, Freeman, San Francisco.
[39] Weibel, E. R. (1991). *Amer. J. Physiol.*, *261*, 361.
[40] Weibel, E. R. (1994). "Design of biological organisms and fractal geometry", *Fractal in Biology and Medicine*, Birkhäuser, Basel e a., 68-85.
[41] Isaeva, V. V., Shukalyuk, A. I., Trofimova, A. V., Korn, O. M. & Rybakov, A. V. (2001). *Crust. Res.*, *30*, 134.
[42] Isaeva, V. V., Shukalyuk, A. I., Korn, O. M. & Rybakov, A. V. (2004). *Crust. Res.*, *33*, 61.

[43] Isaeva, V. V., Akhmadieva, A. V. & Shukalyuk, A. I. (2012). *J. Marine Biol. Assoc. UK, 92*, 457.
[44] Shukalyuk, A., Isaeva, V., Kizilova, E. & Baiborodin, S. (2005). *Invert. Reprod. Devel.*, *48*, 41.
[45] Shukalyuk, A., Golovnina, K., Baiborodin, S., Gunbin, K., Blinov, A. & Isaeva, V. (2007). *Cell Biol. Internat.*, *31*, 97.
[46] Shukalyuk, A. I. & Isaeva, V. V. (2012). "Molecular and subcellular gametogenic machinery of stem and germline cells across Metazoa", *Current Frontiers and Perspectives in Cell Biology*, InTech, Rieka, 279.
[47] Isaeva, V. V., Yu. A., Karetin, A. V. Chernyshev, & Yu, D. Shkuratov, *Fractals and Chaos in Biological Morphogenesis*, Dalnauka Vladivostok (2004, in Russian)
[48] Davies, J. A. (2005). "Why a book on branching, and why now?" *Branching Morphogenesis,* Eurekah.com and Springer Science+Business Media, New York, 1.
[49] Fleury, V., Watanabe, T., Nguyen, T. H., Unbekandt, M., Warburton, D., Dejmek, M., Nguyen, M. B., Lindner, A. & Schwartz, L. (2005). "Physical mechanisms of branching morphogenesis in animals. From viscous fingering to cartilage rings", *Branching Morphogenesis,* Eurekah.com and Springer Science+Business Media, New York, 202-234.
[50] Damiani, G. (1994). "Evolutionary meaning, functions and morphogenesis of branching structures in biology", *Fractals in Biology and Medicine*, Birkhäuser Verlag, Basel, 104-115.
[51] Goldberger, A. L. (1997). *Perspect. Biol. Med.*, *40*, 543.
[52] Goldberger, A. L. (2006). *Proc. Amer. Thorac. Soc.*, *3*, 467.
[53] West, G. B., Brown, J. H. & Enquist, B. J. (1999). *Science*, *284*, 1677.
[54] Metzger, R. J. & Krasnow, M. A. (1999). *Science*, *284*, 1635.
[55] Minelli, A. (2003). *The Development of Animal Form: Ontogeny, Morphology, and Evolution*, Cambridge University Press, Cambridge.
[56] Warburton, D., Schwarz, M., Tefft, D., Flores-Delgado, G., Anderson, K. D. & Cardoso, W. V. (2000). *Mech. Devel.*, *92*, 55.
[57] Chernyshev, A. V. & Isaeva, V. V. (2002). *Russ. J. Mar. Biol.*, *28*, 347.
[58] Southward, A. J. (1955). *J. Mar. Biol. Assoc. U.K. 34*, 201.
[59] Thiel, M. E. (1959). "Semaeostomae. Physiologie", *Klassen und Ordnungen des Tierreichs. Bd. 2, Abt. 2: Scyphomedusae*, Ajad. Verlag, Leipzig, 869.
[60] Dawson, M. N. & Jacobs, D. K. (2001). *Biol. Bull.*, *200*, 92.
[61] Dawson, M. N., Martin, L. E. (2001). Hydrobiol., *32*, 259.
[62] Kramp, P. L. (1942). *Medd. Grønland*, *81*, 1.
[63] Seaver, E. (2003). *Int. J. Dev. Biol.*, *47*, 583
[64] Shankland, M. (2003). "Evolution of body axis segmentation in the bilaterian radiation", *The New Panorama of Animal Evolution*, PENSOFT Publ., Sofia, Moscow, 187.
[65] Wilkins, A. S. (2002). *The Evolution of Developmental Pathways*, Sinauer, Sunderland, MA
[66] Baum, B. (2006). *Curr. Biol.*, *16*, R502.
[67] Sará, M. (1999). *Riv. Biol./Biol. Forum*, *92*, 29.
[68] Beloussov, L. V. (1998). *The Dynamic Architecture of a Developing Organism. An Interdisciplinary Approach to the Development of Organisms*. Kluwer, Dordrecht.

[69] Ingber, D. E. (2005). *Proc. Natl. Acad. Sci.*, USA, *102*, 11571.

[70] Nelson, C. M., Jean, R. P., Tan, J. L., Liu, W. F., Sniadecki, N. J., Spector, A. A. & Chen, C. S. (2005). *Proc. Natl. Acad. Sci.*, USA, *102*, 11594

[71] Harris, A. K. (1986). "Cell traction in relationship to morphogenesis and malignancy", *Developmental Biology*, Vol. *3*, Plenum Publishing Corporation, NewYork, London, 339.

[72] Harris, A. K., Stopak, D. & Warner, P. (1984). *J. Embryol. Exp. Morphol.*, *80*, 1-20

[73] Isaeva, V. V. (2012). *Biology Bull.*, *39*, 110.

[74] Farge, E. (2003). *Curr. Biol.*, *13*, 1365.

[75] Johnson, B. R. & Lam, S. K. (2010). *BioScience*, *60*, 879.

[76] Thomas, R. D. K. & Reif, W. E. (1993). *Evolution*, *47*, 341.

Index

A

B

C

D

E

F

G

H

I

N

O

P

R

S

T

U

V

W

Y

Z